OT 32
Operator Theory: Advances and Applications
Vol. 32

Birkhäuser Verlag
Basel · Boston · Berlin

Topics in Operator Theory

Constantin Apostol Memorial Issue

Edited by

I. Gohberg

1988

Birkhäuser Verlag
Basel · Boston · Berlin

MATH
sep lae

Volume Editorial Office:
Raymond and Beverly Sackler Faculty of Exact Sciences
School of Mathematical Sciences
Tel Aviv University
Tel Aviv, Israel

CIP-Titelaufnahme der Deutschen Bibliothek

Topics in operator theory: Constantin Apostol memorial issue
/ ed. by I. Gohberg. – Basel ; Boston ; Berlin : Birkhäuser, 1988
(Operator theory ; Vol. 32)
ISBN 3-7643-2232-2 (Basel …) Pb.
ISBN 0-8176-2232-2 (Boston) Pb.
NE: Gochberg, Izrail' [Hrsg.]; Apostol, Constantin: Festschrift; GT

sd
2/3/89
JJ

Printed in Germany
ISBN 3-7643-2232-2
ISBN 0-8176-2232-2

TABLE OF CONTENTS

	From the Editor......................	VII
	D.A. Herrero: In Memory of Constantin Apostol....................	1
	List of Publications of Constantin Apostol....................	7
	C. Apostol: On the Spectral Equivalence of Operators..............	15
BEN-ARTZI, A. GOHBERG, I.	Fredholm properties of band matrices and dichotomy................	37
BERCOVICI, H. CONWAY, J.B.	A note on the algebra generated by a subnormal operator...............	53
CEAUŞESCU, Z. SUCIU, I.	Extreme points in the set of contractive intertwining dilations....	57
DAVIDSON, K.R. FEEMAN, T.G. SHIELDS, A.L.	Extreme points in quotients of operator algebras.................	67
FOIAS, C. TANNENBAUM, A.	On the four block problem, I.........	93
HERRERO, D.A. TAYLOR, T.J. WANG, Z.Y.	Variation of the point spectrum under compact perturbations...........	113
LARSON, D.R. SOLEL, B.	Bimodules of nest subalgebras of von Neumann algebras	159
RAN, A.C.M. RODMAN, L.	Stability of invariant Lagrangian subspaces I................	181
SALINAS, N.	Products of kernel functions and module tensor products............	219
VASILESCU, F.-H.	Spectral capacities in quotient Fréchet spaces...............	243
VOICULESCU, D.	A note on quasidiagonal operators.....	265

CONSTANTIN APOSTOL
(1936–1987)

FROM THE EDITOR

Autumn 1985 Constantin Apostol became interested in the problem of local and global spectral equivalence. He soon produced a manuscript with interesting results. Who could imagine that this manuscript was to be a draft of his last paper, and it is with this paper that this Memorial issue begins. The final touches in the preparation of this paper for publication were done by Domingo Herrero; the author was already in hospital.

Constantin Apostol will be remembered as an outstanding expert in operator theory, in which area he made very important and deep contributions. His friends and colleagues dedicate this volume to his memory.

Hearty thanks are due to Domingo Herrero for his assistance in preparing this volume.

I. *Gohberg*

Operator Theory:
Advances and Applications, Vol. 32

IN MEMORY OF CONSTANTIN APOSTOL

Domingo A. Herrero

Constantin Apostol was born on February 1st, 1936, in Iaşi, in the region of Moldavia, Romania. His father died when he was only one year old, and he was raised by his mother, in poverty conditions bordering hunger, in a Romania ravaged by the war. In spite of the material difficulties, his mother gave him a strict education; the best she could afford. He was a gifted child, with a passion for reading, especially Geography, History, travels, novels.

In 1961, he received his Master in Arts degree from the University of Bucharest, and in 1968, he received his Ph. D. degree from the same institution. The same year, he was awarded the Romanian Academy Prize "Gheorghe Ţiţeica" for research in Operator Theory. By this time, Bucharest was already a well-known center in that area. Constantin was part of the Romanian School of Operator Theory of the 70's (and a very important part, indeed!) Some day, some historian of Mathematics will analyze the rising and evolution of this outstanding group of mathematicians that suddenly appear in an Eastern European country, among socio-economico - political conditions far from optimal for the development of any kind of science.

We live in an age of massification. People go to a basketball game to see a group of ten individuals whose average height surpasses the two meter mark, and everybody considers that there is nothing special about this group. Sometimes, I think that in the same way, the Operator Theory community (all around the World) lived throughout the 70's expecting every year a flush of exceptionally good results coming from Romania. That was the "natural", the "obvious" thing to expect...

In 1968, Paul R. Halmos introduced the notion of "quasi-triangular" (Hilbert space) operator, and two years later he published that wonderful survey article that has influenced so much the research in this area of Mathematics: "Ten problems in Hilbert space" (Bull. Amer. Math. Soc. 76 (1970), 887-933).

Around 1972, Constantin began to work on the spectral characterization of quasitriangularity, and wrote his article "Quasitriangularity in Hilbert space". It was not the solution of the problem but, clearly, it was pointing in the right direction. Shortly afterwards, the joint work of Constantin Apostol, Ciprian Foiaş and Dan Voiculescu produced the desired result: an operator T is quasitriangular if and only if index$(z - T)$ is non-negative for all z such that $z - T$ is semi-Fredholm.

The series of articles leading to this fundamental result contain an incredible number of "approximation tricks", that will strongly contribute to produce a whole body of work in the area of "non-commutative approximation" in the years to come.

1973 was a key year in Operator Theory: it was the year of the Apostol-Foiaş-Voiculescu Theorem, and the year of the Brown-Douglas-Fillmore Theorem, two cornerstones of the theory!

A year later, in 1974, the joint work of C. Apostol, C. Foiaş and D. Voiculescu solved one of the most difficult problems of Halmos: the spectral characterization of the norm-closure of the set of all nilpotent operators. In 1975, Apostol and Voiculescu extended the spectral characterization of quasitriangularity to operators on c_o, ℓ^p $(1 \le p \le \infty)$, and several other Banach spaces.

1976 was another exceptional year. Apostol introduced his "triangular model" of a Hilbert space operator, and showed that this model is a universal tool to solve problems related to the modification of the behavior by means of compact perturbations. In 1976, Voiculescu's Theorem was recent news; Apostol made one of the first nontrivial applications of this powerful tool in order to characterize the inner derivations with closed range: the inner derivation induced by an operator T has closed range if and only if T is similar to $A \oplus (B \otimes 1)$, where A and B are operators acting on finite dimensional spaces. He also wrote two highly interesting articles with Kevin Clancey on one-side resolvents and generalized inverses of operators. The same year, "The Romanian Troika" proved that the only strongly reductive operators are "the obvious ones": normal operators whose spectra have empty interior and do not separate the plane. (C. Apostol had already done important work on the extension of the ideas of N. Aronszajn and K. T. Smith in connection with the existence of invariant subspaces. The characterization of the strongly reductive operators belongs to this circle of ideas.)

Some years later, in our Seminar at the Arizona State University, Constantin would dismiss his beautiful idea of a "triangular model" by saying that: "As you can see, the whole idea is very simple. It is incredible that nobody had though of it long time ago...".

In 1977, Constantin Apostol and Bernard B. Morrel answered another question of Halmos, by providing a spectral characterization of the closure of the set of all those operators whose spectra lie in a fixed subset of the complex plane. Even more important than the result itself, the paper contains the notion of approximation of operators by certain "simple models" (the "Apostol-Morrel simple models"). These models, Voiculescu's Theorem and Apostol's "simple idea" (of a "triangular model") are some of the main bricks for the construction of our most important joint work: the Apostol - Herrero - Voiculescu spectral characterization of the closure of the similarity orbit of a Hilbert space operator. (This result is the core of the monograph "Approximation of Hilbert space operators", written with the cooperation of Lawrence A. Fialkow.)

From 1965 to 1975, Constantin Apostol was a member of the Institute of Mathematics of Buchárest and, from 1975 to 1982, he was a Senior Research Scientist of the new National Institute for Scientific and Technical Creation (INCREST) of Bucharest.

In 1982, dissapointed with the situation in Romania, he defected to the USA. After a year as a Visiting Professor at the University of Nebraska (Lincoln), he was appointed Full-Professor at Arizona State University. Two years later, Leiba Rodman joined our group. With Constantin Apostol and Leiba Rodman at ASU, and a good number of visitors (especially Carl Pearcy and Kenneth R. Davidson, who stayed there for a whole semester), the research life at ASU was a continuous exhilaration... A sort of uninterrupted Seminar on Operator Theory for several years in a row!

The communication was not always easy. Constantin came from the unique atmosphere of the INCREST, surrounded by a good number of the best operator theorists in the world, used to throw and catch challenging questions every day. No matter how serious his reasons to leave Romania, it was apparent that he missed that atmosphere. I, on the other hand, had spent a large part of my mathematical career in obscure corners of South America, with practically nobody to discuss my favorite problems, far from the big centers, without fluent communication with the rest of the world. Each situation produces, I suppose, some kind of professional deformation (of opposite signs).

The kinds of problems that attracted Constantin were very much the same that attracted me. To hear him in the Seminar was a real pleasure. There, a whole structure was constructed on the basis of a few, simple, geometric ideas. Somehow, however, he failed to transmit these simple underlying ideas to the paper. I found it difficult to reach the core of many of his articles. The geometric intuition is there, but (many times) it has been buried under several layers of unintuitive analytic lemmas.

To discuss and analyze the flow of ideas, problems, points of view on mathematical subjects, etc., coming from Constantin, would have required a whole team, not just one person. It was transparent, most of the time, that I was not enough challenge for him for the discussion of mathematical ideas. But, anyway, we felt that we were having a reasonably good time working together.

After the book was ready, he immediately began to apply some of our tools to work on stably invariant subspaces with Ciprian Foiaş and Norberto Salinas. We also wrote another paper on a recalcitrant question: the still open part of "the closure of a similarity orbit problem". The work with Filakow on "elementary operators" is also connected with the ideas in the book.

As a continuation of his work with Frank Gilfeather in Nebraska, they completely classified the isomorphisms of the "discrete" nest algebras, modulo the compact operators. This article was followed by another one, in cooperation with K. R. Davidson, on the analogous problem for general nest algebras.

But the bulk of his work at ASU was done in cooperation with Hari Bercovici, Ciprian Foiaş and Carl Pearcy. Sprouting from the fertile ideas and techniques invented by Scott Brown for the solution of the invariant subspace problem for subnormal operators, they (with Scott Brown, Béla Sz.-Nagy, Allen L. Shields, and a large number of Pearcy's Ph. D. students) created a whole new body of theory about the structure of a Hilbert space contraction whose spectrum includes the unit circle. During the NSF - CBMS Conference held at ASU in May, 1984, Carl Pearcy (the main speaker of the Conference) emphasized the role of Constantin in these developments by remarking that "The $\mathbb{A}$ (of the $\mathbb{A}_{\aleph_0}$ class of contractions) is for Apostol". It is difficult to single out a particular result to illustrate this enormous body of discipline, but here is one that is dear to my heart:

Let B be the Bergman shift operator on $A^2(\mathbb{D})$; given an arbitrary strict contraction A, there exist two invariant subspaces, M(A) and N(A), with $M(A) \subset N(A)$, such that

$$B = \begin{pmatrix} * & * & * \\ 0 & A & * \\ 0 & 0 & * \end{pmatrix} \begin{matrix} M(A) \\ N(A) \ominus M(A) \\ A^2(\mathbb{D}) \ominus N(A) \end{matrix} .$$

Hundreds of pages of explanations could not give a clearer picture of the incredible pathology of the invariant subspace lattice of the Bergman shift! (More than one operator theorist used to imagine it "pretty much like the unilateral shift".)

Constantin had a quiet, but good sense of humour. He was always ready to enjoy a good joke and a good beer.

He was a gregarious mathematician (as one can easily check from his long list of joint articles), and very generous with his ideas. His last article, which is to appear in this issue, was originated by some problems raised by Leiba Rodman in our

Seminar at ASU.

Unfortunately, there was a cloud in the horizon. Constantin suffered from several ailments. It was clear to his friends that he was (perhaps subconsciously?) concentrating all his energies almost exclusively on Mathematics and on the care of his wife Valentina and his daughter Catalina whom, after serious efforts, he had finally managed to bring with him into the USA. Since he was a chain smoker, we attributed those ailments to minor respiratory and circulatory problems. But, from the beginning on the Fall Semester, 1986, it was clear that something wrong was going on. He looked and felt more and more tired every day. In the first week of November, his health began to deteriorate very rapidly. He was hospitalized and diagnosed lung cancer. He died in the hospital on January 21st, 1987.

Operator Theory:
Advances and Applications, Vol. 32

CONSTANTIN APOSTOL - LIST OF PUBLICATIONS

1. Teorema de existență și unicitate pentru ecuații cu diferențiale în spații local convexe, An. Univ. Buc. 13 (1964), 45-53.

2. Propriétés spectrales des couples d'opérateurs dans les espaces de Banach, Bull. Soc. Roum. Math. 9 (1965), 159-165.

3. Propriétés des certains opérateurs bornés des espaces de Hilbert, Rev. Roum. Math. Pures et Appl. 10 (1965), 643-644.

4. Sur la partie normale d'un ensemble d'opérateurs de l'espace de Hilbert, Acta Math. Acad. Sci. Hung. 17 (1966), 1-4.

5. Asupra unei teoreme de existență și unicitate, An. Univ. Buc. 15 (1966), 137-141.

6. Sur l'équivalence asimptotique des opérateurs, Rev. Roum. Math. Pures et Appl. 12 (1967), 601-606.

7. Some properties of spectral maximal spaces and decomposable operators, Rev. Roum. Math. Pures et Appl. 12 (1967), 607-610.

8. Propriétés des certains opérateurs bornés des espaces de Hilbert, II, Rev. Roum. Math. Pures et Appl. 12 (1967), 759-762.

9. Some spectral properties of a couple of operators on a Banach space, Rev. Roum. Math. Pures et Appl. 12 (1967), 1005-1010.

10. Sur les opérateurs scalaires généralisés, Bull. Soc. Math. France 91 (1967), 57-61.

11. Sur les prolongements des représentations des idéaux d'une algèbre, Matematica 9 (1967), 5-7.

12. Restrictions and quotients of decomposable operators, Rev. Roum. Math. Pures et Appl. 13 (1968), 147-150.

13. On the roots of spectral operator-valued analytic functions, Rev. Roum. Math. Pures et Appl. 13 (1968), 587-589.

14. On the roots of spectral operators, Proc. Amer. Math. Soc. 19 (1968), 811-814.

15. Spectral decompositions and functional calculus, Rev. Roum. Math. Pures et Appl. 13 (1968), 1481-1528.

16. A theorem on invariant subspaces, Bull. Acad. Polon. Sci., Série Sci. Math. Astr. Phys. 16 (1968), 181-183.

17. Roots of scalar operator-valued analytic functions and their functional calculus, J. Sci. Hiroshima Univ., Ser. A-I, Math. 32 (1968), 173-180.

18. On the roots of generalized spectral operator-valued analytic functions, Glas. Math. 3 (1968), 347-352.

19. Remarks on the perturbation and a topology for operators, J. Funct. Anal. 2 (1968), 395-408.

20. Teorie spectrală şi calcul funcţional, Studii Cerc. Mat. 20 (1968), 635-667.

21. (with F. Gîndac) Proprietăţi algebreci ale rădăcinilor operatorilor spectrali, Studii Cerc. Mat. 20 (1968), 1115-1117.

22. Functional calculus in locally convex algebras, J. Sci. Hiroshima Univ., Ser. A-I, Math. 34 (1970), 1-12.

23. (with P. Mankiewicz) An example of a proper closed subspace dense in a scale of Hilbert spaces, Bull. Acad. Polon. Sci., Sér. Sci. Math. Astr. Phys. 18 (1970), 747-749.

24. b*-algebras and their representation, J. London Math. Soc. (2) 3 (1971), 30-38.

25. On the growth of the resolvent, perturbation and invariant subspaces, Rev. Roum. Math. Pures et Appl. 16 (1971), 167-172.

26. Products of contractions in Hilbert space, Acta Sci. (Szeged) 33 (1972), 91-94.

27. Decomposable multiplication operators, Rev. Roum. Math. Pures et Appl. 17 (1972), 323-333.

28. Hypercommutativity and invariant subspaces, Rev. Roum. Math. Pures et Appl. 17 (1972), 335-339.

29. On the left essential spectrum and non-cyclic operators in Banach spaces, Rev. Roum. Math. Pures et Appl. 17 (1972), 1141-1147.

30. Commutators on ℓ^p-spaces, Rev. Roum. Math. Pures et Appl. 17 (1972), 1513-1534.

31. (with C. Foiaş and L. Zsidó) Sur les operateurs non-quasi-triangulaires, C. R. Acad. Sci. Paris, Sér. A, 275 (1972), 501-503.

32. (with C. Foiaş and D. Voiculescu) Structure spectrale des operateurs non-quasi-triangulaires, C. R. Acad. Sci. Paris, Sér. A, 276 (1973), 49-51.

33. Quasitriangularity in Hilbert space, Indiana Univ. Math. J. 22 (1973), 817-825.

34. (with C. Foiaş and L. Zsidó) Some results on non-quasi-triangular operators, Indiana Univ. Math. J. 22 (1973), 1151-1161.

35. (with C. Foiaş and D. Voiculescu) Some results on non-quasitriangular operators, II, Rev. Roum. Math. Pures et Appl. 18 (1973), 159-181.

36. (with C. Foiaş and D. Voiculescu) Some results on non-quasitriangular operators, III, Rev. Roum. Math. Pures et Appl. 18 (1973), 309-324.

37. (with C. Foiaş and D. Voiculescu) Some results on non-quasitriangular operators, IV, Rev. Roum. Math. Pures et Appl. 18 (1973), 487-514.

38. Numerical functions and thin operators, Rev. Roum. Math. Pures et Appl. 18 (1973), 625-631.

39. Commutators on Hilbert space, Rev. Roum. Math. Pures et Appl. 18 (1973), 1013-1024.

40. Commutators on c_o-spaces and on ℓ^∞-spaces, Rev. Roum. Math. Pures et Appl. 18 (1973), 1025-1032.

41. (with C. Foiaş and D. Voiculescu) Some results on non-quasitriangular operators, V, Rev. Roum. Math. Pures et Appl. 18 (1973), 1133-1140.

42. (with L. Zsidó) Ideals in W*-algebras and the function η of A. Brown and C. Pearcy, Rev. Roum. Math. Pures et Appl. 18 (1973), 1151-1170.

43. (with C. Foiaş and D. Voiculescu) Some results on non-quasitriangular operators, VI, Rev. Roum. Math. Pures et Appl. 18 (1973), 1473-1494.

44. On the norm-closure of nilpotents, Rev. Roum. Math. Pures et Appl. 19 (1974), 277-282.

45. (with D. Voiculescu) On a problem of Halmos, Rev. Roum. Math. Pures et Appl. 19 (1974), 283-284.

46. (with C. Foiaş and D. Voiculescu) On the norm-closure of nilpotents, II, Rev. Roum. Math. Pures et Appl. 19 (1974), 549-577.

47. (with N. Salinas) Nilpotent approximation and quasinilpotent operators, Pac. J. Math. 61 (1975), 327-337.

48. (with C.-K. Fong) Invariant subspaces for algebras generated by strongly reductive operators, Duke Math. J. 42 (1975), 495-498.

49. Matrix models for operators, Duke Math. J. 42 (1975), 770-785.

50. Operators quasisimilar to normal operators, Proc. Amer. Math. Soc. 53 (1975), 104-106.

51. Quasitriangularity in Banach space, Rev. Roum. Math. Pures et Appl. 20 (1975), 135-174.

52. (with D. Voiculescu) Quasitriangularity in Banach space, II, Rev. Roum. Math. Pures et Appl. 20 (1975), 171-179.

53. (with C. Foiaş) On the distance to bi-quasitriangular operators, Rev. Roum. Math. Pures et Appl. 20 (1975), 261-265.

54. Quasitriangularity in Banach space, III, Rev. Roum. Math. Pures et Appl. 20 (1975), 389-410.

55. (with K. Clancey) Generalized inverses and spectral theorem, Trans. Amer. Math. Soc. 215 (1976), 293-300.

56. (with J. G. Stampfli) On derivation ranges, Indiana Univ. Math. J. 25 (1976), 857-869.

57. On the operator equation $TX - XT = A$, Proc. Amer. Math. Soc. 59 (1976), 115-118.

58. Local resolvents of operators with one-dimensional self-commutator, Proc. Amer. Math. Soc. 58 (1976), 158-162.

59. (with K. Clancey) On generalized resolvents, Proc. Amer. Math. Soc. 58 (1976), 163-168.

60. (with R. G. Douglas and C. Foiaş) Quasisimilar models for nilpotent operators, Trans. Amer. Math. Soc. 224 (1976), 407-415.

61. On the norm-closure of nilpotents, III, Rev. Roum. Math. Pures et Appl. 21 (1976), 143-153.

62. The correction by compact perturbations of the singular behavior of operators, Rev. Roum. Math. Pures et Appl. 21 (1976), 155-175.

63. Inner derivations with closed range, Rev. Roum. Math. Pures et Appl. 21 (1976), 249-265.

64. (with C. Foiaş and D. Voiculescu) On strongly reductive algebras, Rev. Roum. Math. Pures et Appl. 21 (1976), 633-641.

65. (with C. Foiaş and D. Voiculescu) Strongly reductive operators are normal, Acta Sci. Math. (Szeged) 38 (1976), 261-263.

66. Quasiaffine transforms of quasinilpotent operators, Rev. Roum. Math. Pures et Appl. 21 (1976), 813-816.

67. On the closed range points in the spectrum of operators, Rev. Roum. Math. Pures et Appl. 21 (1976), 971-975.

68. (with C. Pearcy and N. Salinas) Spectra of compact perturbations on operators, Indiana Univ. Math. J. 26 (1977), 345-350.

69. (with C. Foiaş) Operators with closed similarity orbits, Rev. Roum. Math. Pures et Appl. 22 (1977), 13-15.

70. (with B. B. Morrel) On uniform approximation of operators by simple models, Indiana Univ. Math. J. 26 (1977), 427-442.

71. The spectrum and the spectral radius as functions in Banach algebras, Bull. Acad. Polon. Sci., Série Sci. Math. Astr. Phys. 26 (1979), 975-978.

72. (with C. Foiaş and C. Pearcy) That quasinilpotent operators are norm-limits of nilpotent operators, revisited, Proc. Amer. Math. Soc. 73 (1979), 61-64.

73. Ultraweakly closed operator algebras, J. Operator Theory 2 (1979), 49-61.

74. Invariant subspaces for subquasiscalar operators, J. Operator Theory 3 (1980), 159-164.

75. Universal quasinilpotent operators, Rev. Roum. Math. Pures et Appl. 25 (1980), 135-138.

76. Functional calculus and invariant subspaces, J. Operator Theory 4 (1980), 159-190.

77. The spectral flavour of Scott Brown's technique, J. Operator Theory 6 (1981), 3-12.

78. (with M. Martin) A C*-algebra approach to the Cowen-Douglas theory, Topics in Modern Operator Theory, 5-th International Conference on Operator Theory, Timişoara and Herculane (Romania 1980), Operator Theory: Advances and Applications, vol. 2, Birkhäuser-Verlag, Basel-Boston-Stuttgart, 1981, pp. 45-51.

79. (with H. Bercovici, C. Foiaş and C. Pearcy) Quasiaffine transforms of operators, Michigan Math. J. 29 (1982), 243-255.

80. (with B. Chevreau) On M-spectral sets and rationally invariant subspaces, J. Operator Theory 7 (1982), 247-266.

81. (with D. A. Herrero and D. Voiculescu) The closure of the similarity orbit of a Hilbert space operator, Bull. Amer. Math. Soc. 6 (1982), 421-426.

82. On the norm-closure of the similarity orbit of essentially nilpotent operators, Invariant Subspaces and Other Topics, 6-th International Conference on Operator Theory, Timişoara and Herculane (Romania 1981), Operator Theory: Advances and Applications, vol. 6, Birkhäuser-Verlag, Basel-Boston-Stuttgart, 1982, pp. 33-43.

83. Hyperinvariant subspaces for some bilateral weighted shifts, Integral Equations and Operator Theory 7 (1982), 1-9.

84. (with L. A. Fialkow, D. A. Herrero and D. Voiculescu) Approximation of Hilbert Space Operators, Volume II, Research Notee in Math., vol. 102, Pitman Advanced Publ. Program, 1984.

85. (with D. A. Herrero) On closures of similarity orbits of essentially nilpotent operators of order two, Integral Equations and Operator Theory 8 (1985), 437-461.

86. (with H. Bercovici, C. Foiaş and C. Pearcy) Invariant subspaces, dilation theory, and the structure of the predual of a dual algebra, I, J. Funct. Anal. 63 (1985), 369-404.

87. (with H. Bercovici, C. Foiaş and C. Pearcy) Invariant subspaces, dilation theory, and the structure of the predual of a dual algebra, II, Indiana Univ. Math. J. 34 (1985), 845-855.

88. (with C. Foiaş and N. Salinas) On stable invariant subspaces, Integral Equations and Operator Theory 8 (1985), 721-750.

89. The reduced minimum modulus, Michigan Math. J. 32 (1985), 279-294.

90. (with H. Bercovici, C. Foiaş and C. Pearcy) On the theory of the class $\mathbf{A}_{\aleph_0}$ with applications to invariant subspaces and the Bergman shift operator, Advances in Invariant Subspaces and Other Results of Operator Theory, 9-th International Conference on Operator Theory, Timişoara and Herculane (Romania 1984), Operator Theory: Advances and Applications, vol. 17, Birkhäuser-Verlag, Basel-Boston-Stuttgart, 1986, pp. 43-49.

91. (with F. Gilfeather) Isomorphisms of nest algebras, modulo the compact operators, Pac. J. Math. 122 (1986), 263-286.

92. (with L. A. Fialkow) Structural properties of elementary operators, Can. J. Math. 38 (1986), 1485-1524.

93. (with K. R. Davidson) Isomorphisms modulo the compact operators of nest algebras, II,

94. On a spectral equivalence of operators, Integral Equations and Operator Theory (1988), - .

Operator Theory:
Advances and Applications, Vol. 32

ON A SPECTRAL EQUIVALENCE OF OPERATORS

Constantin Apostol[1]

INTRODUCTION

Let H be a complex, separable Hilbert space and let L(H) denote the algebra of all bounded, linear operators acting in H. Let F_1, F_2 be L(H)-valued analytic functions whose domains of definition contain an open set $G \subset \mathbf{C}$. As in [9] we shall say that F_1 and F_2 are equivalent on G if

$$F_1(\lambda) = A(\lambda)F_2(\lambda)B(\lambda), \quad \lambda \in G,$$

for some analytic functions A, B, defined in G, whose values are invertible operators. Results on the general theory of this type of equivalence can be found in [9], [10], [12], [14] and their references. Throughout this paper we shall restrict to the study of the equivalence on some open set G of the particular functions

$$F_1(\lambda) = T - \lambda, \quad F_2(\lambda) = S - \lambda, \quad \lambda \in \mathbf{C},$$

where T and S belong to L(H). Because T and S uniquely determine F_1 and F_2 we shall say that T and S are equivalent on G. We shall denote this relation by $T \overset{G}{\sim} S$. If $\mu \in \mathbf{C}$ is given, we shall say that T and S are equivalent at μ, abbreviated as $T \overset{\mu}{\sim} S$, if $T \overset{G}{\sim} S$ for some $G \ni \mu$. Very little is known about the properties shared by T and S when $T \overset{G}{\sim} S$. Certainly no information can be derived from the relation $T \overset{G}{\sim} S$, when $G \subset \rho(T) \cap \rho(S)$, because $T - \lambda = A(\lambda)(S-\lambda)B(\lambda)$, $\lambda \in G$, where $A(\lambda) = I$, $B(\lambda) = (S-\lambda)^{-1}(T-\lambda)$. On the other hand if $T \overset{G}{\sim} S$ and $G \supset \sigma(T) \cup \sigma(S)$ then T is necessarily similar to S(i.e. $T = R^{-1}ST$ for some

(1) This research was partially supported by a Grant of the National Science Foundation.

invertible R). This result belongs to Kaashoek, van der Mee and Rodman ([12], Corollary 0.2) and answers a question asked in [9]. The following question of Gohberg, Kaashoek and Lay also appear in [9]: Does the equivalence of T and S at every $\mu \in G$, imply $T \overset{G}{\sim} S$ for an arbitrary G. In view of the results of [12], we can ask if in particular the implication

$$T \overset{\mu}{\sim} S, \ (\forall)\mu \in \mathbf{C} \quad \Rightarrow T \text{ is similar to } S$$

holds true. We shall show that this implication holds true in the particular case when both T and S are spectral operators in Dunford's sense (see Theorem 3.4). Our efforts will concentrate on the role played by the position of G with respect to $\sigma(T) \cup \sigma(S)$ when $T \overset{G}{\sim} S$. A consequence of Lemma 2.2 is that at least for some (non-trivially positioned) sets G, the answer to the question of [9], is affirmative. Finally, we mention that Theorem 4.1 suggest that the inclusion $G \supset \sigma(T) \cup \sigma(S)$, when $T \overset{G}{\sim} S$, might be non-necessary for the similarity of T and S.
The author expresses his gratitude to J. Froelich, I. Gohberg and L. Rodman for discussing the subject.

1. Notation and Preliminaries

Throughout the paper H will denote a fixed complex, separable, infinite dimensional Hilbert space, with the inner product $\langle \cdot , \cdot \rangle$, L(H) will denote the algebra of all bounded linear operators and Inv(H) will denote the group of invertible elements in L(H). If $T \in L(H)$ and if $\tilde{T}$ denotes the image of T in the Calkin algebra, then the essential spectrum, the left essential spectrum and the right essential spectrum of T will be defined by

$$\sigma_e(T) = \sigma(\tilde{T}), \ \sigma_{\ell e}(T) = \sigma_\ell(\tilde{T}), \ \sigma_{re}(T) = \sigma_r(\tilde{T}).$$

Let $\mu \in \mathbf{C}$. We shall say that μ is T-regular if the function $\lambda \to P_{\ker(T-\lambda)}$ is norm continuous at μ. If μ is not T-regular, we call it T-singular. As in [4], define the reduced minimum modulus of T by

$$\gamma(T) = \begin{cases} \inf\{\|Tx\| : \operatorname{dist}(x,\ker T) = 1\} & \text{if } T \neq 0 \\ 0 & \text{if } T = 0 . \end{cases}$$

For a non-zero T, the range of T is closed iff $\gamma(T) > 0$. In any case, $\lim\limits_{\lambda\to\mu} \gamma(T-\lambda)$ exists for each $\mu \in \mathbf{C}$. In the sequel we shall need to identify the following sets

$$\sigma_{c.r.}(T) = \{\lambda \in \sigma(T) : (T-\lambda)H = ((T-\lambda)H)^{-}\}$$

$$\sigma^{r}_{c.r.}(T) = \{\lambda \in \sigma_{c.r.}(T) : \lambda \text{ is T-regular}\},$$

$$\sigma^{s}_{c.r.}(T) = \{\lambda \in \sigma_{c.r.}(T) : \lambda \text{ is T-singular}\},$$

$$\sigma_{\gamma}(T) = \{\mu \in \mathbf{C} : \lim_{\lambda\to\mu} \gamma(T-\lambda) = 0\},$$

$$\rho_{\gamma}(T) = \mathbf{C} \setminus \sigma_{\gamma}(T).$$

If T has closed range and at least one of dimkerT and dimkerT* is finite we call T a semi-Fredholm operator in which case its index is (possibly infinite)

$$\operatorname{ind}T = \dim\ker T - \dim\ker T^{*}.$$

If T is semi-Fredholm and ind T is finite we shall say that T is a Fredholm operator. We shall denote by $\rho_{s-F}(T)$ the semi-Fredholm domain of T, i.e.

$$\rho_{s-F}(T) = \{\lambda \in \mathbf{C} : T - \lambda \text{ is semi-Fredholm}\}$$

and $\rho^{r}_{s-F}(T)$, $\rho^{s}_{s-F}(T)$ will denote the corresponding T-regular or T-singular parts. The symbols $\rho_{F}(T)$, $\rho^{r}_{F}(T)$, $\rho^{s}_{F}(T)$ will denote the Fredholm analogues. For every closed set $\sigma \subset \mathbf{C}$, the symbol $H_T(\sigma)$ will denote the set of all $x \in H$, such that the equation

$$(\lambda-T)f(\lambda) = x, \quad \lambda \in \mathbf{C} \setminus \sigma,$$

has at least one H-valued analytic solution, defined in $C \setminus \sigma$. Any f as above can be uniquely extended to $\rho(T) \cup (\mathbf{C} \setminus \sigma)$, thus we have $H_T(\sigma) = H_T(\sigma \cap \sigma(T))$. Let $G \subset \mathbf{C}$ be open. We shall say that G is a set of analytic uniqueness for T (see [15], IV, Definition 3.2), if the equation

$$(T-\lambda)f(\lambda) = 0, \quad \lambda \in G$$

has only the trivial H-valued analytic solution, defined in G. A sufficient condition for G to be a set of analytic uniqueness for T is, that each component of G, meets the resolvent set $\rho(T)$. Further we denote by $C(T)$ the family of all closed sets $\sigma \subset \mathbf{C}$, such that $C \setminus \sigma$ is a set of analytic uniqueness for T. If $\sigma \in C(T)$, then for every $x \in H_T(\sigma)$, the equation

$$(\lambda-T)x_T(\lambda) = x, \quad \lambda \in \mathbf{C} \setminus (\sigma \cap \sigma(T))$$

has a unique H-valued analytic solution, defined in $\mathbf{C} \setminus (\sigma \cap \sigma(T))$. Moreover if Γ is an admissible contour (in the sense of [8], VII. 3.9), surrounding $\sigma \cap \sigma(T)$, then we have

$$\frac{1}{2\pi i} \int_\Gamma x_T(\lambda)d\lambda = \frac{1}{2\pi i} \int_{|\lambda|=a} (\lambda-T)^{-1}x\,d\lambda = x \quad (a>0, \text{ large}).$$

(In fact the uniqueness of x_T does not play a role in the evaluation of the above integral.)

Now recall some definitions and results from [6]. We shall say that T has the single valued extension property, abbreviated as s.v.e.p. if every open set $G \subset \mathbf{C}$ is a set of analytic uniqueness for T. If T has the s.v.e.p. then for each $x \in H$, the equation

$$(\lambda-T)x_T(\lambda) = x, \quad \lambda \in \rho_T(x),$$

has a unique H-valued analytic solution defined in a (unique) maximal open set $\rho_T(x)$. If $\sigma \subset \mathbf{C}$ is closed, then putting $\sigma_T(x) = \mathbf{C} \setminus \rho_T(x)$, we have

$$H_T(\sigma) = \{x \in H : \sigma_T(x) \subset \sigma\}.$$

A spectral maximal space of T will be an invariant (closed) subspace of T, X, which contains any invariant subspace of T, Y, such that $\sigma(T|Y) \subset \sigma(T|X)$. The operator T will be called decomposable if for any open covering $\{G_k\}_{k=1}^n$ of $\sigma(T)$, there exist spectral maximal spaces of T, $\{X_k\}_{k=1}^n$, such that

$$\sigma(T|X_k) \subset G_k, \quad \sum_{k=1}^n X_k = X.$$

If T is decomposable then T has the s.v.e.p., and for every closed set $\sigma \subset \mathbf{C}$, $H_T(\sigma)$ is a spectral maximal space of T and $\sigma(T|H_T(\sigma)) \subset \sigma \cap \sigma(T)$.

Let $G \subset \mathbf{C}$ be open and let $Ho\ell(G, Inv(H))$ denote the set of all $Inv(H)$-valued analytic functions defined in G. If T, $S \in L(H)$ are given, then as mentioned in the introduction, we shall say that T and S are equivalent on G, or $T \overset{G}{\sim} S$ if the relation

$$(*) T - \lambda = A(\lambda)(S-\lambda)B(\lambda), \quad \lambda \in G,$$

holds true for some $A, B \in Ho\ell(G, Inv(H))$. If $\mu \in \mathbf{C}$ is given then we say that T and S are equivalent at ν, or $T \overset{\mu}{\sim} S$, if $T \overset{G}{\sim} S$ for some open set G containing μ. It is obvious that both "$\overset{G}{\sim}$" and "$\overset{\mu}{\sim}$" are equivalence relations in L(H).

For every $T \in L(H)$, $\sigma \subset \mathbf{C}$, T^* will denote the Hilbert space adjoint of T and s^* will be the complex conjugate of σ. Let $G \subset \mathbf{C}$ be open and let F be an L(H)-valued analytic function defined in G. Then the equation

$$F^*(\lambda) = F(\bar{\lambda})^* \quad , \quad \lambda \in G^*,$$

defines an analytic function in G^*. If $\sigma \in C(T)$ and $\sigma \subset G$, we shall define the linear map

$$T_\sigma(F) : H_T(\sigma) \to H,$$

by the equation

$$T_\sigma(F)x = \frac{1}{2\pi i} \int_\Gamma F(\lambda)x_T(\lambda)d\lambda, \quad x \in H_T(\sigma),$$

where $\Gamma \subset G$ is an admissible contour surrounding $\sigma \cap \sigma(T)$, and $\Gamma = \partial G_\Gamma$, $\sigma \cap \sigma(T) \subset G_\Gamma \subset G$.

LEMMA 1.1. *Let* $T, S \in L(H)$, $G = \text{int}G \subset \mathbf{C}$ *and suppose* $T \overset{G}{\sim} S$. *Then we have*

(*i*) $T^* \overset{G^*}{\sim} S^*$,

(*ii*) $G \cap \sigma(T) = G \cap \sigma(S)$,

(*iii*) $G \cap \sigma_\gamma(T) = G \cap \sigma_\gamma(S)$, $G \cap \sigma^r_{c.r.}(T) = G \cap \sigma^r_{c.r.}(S)$ $G \cap \sigma^s_{c.r.}(T) = G \cap \sigma^s_{c.r.}(S)$,

(*iv*) $\dim\ker(T-\lambda) = \dim\ker(S-\lambda)$, $\lambda \in G$.

PROOF. Let $A, B \in \mathcal{H}o\ell(G, \text{Inv}(H))$ verify (*). From the definition of γ, it follows

$$\gamma(S-\lambda) \leq \|A^{-1}(\lambda)\| \; \|B^{-1}(\lambda)\|\gamma(t-\lambda),$$
$$\gamma(T-\lambda) \leq \|A(\lambda)\| \; \|B(\lambda)\|\gamma(S-\lambda),$$

thus $\lim_{\lambda\to\mu} \gamma(T-\lambda) = 0$ iff $\lim_{\lambda\to\mu} \gamma(S-\lambda) = 0$ for $\mu \in G$, and in particular $G \cap \sigma_\gamma(T) = G \cap \sigma_\gamma(S)$. The rest of the proof is an easy exercise. □

LEMMA 1.2. *Let* $T, S \in L(H)$, A, B, $\in \mathcal{H}o\ell(G,\text{Inv}(H))$ *verify* (*) *and let* $\sigma \in C(T) \cap C(S)$, $\sigma \subset G$. *Then we have*

(*i*) $T_\sigma(B)H_T(\sigma) = H_S(\sigma)$, $S_\sigma(B^{-1})H_S(G) = H_T(\sigma)$,

(*ii*) $S_\sigma(B^{-1})T_\sigma(B) = I|H_T(\sigma)$,

$T_\sigma(B)S_\sigma(B^{-1}) = I|H_S(\sigma)$,

(*iii*) $ST_\sigma(B) = T_\sigma(B)(T|H_T(\sigma))$,

$TS_\sigma(B^{-1}) = S_\sigma(B^{-1})(S|H_S(\sigma))$

PROOF. Because the above relations are symmetric, it suffices to prove that we have

$$T_\sigma(B)H_T(\sigma) \subset H_S(\sigma),\ S_\sigma(B^{-1})T_\sigma(B) = I|H_T(\sigma),$$
$$ST_\sigma(B) = T_\sigma(B)(T|H_T(\sigma)).$$

Let $x \in H_T(\sigma)$ and put $y = T_\sigma(B)x$. Let $\Gamma \subset G$ be an admissible contour surrounding $\sigma \cap \sigma(T)$ and for each $\lambda \in \mathbf{C} \setminus (\sigma \cap \sigma(T))$, choose an admissible contour $\Gamma' \subset G$, leaving both Γ and λ outside. It is obvious that the integral

$$f(\lambda) = \frac{1}{2\pi i} \int_{\Gamma'} \frac{B(\zeta)x_T(\zeta)}{\lambda - \zeta} d\zeta$$

does not depend on Γ' and defines an analytic function. Using the relations

$$(\lambda - S)f(\lambda) = \frac{1}{2\pi i} \int_{\Gamma'} B(\zeta)x_T(\zeta)d\xi + \frac{1}{2\pi i} \int_{\Gamma'} \frac{(\zeta - S)B(\zeta)x_T(\zeta)}{\lambda - \zeta} d\zeta$$
$$= y + \frac{1}{2\pi i} \int_{\Gamma'} \frac{A^{-1}(\zeta)x}{\lambda - \zeta} d\zeta = y,$$

we derive $y \in H_S(\sigma)$ and $y_S = f$. On the other hand this implies

$$S_\sigma(B^{-1})T_\sigma(B)x = \frac{1}{2\pi i} \int_{\Gamma} B^{-1}(\lambda)\Big(\frac{1}{2\pi i} \int_{\Gamma'} \frac{B(\zeta)x_T(\zeta)}{\lambda - \zeta} d\zeta\Big)d\lambda =$$
$$= \frac{1}{2\pi i} \int_{\Gamma'} \Big(\frac{1}{2\pi i} \int_{\Gamma} \frac{B^{-1}(\lambda)B(\zeta)x_T(\zeta)}{\lambda - \zeta} d\lambda\Big)d\zeta =$$
$$= \frac{1}{2\pi i} \int_{\Gamma'} B^{-1}(\zeta)B(\zeta)x_T(\zeta)d\zeta = \frac{1}{2\pi i} \int_{\Gamma'} x_T(\zeta)d\zeta = x.$$

Because obviously $(Tx)_T(\lambda) = Tx_T(\lambda)$, $\lambda \in \mathbf{C} \setminus (\sigma \cap \sigma(T))$, we also infer

$$T_\sigma(B)Tx = \frac{1}{2\pi i}\int_\Gamma B(\lambda)Tx_T(\lambda)d\lambda =$$

$$= \frac{1}{2\pi i}\int_\Gamma \lambda B(\lambda)x_T(\lambda)d\lambda - \frac{1}{2\pi i}\int_\Gamma B(\lambda)xd\lambda =$$

$$= ST_\sigma(B)x + \frac{1}{2\pi i}\int_\Gamma (\lambda-S)B(\lambda)x_T(\lambda)d\lambda - \frac{1}{2\pi i}\int_\Gamma B(\lambda)xd\lambda$$

$$= ST_\sigma(B)x + \frac{1}{2\pi i}\int_\Gamma (A^{-1}(\lambda) - B(\lambda))xd\lambda = ST_\sigma(B)x. \qquad \square$$

COROLLARY 1.3. *Let* T, S, G, σ, A, B *be as in Lemma 1.2 and let* X *be an invariant subspace of* T *such that* $\sigma(T|X) \subset \sigma$. *Then* $X \subset H_T(\sigma)$ *and* $T_\sigma(B)|X$ *is bounded.*

PROOF. The inclusion $X \subset H_T(\sigma)$ is obvious. If we put

$$f(\lambda) = \begin{cases} (\lambda - (T|X))^{-1}, & \lambda \in \rho(T|X) \\ (\lambda - T)^{-1}, & \lambda \in \rho(T) \setminus \rho(T|X) \end{cases}$$

then for every $x \in H_T(\sigma)$ we have

$$\|T_\sigma(B)x\| = \frac{1}{2\pi}\left\|\int_\Gamma B(x)F(\lambda)d\lambda\right\| \leq \frac{1}{2\pi}\int_\Gamma \|B(\lambda)F(\lambda)\|\,|d\lambda|\,\|x\|,$$

where $\Gamma \subset G$ is an admissible contour surrounding $\sigma \cap \sigma(T)$.

REMARK 1.4. Lemma 1.2 and Corollary 1.3 are valid in Banach spaces.

2. The Equivalence Class $\{T\}_G$.

Throughout this section $T \in L(H)$ will be a fixed operator and $G \subset \mathbf{C}$ will be a fixed open set. We shall put

$$\{T\}_G = \{S \in L(H) : T \overset{G}{\sim} S\}.$$

LEMMA 2.1. *For every* $S \in \{T\}_G$ *and* $\lambda \in G$ *we have*

(*i*) $\dim\ker(T-\lambda)^n = \dim\ker(S-\lambda)^n$, $n \geq 1$, (*ii*) $\dim\ker(T-\lambda)^{*n} = \dim\ker(S-\lambda)^{*n}$, $n \geq 1$.

PROOF. Let us put

$$\sigma = \{\lambda\}, \quad X = \bigcup_{n=1}^{\infty} \ker(T-\lambda), \quad Y = \bigcup_{n=1}^{\infty} \mathrm{ler}(S-\lambda)^n.$$

Since obviously $\sigma \in C(T) \cap C(S)$ and $X \subset H_T(\sigma)$, $Y \subset H_S(\sigma)$, defining $T_\sigma(B)$ and $S_\sigma(B^{-1})$, where A and B verify(*), we can derive (i) from Lemma 1.2. To conclude the proof we observe that (ii) is a consequence of (i) and Lemma 1.1. □

LEMMA 2.2. $S \in L(H)$

$$G^- \subset \rho_r(T) \cap \rho_r(S), \ \dim\ker(T-\lambda) = \dim\ker(s-\lambda), \ \lambda \in G.$$

$S \in \{T\}_G$.

PROOF. It suffices to prove that for each connected component G_0 of G, we can find $B_0 \in \mathrm{Hol}(G_0, \mathrm{Inv}(H))$ such that

$$(\lambda-T) = (\lambda-S)B_0(\lambda), \quad \lambda \in G_0.$$

Since $G_0^- \subset \rho_r(T) \cap \rho_r(S)$, using [5], IX, Proposition 9.17, we can find right resolvents F_1, F_2 for T, respectively S, both defined in G_0. By [5], IX, Lemma 9.6 and Lemma 9.7 we may suppose that

$$\mathrm{ran}\, F_1(\lambda) = H_1, \ \mathrm{ran}\, F_2(\lambda) = H_2, \ \lambda \in G_0,$$

$$H_1 + \ker(T-\lambda) = H_2 + \ker(S-\lambda) = H, \ \lambda \in G_0,$$

$$H_1 \cap \ker(T-\lambda) = H_2 \cap \ker(S-\lambda) = \{0\}, \ \lambda/\in G_0.$$

Since $\dim\ker(T-\lambda) = \dim\ker(S-\lambda)$, $\lambda \in G_0$ we derive $\dim H_1^{\perp} = \dim H_2^{\perp}$, thus we can find a partial isometry $U \in L(H)$ such that $UH = H_2^{\perp}$ and $U^*H = H_1^{\perp}$. If we put

$$P_1(\lambda) = I - F_1(\lambda)(\lambda-T), \ P_2(\lambda) = I - F_2(\lambda)(\lambda-S),$$

then $P_1(\lambda)$ and $P_2(\lambda)$ will be projections onto $\ker(T-\lambda)$, resp. $\ker(S-\lambda)$ and $P_2(\lambda)UP_1(\lambda)$ will map $P_1(\lambda)H$ injectively onto

$P_2(\lambda)H$. Now defining

$$B_0(\lambda) = F_2(\lambda)(\lambda - T) + P_2(\lambda)UP_1(\lambda), \quad \lambda \in G_0,$$

it is easy to check that $B_0 \in \mathrm{Hol}(G_0, \mathrm{Inv}(H))$ and

$$\lambda - T = (\lambda - S)B_0(\lambda), \quad \lambda \in G_0. \qquad \square$$

LEMMA 2.3. *Suppose that* T *has a matrix representation of the form*

$$T = \begin{pmatrix} T_{11} & T_{12} \\ 0 & T_{22} \end{pmatrix}, \quad H = H_1 \oplus H_2,$$

and $G \subset \rho_r(T_{11})$. *Then* $T \overset{G}{\sim} T_{11} \oplus T_{22}$.

PROOF. By the Corollary of [1], Theorem 1, we can find an analytic function F_1, defined in G and valued in $L(H_1)$, such that $(\lambda - T_{11})F_1(\lambda) = I_1$, $\lambda \in G$. Now putting

$$F(\lambda) = \begin{pmatrix} I_1 & F_1(\lambda)T_{12} \\ 0 & I_2 \end{pmatrix}, \quad \lambda \in G,$$

we can easily check that $F \in \mathrm{Hol}(G, \mathrm{Inv}(H))$ and

$$(T - \lambda)F(\lambda) = (T_{11} - \lambda) \oplus (T_{22} - \lambda), \quad \lambda \in G. \qquad \square$$

COROLLARY 2.4. *Suppose that* T *has a matrix representation of the form*

$$T = \begin{pmatrix} T_{11} & * & * \\ 0 & T_{22} & * \\ 0 & 0 & T_{33} \end{pmatrix}, \quad H = H_1 \oplus H_2 \oplus H_3,$$

and $G \subset \rho_r(T_{11}) \cap \rho_\ell(T_{22})$. *Then* $T \overset{G}{\sim} T_{11} \oplus T_{22} \oplus T_{33}$.

PROOF. We apply Lemma 2.2 two times, combined with Lemma 1.1. $\square$

THEOREM 2.5. *Let* $S \in L(H)$ *and suppose* $G^- \subset \rho_\gamma(T) \cap \rho_\gamma(S)$. *Then the following are equivalent*

(*i*) $T \overset{G}{\sim} S$

(*ii*) $\dim\ker(T-\lambda) = \dim\ker(S-\lambda)$, $\dim\ker(T-\lambda)^* = \dim\ker(S-\lambda)^*$, $\lambda \in G$.

PROOF. It suffices to prove the theorem when G is connected and $G \subset \sigma(T) \cap \sigma(S)$. Since by [4], Proposition 2.4 we also have $G \subset \sigma^r_{c.r.}(T) \cap \sigma^r_{c.r.}(S)$, applying [4], Theorem 1.5 we may suppose that T and S have the matrix representations

$$T = \begin{pmatrix} T_{11} & * \\ 0 & T_{22} \end{pmatrix}, \quad H = H_1 \oplus H_2,$$

$$S = \begin{pmatrix} S_{11} & * \\ 0 & S_{22} \end{pmatrix}, \quad H = H_1' \oplus H_2',$$

such that

$$G^- \subset \rho_r(T_{11}) \cap \rho_r(S_{11}) \cap \rho_\ell(T_{22}) \cap \rho_\ell(S_{22}),$$

$$\dim\ker(T_{11}-\lambda) = \dim\ker(S_{11}-\lambda), \qquad \lambda \in G,$$

$$\dim\ker(T_{22}-\lambda)^* = \dim\ker(S_{22}-\lambda)^*, \qquad \lambda \in G.$$

If either $G \subset \rho(T_{11}) \cap \rho(S_{11})$ or $G \subset \rho(T_{22}) \cap \rho(S_{22})$ we apply Lemma 2.2, and we apply Lemma 2.3 and then Lemma 2.2 otherwise.□

THEOREM 2.6. *Let* $S \in L(H)$ *and suppose* $G^- \subset \rho_{s-F}(T) \cap \rho_{s-F}(S)$. *Then* $S \in \{T\}_G$ *if and only if*

(*i*) $\dim\ker(T-\lambda)^n = \dim\ker(S-\lambda)^n$, $n \geq 1$,

(*ii*) $\dim\ker(T-\lambda)^{*n} = \dim\ker(S-\lambda)^{*n}$, $n \geq 1$.

PROOF. Let $S \in L(H)$ and assume that (i) and (ii) hold true. Since under our assumptions both $\rho^s_{s-F}(T) \cap G$ and $\rho^s_{s-F}(S) \cap G$ must be finite sets, applying [11], III, Theorem 3.38 (x) we may suppose that

$$T = T_1 \oplus T_2, \; H = H_1 \oplus H_2, \; S = S_1 \oplus S_2, \; H = H_1' + H_2',$$

where $G^- \subset \rho_\gamma(T_1) \cap \rho_\gamma(S_1)$, $\dim H_2 + \dim H_2' < \infty$ and $\sigma(T_2) \cup \sigma(S_2) \subset G$. It is an easy exercise to derive that (i) and (ii) imply $T_1 \overset{G}{\sim} S_1$ (via Theorem 2.5) and

$$\dim(T_2-\lambda)^n = \dim(S_2-\lambda)^n, \quad \lambda \in G.$$

Because H_2 and H_2' are finite dimensional spaces, the last relation, in turn implies $\dim H_2 = \dim H_2'$ as well as the fact that T_2 and S_2 are similar. Since the conditions (i) and (ii) are necessary by Lemma 2.1, the proof is concluded. □

PROPOSITION 2.7. *Let* $S \in L(H)$ *and suppose that*

$$T = T_1 \oplus T_2, \; H = H_1 \oplus H_2, \; \sigma(T_1) \subset G, \; G \cap \sigma(T_2) = \phi.$$

If $S \in \{T\}_G$, *then* S *is similar to* S', *where*

$$S' = S_1 \oplus S_2, \; H = H_1' \oplus H_2', \; G \cap \sigma(S_2) = \phi,$$

and S_1 *is similar to* T_1. *Moreover if* $\dim H_2 = \dim H_2'$ *and* S' *is as above, then we have* $S' \in \{T\}_G$.

PROOF. If $S \in \{T\}_G$, then applying Lemma 1.1, we may suppose (via a similarity) that we have

$$S = S_1 \oplus S_2, \; H = H_1' \oplus H_2', \; \sigma(S_1) = \sigma(T_1), \; G \cap \sigma(T_2) = \phi.$$

If we put $\sigma = \sigma(T_1)$, then each connected component of $\mathbf{C} \setminus \sigma$ will intersect both $\rho(T)$ and $\rho(S)$, thus $\sigma \in \mathcal{C}(T) \cap \mathcal{C}(S)$, and moreover we have

$$H_T(\sigma) = H_1 \oplus \{0\}, \; H_S(\sigma) = H_1' \oplus \{0\}.$$

Now using Lemma 1.2 and Corollary 1.3 we easily derive that S_1 is similar to T_1. If $\dim H_2 = \dim H_2'$ and S' is as in the statement, we may suppose that we have $H_1' = H_1$, $H_2' = H_2$, $S_1 = T_1$ and consequently $S' = T_1 \oplus S_2$. If we define B by the equation

$$B(\lambda) = I_1 \oplus (S_2-\lambda)^{-1}(T_2-\lambda), \quad \lambda \in G$$

we obviously have

$$B \in Hol(G, Inv(H)), \quad T - \lambda = (S-\lambda)B(\lambda), \quad \lambda \in G. \qquad \square$$

REMARK 2.8. Suppose that T is as in Proposition 2.7 and $S \in \{T\}$. J. Froelich proved that it is possible to have $\dim H_2 = 0 \neq \dim H_2'$? (Private communication.) We conclude this section observing that Lemma 2.1 holds true in Banach spaces (of course (ii) involves Banach space adjoints).

3. The Equivalence Class $\{T\}_\mu$.

Throughout this section $T \in L(H)$ will be fixed. For every $\mu \in \mathbf{C}$ we shall put

$$\{T\}_\mu = \{S \in L(H) : T \overset{\mu}{\sim} S\}$$

THEOREM 3.1. *Let* $\mu \in \rho_\gamma(T) \cup \rho^s_{s-F}(T)$. *Then* $S \in \{T\}_\mu$ *iff* $\mu \in \rho_\gamma(S) \cup \rho^s_{s-F}(T)$ *and*

$$\dim\ker(T-\mu)^n = \dim\ker(S-\mu)^n, \ \dim\ker(T-\mu)^{*n} =$$

$$= \dim\ker(S-\mu)^{*n}, \ n \geq 1.$$

PROOF. Suppose that $\mu \in \rho_\gamma(T)$ and S verifies the hypothesis of our theorem. Then by Lemma 1.1, $\mu \in \rho_\gamma(S)$ and we can find an open, connected set $G \ni \mu$ such that $G^- \subset \rho_\gamma(T) \cap \rho_\gamma(S)$. Further applying Theorem 2.5, we deduce $T \overset{G}{\sim} S$, and in particular $T \overset{\mu}{\sim} S$. If $\mu \in \rho^s_{s-F}(T)$, we argue as in Theorem 2.6, to deduce again $T \overset{\mu}{\sim} S$. Because the part "only if" is an easy consequence of Lemma 1.1 and Lemma 2.1, the proof is concluded. $\square$

Let $\varepsilon \geq 0$ be given and put

$$\sigma(\mu,\varepsilon) = \{\lambda \in \mathbf{C} : |\lambda - \mu| \leq \varepsilon\}.$$

Recall that if T is decomposable then $H_T(\sigma(\mu,\varepsilon))$ is a spectral maximal space of T and $\sigma(T|H_T(\sigma(\mu,\varepsilon))) \subset \sigma(\mu,\varepsilon) \cap \sigma(T)$ (see [6], Ch. 2).

PROPOSITION 3.2. *Let* $S \in \{T\}_\mu$ *and assume that both* T *and* S *are decomposable. Then there exists* $\varepsilon > 0$ *such that* $T|H_T(\sigma(\mu,\varepsilon))$ *is similar to* $S|H_S(\sigma(\mu,\varepsilon))$.

PROOF. Let $S \in \{T\}_\mu$ and let G be an open set such that $G \ni \mu$, $T \overset{G}{\sim} S$. Let $\varepsilon > 0$ be small enough to have $\sigma(\mu,\varepsilon) \subset G$. Taking $\sigma = \sigma(\mu,\varepsilon)$ and applying Lemma 1.2 and Corollary 1.3, we readily derive that $T|H_T(\sigma(\mu,\varepsilon))$ is similar to $S|H_S(\sigma(\mu,\varepsilon))$. □

LEMMA 3.3. *Let* $S \in L(H)$ *and assume that both* T *and* S *are decomposable,* $T|H_T(\sigma(\mu,\varepsilon))$ *is similar to* $S|H_S(\sigma(\mu,\varepsilon))$ *and* $\dim H_T(\sigma(\mu,\varepsilon))^\perp = \dim H_S(\sigma(\mu,\varepsilon))^\perp$. *Then* $S \in \{T\}_\mu$.

PROOF. Let us put

$$G = \{\lambda \in \mathbf{C} : |\lambda - \mu| < \varepsilon/2\}.$$

Applying [2], Proposition 3.3, we easily derive that T and S have matrix representations of the form

$$T = \begin{pmatrix} T_1 & * \\ 0 & T_2 \end{pmatrix}, \quad S = \begin{pmatrix} S_1 & * \\ 0 & S_2 \end{pmatrix}$$

where $T_1 = T|H_T(\sigma(\mu,E))$, $S_1 = S|H_S(\sigma(\mu,\varepsilon))$ and $G \cap (\sigma(T_2) \cup \sigma(S_2)) = \phi$. Further applying Lemma 1.1, (i) and Lemma 2.2, we know that we have

$$T \overset{G}{\sim} T_1 \oplus T_2, \quad S \overset{G}{\sim} S_1 \oplus S_2$$

and the proof will be concluded arguing as in the second part of the proof of Proposition 2.7. □

THEOREM 3.4. *Let* $S \in L(H)$ *and suppose that both* T *and* S *are spectral operators in Dunford's sense* ([7]). *Then the following are equivalent:*

(*i*) T *is similar* to S,

(*ii*) $T \overset{\mu}{\sim} S$ *for every* $\mu \in \mathbf{C}$.

PROOF. The implication (i) => (ii) is trivial thus assume that (ii) holds true. Applying [6], Ch. 2, 1.6 iii) we deduce that T and S are decomposable. Let $\{\sigma(\mu_k, \varepsilon_k)\}_{k=1}^{n}$ be a finite covering of $\sigma(T) \cup \sigma(S)$, such that $T|H_T(\sigma(\mu_k, \varepsilon_k))$ is similar to $S|H_S(\sigma(\mu_k, \varepsilon_k))$ (see Lemma 3.2). If E_1, E_2 denote the spectral measures of T, resp. S and if σ_k is a Borel set $\sigma_k \subset \sigma(\mu_k, \varepsilon_k)$ then it is an easy exercise to prove that $T|E_1(\sigma_k)H$ must be similar to $S|E_2(\sigma_k)H$. Now choosing $\sigma_k \subset \sigma(\mu, \varepsilon_k)$ such that $\bigcup_{k=1}^{n} \sigma_k = \sigma(T) \cup \sigma(S)$ and $\sigma_k \cap \sigma_j = \phi$, for $j \neq k$, we can construct piecewise the needed similarity. □

PROPOSITION 3.5. *Let* $S \in L(H)$. *Then the following are equivalent:*

(*i*) $T \overset{\mu}{\sim} S$ *for every* $\mu \in \mathbf{C}$,

(*ii*) $T \overset{\mu}{\sim} S$ *for every* $\mu \in \sigma_\gamma(T) \cup \sigma_\gamma(S)$.

PROOF. Suppose that (ii) holds true. Then by Lemma 1.1 we derive $\sigma_\gamma(T) = \sigma_\gamma(S)$. Let $\mu_0 \in \rho_\gamma(T)$, let G_0 denote the component of $\rho_\gamma(T)$ that contains μ_0 and let $\mu \in \partial G_0$. Since $T \overset{\mu}{\sim} S$ we can find an open set G such that $T \overset{G}{\sim} S$ and we can pick $\mu_1 \in G_0 \cap G$. Because by [4], Proposition 2.4 we have

$$\dim\ker(T-\lambda) = \dim\ker(T-\mu_1), \quad \lambda \in G_0,$$

$$\dim\ker(S-\lambda) = \dim\ker(S-\mu_1), \quad \lambda \in G_0,$$

using Lemma 2.1 we derive

$$\dim\ker(T-\lambda) = \dim\ker(S-\lambda), \quad \lambda \in G_0.$$

We analogously show that we have

$$\dim\ker(T-\lambda)^* = \dim\ker(S-\lambda)^*, \quad \lambda \in G_0$$

and then (i) is a consequence of Theorem 2.5. The implication

(i) => (ii) is trivial. □

REMARK 3.6. The only positive answer to the question mentioned in introduction is contained in Theorem 3.4. Note that even this particular result is obtained assuming (see also Proposition 3.2 and Lemma 3.2) that both T and S share "good" spectral properties. Excepting Lemma 3.3 and Proposition 3.5, the results of this section are valied in Banach spaces.

4. A Conjecture

Throughout this section $T, S \in L(H)$ will be fixed operators and G will denote an open set containing $\sigma_\gamma(T) \cup \sigma_\gamma(S)$. We shall assume that $T \overset{G}{\sim} S$ and this equivalence on G is implemented by

$$T - \lambda = A(\lambda)(S-\lambda)B(\lambda), \quad \lambda \in G,$$

for some $A, B \in Hol(G, Inv(H))$. In view of the results contained in §2 and §3, it seems reasonable to conjecture that T <u>is similar to</u> S. Using Proposition 3.5 we observe that we have $T \overset{\mu}{\sim} S$, for every $\mu \in \mathbf{C}$, thus if our conjecture is false, then the question of Gohberg, Kaashoek and Lay ([9]) has a negative answer. In the remainder of this section we shall support the above conjecture by proving its validity for a particular but non-trivial choice of T. More precisely we have

THEOREM 4.1. *If T is a forward shift of finite multiplicity then T is similar to S.*

The proof of the theorem will be given at the end after the presentation of some prelimiaries. Unfortunately our proof gives little hope for a significant generalization.

Let $r \geq 0$ *and put*

$$\Gamma_r = \{\lambda \in \mathbf{C} : |\lambda| = r\}.$$

If $0 \leq a < b < \infty$ *lets also put*

$$G_{a,b} = \{\lambda \in \mathbf{C} : a < |\lambda| < b\}$$

LEMMA 4.2. *Suppose that* $\|T\| \leq 1$ *and* $G \supset G^{-}_{a,b}$, *for some* $0 < a < 1 < b$. *Then* S *is power bounded.*

PROOF. If A and B have the series expansions

$$A(\lambda) = \sum_{k=-\infty}^{\infty} \lambda^k A_k \quad , \quad B(\lambda) = \sum_{k=-\infty}^{\infty} \lambda^k B_k$$

and if we put $r = \max\{a, b^{-1}\}$, we can find $M > 0$ such that

$$\|A_k\| \leq Mr^{|k|} \quad , \quad \|B_k\| \leq Mr^{|k|}.$$

On the other hand because $\|T\| \leq 1$ and $T \overset{G}{\sim} S$ we easily derive $|S|_{sp} \leq 1$ (see Proposition 3.5) and

$$S^n = \frac{1}{2\pi i} \int_{\Gamma_r} \lambda^n B(\lambda)(\lambda - T)^{-1} A(\lambda) d\lambda = \sum_{k=0}^{\infty} \sum_{m=-\infty}^{\infty} B_{k-n-m} T^k A_m,$$

$n \geq 0$. This implies

$$\|S^n\| \leq \sum_{k=0}^{\infty} \sum_{m=-\infty}^{\infty} \|B_{k-n-m}\| \|A_m\| \leq M^2 \sum_{k=0}^{\infty} \sum_{m=-\infty}^{\infty} r^{|k-n-m|+|m|} \leq$$

$$\leq M^2 \sum_{k=0}^{\infty} \left(\frac{1+r^2}{1-r^2} + |n-k|\right) r^{|n-k|} \leq 4M^2 \frac{1+r^2}{r(1-r)^2}. \qquad \square$$

Observe that if T is a shift we have $\sigma_\gamma(T) = \sigma_\gamma(S) = \Gamma_1 \subset G \cap G^*$.

LEMMA 4.3. *Suppose that* T *is a forward shift of finite multiplicity and choose* $0 < r < 1$ *such that* $G^{-}_{r,1/r} \subset G$. *Let us put* $\Gamma = \Gamma_r \cup \Gamma_{1/r}$ *and*

$$F(\lambda) = \begin{cases} (\lambda - T^*)^{-1}, & |\lambda| > 1 \\ -\sum_{k=0}^{\infty} \lambda^k T^{k+1}, & |\lambda| < 1 \end{cases} \quad , \quad M = \frac{1}{2\pi i} \int_{\Gamma} A^*(\lambda) F(\lambda) d\lambda.$$

Then M *is a Fredholm operator and* $S^*MT = M$.

PROOF. Define $N \in L(H)$ by the equation

$$N = \frac{1}{2\pi i} \int_{\Gamma} F(\lambda) B^*(\lambda) d\lambda.$$

Using the relations

$$T^* - \lambda = B^*(\lambda)(S^*-\lambda)A^*(\lambda), \quad \widetilde{F(\lambda)} = (\lambda-\tilde{T}^*)^{-1}, \quad \lambda \in \Gamma,$$

and arguing as in the proof of Lemma 1.2, we deduce $\tilde{M}\tilde{N} = \tilde{N}\tilde{M} = I$, thus M is a Fredholm operator. Because $F(\lambda)$ is a right inverse of $\lambda - T^*$ and we have

$$S^*A^*(\lambda) = (S^*-\lambda)A^*(\lambda) + A^*(\lambda)(\lambda-T^*) + A^*(\lambda)T^* =$$

$$= [A^*(\lambda) - B^*(\lambda)^{-1}](\lambda-T^*) + A^*(\lambda)T^*,$$

we deduce

$$S^*MT = \frac{1}{2\pi i}\int_\Gamma S^*A^*(\lambda)F(\lambda)Td\lambda = \frac{1}{2\pi i}\int_\Gamma A^*(\lambda)T^*F(\lambda)Td\lambda.$$

Finally, using the relations

$$T^*F(\lambda)T = T^*TF(\lambda) = F(\lambda), \quad \lambda \in \Gamma_r,$$

$$T^*F(\lambda)T = F(\lambda)T^*T = F(\lambda), \quad \lambda \in \Gamma_{1/r},$$

we get

$$S^*MT = \frac{1}{2\pi i}\int_\Gamma A^*(\lambda)T^*F(\lambda)Td\lambda = \frac{1}{2\pi i}\int_\Gamma A^*(\lambda)F(\lambda)d\lambda = M. \square$$

THE PROOF OF THEOREM 4.1. Let $\operatorname{L|M}_{n\to\infty}$ be a Banach limit, as defined in [8], 11.4.22. Since T is a shift of finite multiplicity and by lemma 4.2, S is power bounded, the equation

$$\langle D^2x,y\rangle = \operatorname{L|M}_{n\to\infty} \langle S^{*n}S^nx,y\rangle, \quad x, \quad y \in H,$$

defines a positive bounded operator $D \in L(H)$. Let M be defined as in Lemma 4.3. Because M is Fredholm and

$$\|M^*x\|^2 = \|T^{*n}M^*S^nx\|^2 \leq \|M\|^2 \langle S^{*n}S^nx,x\rangle, \quad x \in H,$$

we deduce $MM^* \leq \|M\|^2 D^2$, consequently D must be Fredholm. Let $x_0 \in \ker D$. Since we have $T^{*m}M^*S^m = T^{*n+m}M^*S^{n+m} = M^*$ we derive

$$\|M^*S^m x_0\|^2 \leq \|M\|^2 \langle S^{*n+m}S^{n+m}x_0, x_0\rangle$$

and this implies

$$\|M^*S^m x_0\| \leq \|M\|^2 \|Dx_0\|^2 = 0, \quad m \geq 1.$$

But because $\dim\ker M^* < \infty$ and $\sigma_\rho(S) = \phi$ (see Proposition 3.5) we must have $x_0 = 0$; therefore D is invertible. Now using the relation

$$\|DSx\|^2 = \lim_{n\to\infty} \langle S^{*n+1}S^{n+1}x, x\rangle = \|Dx\|^2, \quad x \in H$$

we infer that DSD^{-1} is an isometry of the same index as T (see Lemma 2.1). To conclude the proof we have to show that DSD^{-1} is a pure isometry. To this aim assume to the contrary that DSD^{-1} has a unitary part and consequently $H_{DSD^{-1}}(\Gamma_1) \neq \{0\}$. In particular we have $H_S(\Gamma_1) \neq \{0\}$ and because Γ_1 is a set of analytic uniqueness for both T and S (via $\sigma_\rho(T) = \sigma_\rho(S) = \phi$), Lemma 1.2 implies $H_T(\Gamma_1) \neq \{0\}$, an obvious contradiction. □

REMARK 4.4. Under the hypothesis of Theorem 4.1 we have

$$\sigma_\gamma(T) = \sigma(T) \setminus \rho^r_{s-F}(T) = \sigma(T) \setminus \rho^r_F(T).$$

In general, if T is an arbitrary operator we have

$$\sigma_\gamma(T) \subset \sigma(T) \setminus \rho^r_{s-F}(T) \subset \sigma(T) \setminus \rho^r_F(T).$$

If $T \overset{G}{\sim} S$ and $G \supset (\sigma(T) \setminus \rho^r_F(T)) \cup (\sigma(S) \setminus \rho^r_F(S))$ we in particular have $G \supset \sigma_\gamma(T) \cup \sigma_\gamma(S)$. Is, in this case, T similar to S? Note that the relation

$$\sigma_\gamma(T) = \sigma(T) \setminus \rho^r_F(T)$$

played an important role in Theorem 4.1, when we proved that M is a Fredholm operator. We conclude by observing that Lemma 4.2 is valid in Banach spaces.

REFERENCES

1. Allen, G.R.: Holomorphic vector-valued functions on a domain of holomorphy, J. London Math. Soc., 42 (1967), 509-513.

2. Apostol, C.: Spectral decompositions and functional calculus, Rev. Roum. Math. Pures et Appl., 13 (1968), 1481-1528.

3. Apostol, C.: The correction by compact perturbation of the singular behaviour of operators, Rev. Roum. Math. Pures et Appl., 21 (1976), 155-175.

4. Apostol, C.: The reduced minimum modulus, Michigan Math. J., 32 (1985), 279-284.

5. Apostol, C., Fialkow, L., Herrero, D.A., and Voiculescu, D.: Approximation of Hilbert Space Operators, II, Pitman, London-Boston-Melbourne, 1984.

6. Colojoara, I. and Foias, C.: Theory of Generalized Spectral Operators, Gordon and Breach, New York, 1968.

7. Dunford, N.: Spectral operators, Pacific J. Math., 4 (1954), 321-354.

8. Dunford, N. and Schwartz, J.T.: Linear Operators, I, Interscience Publishers, New York, 1957.

9. Gohberg, I., Kaashoek, M.A., and Lay, D.C.: Spectral classification of operators and operator functions, Bull. Amer. Math. Soc., 82 (1976), 587-589.

10. Gohberg, I., Lancaster, P., and Rodman, L.: Matrix Polynomials, Academic Press, New York, 1982.

11. Herrero, D.A.: Approximation of Hilbert Space Operators,I, Pitman, London-Boston-Melbourne, 1982.

12. Kaashoek, M.A., van der Mee, C.V.M., and Rodman, L.: Analytic operator functions with compact support, I, Int. Eq. Op. Th. 4 (1981), 504-547.

13. Kato, T.: Perturbation Theory for Linear Operators, Springer-Verlag, New York, 1966.

14. Leiterer, J.: Local and global equivalence of meromorphic operator functions, I and II, Math. Nachr. 83 (1978), 7-29 and 84 (1978), 145-170.

15. Vasilescu, F.H.: Analytic Functional Calculus, Editura Academiei & D. Reidel Publ. Comp., 1982.

Department of Mathematics
Arizona State University
Tempe, Arizona 85287, USA

Operator Theory:
Advances and Applications, Vol. 32

FREDHOLM PROPERTIES OF BAND MATRICES AND DICHOTOMY

A. Ben-Artzi and I. Gohberg

Dedicated to the memory of Constantin Apostol.

The invertibility and Fredholm properties of a regular block band matrix are described in terms of the dichotomy of the companion sequence.

1. INTRODUCTION

The following theorem is a corollary of the main results of this paper.

THEOREM 1.1. *Let* $A = (a_{ij})_{i,j=-\infty}^{\infty}$ *be a block matrix, the entries of which are* $r \times r$ *complex matrices with the following properties:*

a) $a_{ij} = 0$ *if* $i < j - m_1$ *or* $i > j + m_2$, *where* m_1 *and* m_2 *are some nonnegative integers, not both zero.*

b) $\sup_{ij} \|a_{ij}\| < +\infty$.

c) $a_{j-m_1,j}$ and $a_{j+m_2,j}$ *are invertible matrices for all* $j = 0, \pm 1, \ldots$.

d) $\sup_j \|a_{j-m_1,j}^{-1}\| < +\infty$.

The matrix A *defines an invertible operator on* $\ell_r^2(\mathbb{Z})$ *if and only if the following sequence of* $(m_1 + m_2)r \times (m_1 + m_2)r$ *matrices*

$$(1.1) \qquad C_n = \begin{pmatrix} 0 & 0 & \cdots & 0 & -a_{n-m_1,n}a_{n+m_2,n}^{-1} \\ I & 0 & \cdots & 0 & -a_{n-m_1+1,n}a_{n+m_2,n}^{-1} \\ 0 & I & \cdots & 0 & -a_{n-m_1+2,n}a_{n+m_2,n}^{-1} \\ \cdot & \cdot & \cdots & \cdot & \cdot \\ 0 & 0 & \cdots & I & -a_{n+m_2-1,n}a_{n+m_2,n}^{-1} \end{pmatrix} \qquad (n = 0, \pm 1, \cdots)$$

where $I = (\delta_{ij})_{i,j=1}^{r}$, *admits a left dichotomy. If* A *is invertible, then* $G = (a_{ij})_{i,j=0}^{\infty}$ *defines in* ℓ_r^2 *a Fredholm operator with*

$$\text{index}(G) = -p + m_1 r$$

where p *is the rank of the dichotomy. Furthermore, for an invertible operator* A *the following estimate holds:*

$$\|A^{-1}\| \leq 2M^2(1-a)^{-1} \sup_j \|a_{j-m_1,j}^{-1}\|$$

where (a, M) is a bound of the dichotomy.

To make this statement clear, it remains to define dichotomy of a sequence of matrices. Let $(A_n)_{n=-\infty}^{\infty}$ be a sequence of $h \times h$ invertible matrices. A sequence $(P_n)_{n=-\infty}^{\infty}$ of projections in $\mathbb{C}^h$, is called a left dichotomy for $(A_n)_{n=-\infty}^{\infty}$ if the following conditions hold:

$$P_n A_n = A_n P_{n+1} \qquad (n = 0, \pm 1, \dots), \tag{1.2}$$

$$\|P_n A_n \cdots A_{n+j-1}\| \leq M a^j \qquad (n = 0, \pm 1, \dots; j = 0, 1, \dots), \tag{1.3}$$

and

$$\|(I_h - P_n) A_{n-1}^{-1} \cdots A_{n-j}^{-1}\| \leq M a^j \qquad (n = 0, \pm 1, \dots; j = 0, 1, \dots), \tag{1.4}$$

where a and M are positive numbers such that $a < 1$, and $I_h = (\delta_{ij})_{i,j=1}^{h}$. Such a pair of numbers (a, M) is called a bound of the dichotomy, and the number rank (P_n), which, by (1.2), does not depend on n, is called the rank of the dichotomy. A description of dichotomy and its properties can be found in [1], which also treats the noninvertible case.

In general, an infinite block matrix $(a_{ij})_{i,j=-\infty}^{\infty}$ or $(a_{ij})_{i,j=0}^{\infty}$ will be considered (when dealing with boundedness, invertibility or Fredholm properties) as an operator acting in $\ell_r^2(\mathbb{Z})$ or ℓ_r^2, respectively. It is easily seen that the theorems in this paper still hold if $\ell_r^2(\mathbb{Z})$ and ℓ_r^2 are replaced by many other Banach spaces, for example, $\ell_r^p(\mathbb{Z})$ and ℓ_r^p $(1 \leq p \leq \infty)$.

This paper is divided into six sections. The first is the introduction. In the second we state the main results, which are Theorems 2.1 and 2.2. The third section contains a theorem on exponential decay of the entries of the inverse of a band matrix far away from the main diagonal. The main part of the paper is contained in Sections 4 and 5. In Section 4 it is shown that dichotomy of the companion sequence is a necessary condition for the invertibility of a band matrix. In Section 5 the proof of Theorem 2.1 is completed. The last section contains the proof of Theorem 2.2, which deals with Fredholm properties, and ends with the proof of Theorem 1.1, stated above. Applications of the results of this paper will appear in [2].

Finally, let us mention that the results of this paper are related to some results of C. de Boor [3]. For the case of a bounded band matrix, our results about the relationship of dichotomy to invertibility and Fredholm properties are more complete.

2. THE MAIN RESULTS

Theorem 1.1 of the introduction will be derived from the main results of this paper, which are Theorems 2.1 and 2.2, stated below.

Let $A = (a_{ij})_{i,j=-\infty}^{\infty}$ be a block matrix, the entries of which are $r \times r$ complex matrices satisfying conditions a), b) and c) of Theorem 1.1. We call such a matrix, a regular (m_1, m_2)-band matrix. For any regular (m_1, m_2)-band matrix we call the sequence of matrices $(C_n)_{n=-\infty}^{\infty}$, defined by the equalities (1.1), the companion sequence of A.

THEOREM 2.1. *Let* $A = (a_{ij})_{i,j=-\infty}^{\infty}$ *be a regular* (m_1, m_2)*-band matrix. Then* A *defines an invertible operator on* $\ell_r^2(\mathbb{Z})$ *if and only if its companion sequence admits a left dichotomy* $(P_n)_{n=-\infty}^{\infty}$ *such that*

$$K_1 = \sup_n \|(a_{n-m_1,n}^{-1}, \dots, 0)(I - P_n)\| < +\infty \tag{2.1}$$

and

$$K_2 = \sup_n \|(0, \dots, 0, a_{n+m_2,n}^{-1})P_{n+1}\| < +\infty. \tag{2.2}$$

Moreover, if A *defines an invertible operator on* $\ell_r^2(\mathbb{Z})$ *then a bound of the dichotomy is given by*

$$(a, M) = \left(1 - \frac{1}{16m^2\|A^{-1}\|\|A\|}, 16m^2\|A^{-1}\|\|A\|\right), \tag{2.3}$$

where $m = \max(m_1, m_2)$*, and the following estimate holds:*

$$\max(K_1, K_2) \le \|A^{-1}\| \le M(1-a)^{-1}(K_1 + K_2), \tag{2.4}$$

for any bound (a, M) *of the dichotomy.*

In the following statement, the existence of a dichotomy follows from the previous result.

THEOREM 2.2. *Let* $A = (a_{ij})_{i,j=-\infty}^{\infty}$ *be a regular* (m_1, m_2)*-band matrix which defines an invertible operator in* $\ell_r^2(\mathbb{Z})$*. Then the matrix* $G = (a_{ij})_{i,j=0}^{\infty}$ *defines a Fredholm operator in* ℓ_r^2 *and*

$$\operatorname{index}(G) = -p + m_1 r \tag{2.5}$$

where p *is the rank of the dichotomy of the companion sequence of* A.

3. A THEOREM ON INVERSES OF BAND MATRICES

THEOREM 3.1. *Let* $A = (a_{ij})_{i,j=-\infty}^{\infty}$ *be a block matrix whose entries* a_{ij} $(i, j = 0, \pm 1, \dots)$ *are* $r \times r$ *complex matrices, with* $a_{ij} = 0$ *if* $|i - j| > m$*, where* m *is some fixed positive integer. Assume that* A *defines a bounded invertible operator on* $\ell_r^2(\mathbb{Z})$*, and let* $A^{-1} = (b_{ij})_{i,j=-\infty}^{\infty}$*. Then*

$$\|b_{ij}\| \le 2\|A^{-1}\|a^{|i-j|} \qquad (i, j = 0, \pm 1, \dots) \tag{3.1}$$

for

$$a = 1 - \frac{1}{16m^2\|A^{-1}\|\|A\|}. \tag{3.2}$$

PROOF. We define $2m+1$ operators in $\ell_r^2(\mathbb{Z})$, which have at most one nonzero diagonal, as follows

$$A_k = (\delta_{k,j-i}a_{i,j})_{i,j=-\infty}^{\infty} \qquad (k = -m, \ldots, m). \tag{3.3}$$

Consider now the rational operator function given by

$$A(z) = \sum_{k=-m}^{m} z^k A_k \qquad (z \in \mathbb{C}).$$

This rational function is analytic at every complex number z, except possibly $z = 0$. It is clear from the band structure of A and the definition (3.3) of the coefficients $(A_k)_{k=-m}^{m}$ that

$$A(z) = (z^{j-i}a_{ij})_{i,j=-\infty}^{\infty} \qquad (z \neq 0).$$

As a consequence of this equality we obtain

$$A(z) = [((z/|z|)^{-i}\delta_{i,j})_{ij=-\infty}^{\infty}]A(|z|)[((z/|z|)^{j}\delta_{i,j})_{ij=-\infty}^{\infty}]. \tag{3.4}$$

Since $A(1) = A$, the previous equality implies that $A(z)$ is invertible for every z such that $|z| = 1$. Thus, there exists an annulus $\Omega = \{z: \rho_1 \leq |z| \leq \rho_2\}$, with $0 < \rho_1 < 1 < \rho_2$, such that $A(z)$ is invertible for $z \in \Omega$.

Since $A(z)$ is analytic on Ω, so is $A(z)^{-1}$. Let the series

$$A(z)^{-1} = \sum_{k=-\infty}^{\infty} z^k B_k \qquad (\rho_1 \leq |z| \leq \rho_2)$$

be the Laurent expansion of $A(z)^{-1}$ in Ω. The coefficients B_k $(k = 0, \pm1, \ldots)$ are operators in $\ell_r^2(\mathbb{Z})$, which are given by the formulas

$$B_k = \frac{1}{2\pi i}\int_{|z|=\rho} z^{-1-k}A(z)^{-1}dz$$

where ρ is an arbitrary number satisfying $\rho_1 \leq \rho \leq \rho_2$. Let $L = \sup_{z\in\Omega}\|A(z)^{-1}\|$. Then from the previous integral representation we obtain the estimate $\|B_k\| \leq L\rho^{-k}$ $(k = 0, \pm1, \ldots)$. Letting $\rho = \rho_1$ and $\rho = \rho_2$ in the last inequality, we obtain

$$\|B_k\| \leq L(\max(\rho_1, \rho_2^{-1}))^{|k|} \qquad (k = 0, \pm1, \ldots).$$

On the other hand, equality (3.4) implies that for $|z| = 1$,

$$A(z) = [(z^{-i}\delta_{i,j})_{ij=-\infty}^{\infty}]A[(z^{j}\delta_{i,j})_{ij=-\infty}^{\infty}].$$

Therefore

$$A(z)^{-1} = (z^{j-i} b_{ij})_{i,j=-\infty}^{\infty} \qquad (|z| = 1).$$

It follows immediately that

$$B_k = (\delta_{k,j-i} b_{ij})_{i,j=-\infty}^{\infty} \qquad (k = 0, \pm 1, \ldots),$$

and hence

$$\|b_{ij}\| \le L\big(\max(\rho_1, \rho_2^{-1})\big)^{|j-i|} \qquad (i, j = 0, \pm 1, \ldots).$$

We now proceed to obtain the estimate (3.1), with the number a given by (3.2). Let h be any integer such that $1 \le h \le m$. Since $0 < a < 1$, it is clear that

$$1 - a^h = (1 + a + \cdots + a^{h-1})(1 - a) < h(1 - a) \le \frac{1}{16m\|A^{-1}\|\|A\|}.$$

This inequality implies that $\frac{1}{2} < a^h$, and that

$$a^{-h} - 1 = a^{-h}(1 - a^h) < 2(1 - a^h) < \frac{1}{8m\|A^{-1}\|\|A\|}.$$

The last two inequalities imply that

$$|a^{\pm k} - 1| < \frac{1}{8m\|A^{-1}\|\|A\|} \qquad (k = -m, \ldots, m).$$

Now let z be a complex number in the annulus $\{z : a \le |z| \le a^{-1}\}$. By the previous inequality it follows that

$$\big||z|^{\pm k} - 1\big| < \frac{1}{8m\|A^{-1}\|\|A\|} \qquad (k = -m, \ldots, m).$$

Therefore

$$\|A(|z|) - A\| = \|A(|z|) - A(1)\| = \left\| \sum_{k=-m}^{m} (|z|^k - 1) A_k \right\| \le$$

$$\le \frac{1}{8m\|A^{-1}\|\|A\|}(2m + 1)\|A\| \le \frac{1}{2\|A^{-1}\|}.$$

Here we have used the inequalities $\|A_k\| \le \|A\|$ $(k = -m, \ldots, m)$, which follow from (3.3). The last inequality implies that $A(|z|)$ is invertible and that $\|A(|z|)^{-1}\| \le 2\|A^{-1}\|$. On the other hand, equality (3.4) implies that $A(z)$ is unitarily equivalent to $A(|z|)$. Therefore, $A(z)$ is invertible and $\|A(z)^{-1}\| \le 2\|A^{-1}\|$.

Thus, we have shown that $A(z)$ is invertible for $z \in \Omega = \{z : a \le |z| \le a^{-1}\}$ and that

$$\sup_{z \in \Omega} \|A(z)^{-1}\| \le 2\|A^{-1}\|.$$

This allows us to specify ρ_1 and ρ_2 in the following way: $\rho_1 = a$ and $\rho_2 = a^{-1}$, and hence

$$\|b_{ij}\| \leq \sup_{z\in\Omega} \|A(z)^{-1}\| a^{|j-i|} \qquad (i,j = 0, \pm 1, \ldots).$$

Inequality (3.1) follows from the last two inequalities. □

4. DICHOTOMY

In this section, we prove the following result.

THEOREM 4.1. *Let $A = (a_{ij})_{i,j=-\infty}^{\infty}$ be a regular (m_1, m_2)-band matrix. If A is invertible, then its companion sequence admits a left dichotomy, and a bound for the dichotomy is given by*

$$(4.1) \qquad (a, M) = \left(1 - \frac{1}{16m^2\|A^{-1}\|\|A\|}, 16m^2\|A^{-1}\|\|A\|\right)$$

where $m = \max(m_1, m_2)$.

Here, and in the rest of the paper, the vectors are row vectors, and they multiply matrices from the left. Thus, if D is a matrix, then $\operatorname{Im}(D)$ is the linear span of the rows of D. If P is a projection, then $\operatorname{Ker} P$ is the space of row vectors x such that $xP = 0$. In general, we will represent a vector $x \in \ell_r^2(\mathbb{Z})$ (respectively, ℓ_r^2) by $x = (x_i)_{i=-\infty}^{\infty}$ (respectively, $x = (x_i)_{i=0}^{\infty}$) where each x_i is a row vector in $\mathbb{C}^r$. In the proof of Theorem 4.1 we will use the following lemma.

LEMMA 4.2. *Let $A = (a_{ij})_{i,j=-\infty}^{\infty}$ be a regular (m_1, m_2)-band matrix which is invertible and let $(b_{ij})_{i,j=-\infty}^{\infty} = A^{-1}$. Then for every integer n, the two subspaces*

$$(4.2) \qquad T_n = \operatorname{span}\{\operatorname{Im}(b_{n+i,n-m_1}, \ldots, b_{n+i,n+m_2-1})\}_{i=0}^{m_1+m_2-1}$$

and

$$(4.3) \qquad S_n = \operatorname{span}\{\operatorname{Im}(b_{n-i,n-m_1}, \ldots, b_{n-i,n+m_2-1})\}_{i=1}^{m_1+m_2}$$

satisfy $T_n \cap S_n = \{0\}$ and $T_n + S_n = \mathbb{C}^{r(m_1+m_2)}$. Additionally, for every vector $v \in \mathbb{C}^{r(m_1+m_2)}$, there exists vectors $u_i \in \mathbb{C}^r$ $(i = n - m_1 - m_2, \ldots, n + m_1 + m_2 - 1)$ such that

$$(4.4) \qquad v = \sum_{i=n-m_1-m_2}^{n+m_1+m_2-1} u_i(b_{i,n-m_1}, \ldots, b_{i,n+m_2-1})$$

and

$$(4.5) \qquad \|u_i\| \leq \|v\|\|A\| \qquad (i = n - m_1 - m_2, \ldots, n + m_1 + m_2 - 1).$$

PROOF. We first prove the second part of the lemma. Let $v = (v_0, \dots, v_{m_1+m_2-1})$ be the block decomposition of v, where $v_i \in \mathbb{C}^r$ $(i = 0, \dots, m_1 + m_2 - 1)$. We define a vector $x = (x_k)_{k=-\infty}^{\infty} \in \ell_r^2(\mathbb{Z})$ via

$$x_k = v_{k-n+m_1} \qquad (k = n - m_1, \dots, n + m_2 - 1) \tag{4.6}$$

and

$$x_k = 0 \qquad (k \neq n - m_1, \dots, n + m_2 - 1). \tag{4.7}$$

We also define $u = xA \in \ell_r^2(\mathbb{Z})$, and let $u = (u_k)_{k=-\infty}^{\infty}$ be its block decomposition. The band property of A and equalities (4.7) imply that $u_k = 0$ if $k \geq n + m_1 + m_2$ or $k < n - m_1 - m_2$. Therefore, we obtain from the equality $x = uA^{-1}$ that

$$x_k = \sum_{i=n-m_1-m_2}^{n+m_1+m_2-1} u_i b_{i,k} \qquad (k = 0, \pm 1, \dots).$$

Thus, in particular,

$$(x_{n-m_1}, \dots, x_{n+m_2-1}) = \sum_{i=n-m_1-m_2}^{n+m_1+m_2-1} u_i (b_{i,n-m_1}, \dots, b_{i,n+m_2-1}).$$

However the vector on the left of this equality is equal to v by (4.6). This proves (4.4). The inequalities (4.5) follows also from $u = xA$.

The existence of vectors $(u_i)_{i=n-m_1-m_2}^{n+m_1+m_2-1}$, satisfying (4.4), proves, in particular, that $T_n + S_n = \mathbb{C}^{r(m_1+m_2)}$.

We now prove that $T_n \cap S_n = \{0\}$. Let $w \in T_n \cap S_n$. By definition, there are vectors $\alpha_i \in \mathbb{C}^r$ $(i = 0, \dots, m_1 + m_2 - 1)$ and $\beta_i \in \mathbb{C}^r$ $(i = 1, \dots, m_1 + m_2)$ such that

$$\begin{aligned} w = & \sum_{i=0}^{m_1+m_2-1} \alpha_i (b_{n+i,n-m_1}, \dots, b_{n+i,n+m_2-1}) = \\ = & \sum_{i=1}^{m_1+m_2} \beta_i (b_{n-i,n-m_1}, \dots, b_{n-i,n+m_2-1}). \end{aligned}$$

Define a vector $x = (x_k)_{k=-\infty}^{\infty} \in \ell_r^2(\mathbb{Z})$ via

$$x_k = \sum_{i=0}^{m_1+m_2-1} \alpha_i b_{n+i,k} \qquad (k = n-1, n-2, \dots)$$

and

$$x_k = \sum_{i=1}^{m_1+m_2} \beta_i b_{n-i,k} \qquad (k = n, n+1, \dots).$$

It is easily verified that

$$x_{k-m_1} a_{k-m_1,k} + \cdots + x_{k+m_2} a_{k+m_2,k} = 0 \qquad (k = 0, \pm 1, \ldots).$$

Since A is invertible, this shows that $x_k = 0$ for each $k = 0, \pm 1, \ldots$. But then $w = (x_{n-m_1}, \ldots, x_{n+m_2-1}) = 0$. □

PROOF OF THEOREM 4.1. Let $B = (b_{ij})_{i,j=-\infty}^{\infty} = A^{-1}$. We define the projection P_n on S_n along T_n, where n is an integer and S_n and T_n are as in Lemma 4.2.

Let n be an arbitrary integer. The equalities

$$b_{k,n-m_1} a_{n-m_1,n} + \cdots + b_{k,n+m_2} a_{n+m_2,n} = 0 \qquad (k \neq n)$$

imply that

$$(4.8) \qquad (b_{k,n-m_1}, \ldots, b_{k,n+m_2-1}) C_n = (b_{k,n-m_1+1}, \ldots, b_{k,n+m_2}) \qquad (k \neq n).$$

Consequently, we have for $j = 0, 1, \ldots$ and $k = n-1, n-2, \ldots$

$$(4.9) \qquad (b_{k,n-m_1}, \ldots, b_{k,n+m_2-1}) C_n \cdots C_{n+j-1} = (b_{k,n-m_1+j}, \ldots, b_{k,n+m_2-1+j}),$$

while for $k = n, n+1, \ldots$

$$(4.10) \qquad (b_{k,n-m_1}, \ldots, b_{k,n+m_2-1}) C_{n-1}^{-1} \cdots C_{n-j}^{-1} = (b_{k,n-m_1-j}, \ldots, b_{k,n+m_2-1-j}).$$

Let v be an arbitrary vector in $\mathbb{C}^{r(m_1+m_2)}$, and let $u_{n-m_1-m_2}, \ldots, u_{n+m_1+m_2-1}$ be as in the preceding lemma. It follows from (4.4) that

$$vP_n = \sum_{i=n-m_1-m_2}^{n-1} u_i (b_{i,n-m_1}, \ldots, b_{i,n+m_2-1}),$$

and

$$v(I - P_n) = \sum_{i=n}^{n+m_1+m_2-1} u_i (b_{i,n-m_1}, \ldots, b_{i,n+m_2-1}).$$

Thus, (4.9) and (4.10) imply that for $j = 0, 1, \ldots$,

$$vP_n C_n \cdots C_{n+j-1} = \sum_{i=n-m_1-m_2}^{n-1} u_i (b_{i,n-m_1+j}, \ldots, b_{i,n+m_2-1+j}),$$

and

$$v(I - P_n) C_{n-1}^{-1} \cdots C_{n-j}^{-1} = \sum_{i=n}^{n+m_1+m_2-1} u_i (b_{i,n-m_1-j}, \ldots, b_{i,n+m_2-1-j}).$$

Taking into account the inequalities (3.1) and (4.5), we obtain the estimates

$$\|vP_nC_n\cdots C_{n+j-1}\| \leq \|v\|\|A\|2\|A^{-1}\|a^{j-m_1+1}(2m)^2 \qquad (j=0,1,\ldots),$$

and

$$\|v(I-P_n)C_{n-1}^{-1}\cdots C_{n-j}^{-1}\| \leq \|v\|\|A\|2\|A^{-1}\|a^{j-m_2+1}(2m)^2 \qquad (j=0,1,\ldots),$$

where a is given by (4.1). It is easily seen, as in Section 3, that $a^{-m} < 2$. Therefore, the above inequalities imply that the dichotomy inequalities (1.3) and (1.4) hold with the bound (4.1).

We will now prove that for every integer n,

(4.11) $$(\operatorname{Im} P_n)C_n \subseteq \operatorname{Im} P_{n+1},$$

and

(4.12) $$(\operatorname{Ker} P_n)C_n \subseteq \operatorname{Ker} P_{n+1}.$$

First note that

$$\sum_{i=1}^{m_1+m_2+1} a_{n-m_1,n+1-i}(b_{n+1-i,n+1-m_1},\ldots,b_{n+1-i,n+m_2}) = 0.$$

Since $a_{n-m_1,n-m_1-m_2}$ is invertible, it follows that

$$\operatorname{Im}(b_{n-m_1-m_2,n+1-m_1},\ldots,b_{n-m_1-m_2,n+m_2}) \subseteq \operatorname{Im} P_{n+1}.$$

By (4.8) this implies that

$$\operatorname{Im}(b_{n-m_1-m_2,n-m_1},\ldots,b_{n-m_1-m_2,n+m_2-1})C_n \subseteq \operatorname{Im} P_{n+1}.$$

On the other hand, (4.8) also implies that for $i = 1,\ldots,m_1+m_2-1$,

$$\operatorname{Im}(b_{n-i,n-m_1},\ldots,b_{n-i,n+m_2-1})C_n = \operatorname{Im}(b_{n-i,n+1-m_1},\ldots,b_{n-i,n+m_2}) \subseteq \operatorname{Im} P_{n+1}.$$

The last two inclusions prove that (4.11) hold.

In a similar way, note that

$$\sum_{i=-1}^{m_1+m_2-1} a_{n+m_2,n+1+i}(b_{n+1+i,n+1-m_1},\ldots,b_{n+1+i,n+m_2-1},b_{n+1+i,n+m_2} - a_{n+m_2,n}^{-1}\delta_{i,-1}) = 0.$$

But $a_{n+m_2,n}$ is invertible and therefore

$$\operatorname{Im}(b_{n,n+1-m_1},\ldots,b_{n,n+m_2-1},b_{n,n+m_2} - a_{n+m_2,n}^{-1}) \subseteq \operatorname{Ker} P_{n+1}.$$

On the other hand, it follows from

$$b_{n,n-m_1}a_{n-m_1,n}+\cdots+b_{n,n+m_2}a_{n+m_2,n}=I,$$

that

$$(b_{n,n-m_1},\ldots,b_{n,n+m_2-1})C_n=(b_{n,n-m_1+1},\ldots,b_{n,n+m_2-1},b_{n+m_2}-a^{-1}_{n+m_2,n}).$$

Therefore,

$$\operatorname{Im}(b_{n,n-m_1},\ldots,b_{n,n+m_2-1})C_n\subseteq \operatorname{Ker}P_{n+1}.$$

However, (4.8) also shows that for $i=1,\ldots,m_1+m_2-1$,

$$\operatorname{Im}(b_{n+i,n-m_1},\ldots,b_{n+i,n+m_2-1})C_n=\operatorname{Im}(b_{n+i,n+1-m_1},\ldots,b_{n+i,n+m_2})\subseteq \operatorname{Ker}P_{n+1}.$$

The last two relations prove (4.12).

Finally, it is clear that (4.11) and (4.12) prove that the commutation relations (1.2) are satisfied. □

We close this section with the following proposition which gives a criterion for a sequence of projections to be a dichotomy of a given sequence of invertible matrices.

PROPOSITION 4.3. *Let $(A_n)_{n=-\infty}^{\infty}$ be a sequence of $h\times h$ invertible matrices. A sequence $(P_n)_{n=-\infty}^{\infty}$ of projections in $\mathbb{C}^h$ is a dichotomy for $(A_n)_{n=-\infty}^{\infty}$ if and only if* rank(P_n), $n=0,\pm1,\ldots$ *is constant and inequalities* (1.3)–(1.4) *are satisfied with* $0<a<1$ *and* $0<M$.

PROOF. It is clear that if $(P_n)_{n=-\infty}^{\infty}$ is a dichotomy, then equalities (1.2) show that rank$(P_n)_{n=-\infty}^{\infty}$ is constant. Conversely assume that $(P_n)_{n=-\infty}^{\infty}$ satisfies the conditions of the proposition. We will denote $p=\operatorname{rank}(P_n)$. We first show that

$$\operatorname{Im}(P_n)=\{x\colon \lim_{j\to+\infty} xA_n\cdots A_{n+j}=0\}\qquad (n=0,\pm1,\ldots). \tag{4.13}$$

Let n be an integer, and denote $N=\{x\colon \lim_{j\to+\infty}xA_n\cdots A_{n+j}=0\}$. It is clear by (1.3) that $\operatorname{Im}(P_n)\subseteq N$. Assume that $\operatorname{Im}(P_n)\neq N$. Then necessarily $p<\dim N$. Let $x_1,\ldots,x_\ell$ be an orthonormal basis of N, where $\ell=\dim N$. By the definition of N, there exists an integer $k>0$ such that $\|x_iA_n\cdots A_{n+k}\|\le\frac{1}{2\ell}$ $(i=1,\ldots,\ell)$, and we may also take k large enough so that $Ma^{k+1}<1$. Let x be an arbitrary vector in N. There exist complex numbers $\alpha_1,\ldots,\alpha_\ell$ such that $x=\sum_{i=1}^{\ell}\alpha_ix_i$. Since $\|x\|^2=\sum_{i=1}^{\ell}|\alpha_i|^2$, then $|\alpha_i|\le\|x\|$ $(i=1,\ldots,\ell)$ and therefore

$$\|xA_n\cdots A_{n+k}\|\le\sum_{i=1}^{\ell}|\alpha_i|\frac{1}{2\ell}\le\frac{\|x\|}{2}. \tag{4.14}$$

However, $\dim NA_n\cdots A_{n+k}=\dim N>p=\operatorname{rank}(P_{n+k+1})$. Therefore, there exists a nonzero vector x in N such that $xA_n\cdots A_{n+k}\in\operatorname{Ker}(P_{n+k+1})$. Let $y=xA_n\cdots A_{n+k}$. Then by (1.4), we obtain

$$\|x\|=\|y(I_h-P_{n+k+1})A^{-1}_{n+k}\cdots A_n^{-1}\|\le Ma^{k+1}\|y\|.$$

Therefore $\|x\| < \|xA_n \cdots A_{n+k}\|$, which contradicts inequality (4.14).

It follows immediately from (4.13) that

$$\text{(4.15)} \qquad \operatorname{Im}(P_n)A_n = \operatorname{Im}(P_{n+1}) \qquad (n = 0, \pm 1, \ldots).$$

In a similar way it can be proved that

$$\operatorname{Ker}(P_n) = \{\lim_{j \to +\infty} xA_{n-1}^{-1} \cdots A_{n-j}^{-1} = 0\} \qquad (n = 0, \pm 1, \ldots).$$

This shows that $\operatorname{Ker}(P_{n+1})A_n^{-1} = \operatorname{Ker}(P_n)$, $(n = 0, \pm 1, \ldots)$, and therefore

$$\operatorname{Ker}(P_n)A_n = \operatorname{Ker}(P_{n+1}) \qquad (n = 0, \pm 1, \ldots).$$

The commutation relations (1.2) from this equality and (4.15). □

5. INVERTIBILITY

This section is devoted to the proof of Theorem 2.1. In the proof we will use Corollary 7.2 of [1], which, for sake of convenience, is stated here as Proposition 5.1.

PROPOSITION 5.1. *If the sequence $(C_n)_{n=-\infty}^{\infty}$ of invertible matrices admits the left dichotomy $(P_n)_{n=-\infty}^{\infty}$, then*

$$\operatorname{Im} P_n = \{x \in \mathbb{C}^r\colon \lim_{k \to +\infty} xC_n \cdots C_{n+k} = 0\} \qquad (n = 0, \pm 1, \ldots),$$

and

$$\operatorname{Ker} P_n = \{x \in \mathbb{C}^r\colon \lim_{k \to +\infty} xC_{n-1}^{-1} \cdots C_{n-k}^{-1} = 0\} \qquad (n = 0, \pm 1, \ldots).$$

In particular, if a dichotomy exists, then it is unique.

We first prove an auxiliary result.

LEMMA 5.2. *Let A be a regular (m_1, m_2)-band matrix such that its companion sequence $(C_n)_{n=-\infty}^{\infty}$ admits a left dichotomy $(P_n)_{n=-\infty}^{\infty}$, and let n be an integer. Then the equation*

$$\text{(5.1)} \qquad xA = (\delta_{n,j}I)_{j=-\infty}^{\infty}$$

admits a unique solution x in $\ell_r^2(\mathbb{Z})$. This solution $x = (x_i)_{i=-\infty}^{\infty}$ is determined by the equalities

$$\text{(5.2)} \qquad (x_{n-m_1+1+j}, \ldots, x_{n+m_2+j}) = (0, \ldots, 0, a_{n+m_2,n}^{-1})P_{n+1}C_{n+1} \cdots C_{n+j},$$

and

$$\text{(5.3)} \qquad (x_{n-m_1-j}, \ldots, x_{n+m_2-1-j}) = (a_{n-m_1,n}^{-1}, 0, \ldots, 0)(I - P_n)C_{n-1}^{-1} \cdots C_{n-j}^{-1},$$

for $j=0,1,\dots$. If $j=0$, then the empty product on the right hand side is taken as the identity.

PROOF. We first prove the uniqueness, in $\ell_r^2(\mathbb{Z})$, of the solution of equation (5.1). Assume that $xA=0$ for some vector $x=(x_i)_{i=-\infty}^{\infty}$. Then

$$x_{k-m_1}a_{k-m_1,n}+\cdots+x_{k+m_2}a_{k+m_2,n}=0 \qquad (k=0,\pm1,\dots).$$

Therefore

$$(x_{k-m_1+1},\dots,x_{k+m_2})=(x_{k-m_1},\dots,x_{k+m_2-1})C_k \qquad (k=0,\pm1,\dots).$$

These equalities imply that for $j=1,2,\dots$,

$$(5.4) \qquad (x_{n-m_1},\dots,x_{n+m_2-1})C_n\cdots C_{n+j-1}=(x_{n+j-m_1},\dots,x_{n+j+m_2-1}),$$

and

$$(5.5) \qquad (x_{n-m_1},\dots,x_{n+m_2-1})C_{n-1}^{-1}\cdots C_{n-j}^{-1}=(x_{n-j-m_1},\dots,x_{n-j+m_2-1}).$$

Let $y=(x_{n-m_1},\dots,x_{n+m_2-1})$. Equalities (5.4) and (5.5) imply that

$$\lim_{j\to+\infty} yC_n\cdots C_{n+j-1}=\lim_{j\to+\infty} yC_{n-1}^{-1}\cdots C_{n-j}^{-1}=0.$$

By the preceding proposition, $y\in \operatorname{Im}P_n\cap\operatorname{Ker}P_n$. Therefore $y=0$, and equalities (5.4) and (5.5) prove that $x=0$.

Next, we construct one solution of (5.1). Define $x_{n-m_1},\dots,x_{n+m_2-1}$ via

$$(5.6) \qquad (x_{n-m_1},\dots,x_{n+m_2-1})=(a_{n-m_1,n}^{-1},0,\dots,0)(I-P_n).$$

The vectors $x_{n+m_2},x_{n+m_2+1},\dots$ are constructed recursively by

$$(5.7) \qquad x_{n+m_2+j}=\left(\delta_{0,j}I-\sum_{k=n-m_1+j}^{n+m_2+j-1} x_k a_{k,n+j}\right)a_{n+m_2+j,n+j}^{-1} \qquad (j=0,1,\dots),$$

while $x_{n-m_1-1},x_{n-m_1-2},\dots$ are defined by

$$(5.8) \qquad x_{n-m_1-j}=-\left(\sum_{k=n-m_1-j+1}^{n+m_2-j} x_k a_{k,n-j}\right)a_{n-m_1-j,n-j}^{-1} \qquad (j=1,2,\dots).$$

It is clear from the above construction that for $k=0,\pm1,\dots$,

$$(5.9) \qquad x_{k-m_1}a_{k-m_1,k}+\cdots+x_{k+m_2}a_{k+m_2,k}=\delta_{n,k}I.$$

In particular, if $k\neq n$, then the right hand side of this equality is zero, and therefore

$$(5.10) \qquad (x_{k-m_1+1},\dots,x_{k+m_2})=(x_{k-m_1},\dots,x_{k+m_2-1})C_k \qquad (k\neq n).$$

Thus for $j = 1, 2, \ldots$,

$$(5.11) \qquad (x_{n-m_1+1+j}, \ldots, x_{n+m_2+j}) = (x_{n-m_1+1}, \ldots, x_{n+m_2})C_{n+1} \cdots C_{n+j},$$

and

$$(5.12) \qquad (x_{n-m_1-j}, \ldots, x_{n+m_2-1-j}) = (x_{n-m_1}, \ldots, x_{n+m_2-1})C_{n-1}^{-1} \cdots C_{n-j}^{-1}.$$

By (5.6), $\operatorname{Im}(x_{n-m_1}, \ldots, x_{n+m_2-1}) \subseteq \operatorname{Ker} P_n$. Therefore the dichotomy inequalities (1.4) and equality (5.12) above imply that

$$(5.13) \qquad \|(x_{n-m_1-j}, \ldots, x_{n+m_2-1-j})\| \leq Ma^j \|(x_{n-m_1}, \ldots, x_{n+m_2-1})\|,$$

for $j = 1, 2, \ldots$, where (a, M) is a bound of the dichotomy.

We will now use the following equality, which will be proved later:

$$(5.14) \qquad (x_{n-m_1+1}, \ldots, x_{n+m_2}) = (0, \ldots, 0, a_{n+m_2,n}^{-1})P_{n+1}.$$

This equality implies that $\operatorname{Im}(x_{n-m_1+1}, \ldots, x_{n+m_2}) \subseteq \operatorname{Im} P_{n+1}$. Thus, the inequalities (1.3) and equalities (5.11) show that

$$\|(x_{n-m_1+1+j}, \ldots, x_{n+m_2+j})\| \leq Ma^j \|(x_{n-m_1+1}, \ldots, x_{n+m_2})\|.$$

This inequality and (5.13) prove that $x \in \ell_r^2(\mathbb{Z})$. By (5.9), x is a solution of (5.1). Moreover, equality (5.2) follows from (5.11) and (5.14), while (5.3) follows from (5.6) and (5.12).

It remains to prove (5.14). It follows from (5.9), with $k = n$, that

$$(x_{n-m_1} - a_{n-m_1,n}^{-1})a_{n-m_1,n} + x_{n-m_1+1}a_{n-m_1+1,n} + \cdots + x_{n+m_2}a_{n+m_2,n} = 0.$$

Therefore,

$$(x_{n-m_1+1}, \ldots, x_{n+m_2}) = (x_{n-m_1} - a_{n-m_1,n}^{-1}, \ldots, x_{n+m_2-1})C_n.$$

However, by (5.6),

$$(x_{n-m_1} - a_{n-m_1,n}^{-1}, x_{n-m_1+1}, \ldots, x_{n+m_2-1}) = -(a_{n-m_1,n}^{-1}, 0, \ldots, 0)P_n.$$

The last two equalities prove that

$$(5.15) \qquad (x_{n-m_1+1}, \ldots, x_{n+m_2}) = -(a_{n-m_1,n}^{-1}, 0, \ldots, 0)P_nC_n.$$

However, the commutation relations (1.2) and the definition (1.1) of the companion sequence imply that

$$\begin{aligned} -(a_{n-m_1,n}^{-1}, 0, \ldots, 0)P_nC_n &= (-a_{n-m_1,n}^{-1}, 0, \ldots, 0)C_nP_{n+1} = \\ &= (0, \ldots, 0, a_{n+m_2,n}^{-1})P_{n+1}. \end{aligned}$$

This equality and (5.15) prove (5.14). □

PROOF OF THEOREM 2.1. If A is invertible, then by Theorem 4.1 the companion sequence $(C_n)_{n=-\infty}^{\infty}$ admits a dichotomy $(P_n)_{n=-\infty}^{\infty}$ with the bound (2.3). Let $B = A^{-1} = (b_{ij})_{i,j=-\infty}^{\infty}$. For every integer n, the preceding lemma shows that

$$(b_{n,n-m_1+1}, \dots, b_{n,n+m_2}) = (0, \dots, 0, a_{n+m_2,n}^{-1})P_{n+1},$$

and

$$(b_{n,n-m_1}, \dots, b_{n,n+m_2-1}) = (a_{n-m_1,n}^{-1}, 0, \dots, 0)(I - P_n).$$

This implies that $\max(K_1, K_2) \le \|A^{-1}\|$.

Conversely, assume that the companion sequence $(C_n)_{n=-\infty}^{\infty}$ admits a dichotomy $(P_n)_{n=-\infty}^{\infty}$, and that conditions (2.1) and (2.2) hold. For every integer n, let $x = (x_k)_{k=-\infty}^{\infty}$ be the solution of equation (5.1) in $\ell_r^2(\mathbb{Z})$, and define $b_{n,k} = x_k$ $(k = 0, \pm 1, \dots)$. Then

$$(5.16) \qquad b_{n,j-m_1} a_{j-m_1,j} + \dots + b_{n,j+m_2} a_{j+m_2,j} = \delta_{n,j} I \qquad (j, n = 0, \pm 1, \dots).$$

Moreover, the representation (5.2)–(5.3), combined with the conditions (2.1)–(2.2) and the dichotomy inequalities (1.3)–(1.4), proves that

$$\|(b_{n,n-m_1+1+j}, \dots, b_{n,n+m_2+j})\| \le K_2 M a^j \qquad (j = 0, 1, \dots),$$

and

$$\|(b_{n,n-m_1-j}, \dots, b_{n,n+m_2-1-j})\| \le K_1 M a^j \qquad (j = 0, 1, \dots).$$

In particular,

$$\|b_{n,k+1-m_1}\| \le K_2 M a^{|k-n|} \qquad (k = n, n+1, \dots),$$

and

$$\|b_{n,k-m_1}\| \le K_1 M a^{|k-n|} \qquad (k = n, n-1, \dots).$$

The last two equalities prove that the matrix $B = (b_{ij})_{i,j=-\infty}^{\infty}$ represents a bounded operator with

$$\|B\| \le M(1-a)^{-1}(K_1 + K_2).$$

By (5.16), $B = A^{-1}$. □

6. INDEX AND DICHOTOMY RANK

In this section we prove Theorems 1.1 and 2.2.

PROOF OF THEOREM 2.2. Define $G_1 = (a_{ij})_{i,j=-\infty}^{-1}$, $G_2 = (a_{ij})_{i=-\infty,j=0}^{-1,\infty}$ and $G_3 = (a_{ij})_{i=0,j=-\infty}^{\infty,-1}$. Then we have a direct sum decomposition of the corresponding operators as follows

$$A = \begin{pmatrix} G_1 & G_2 \\ G_3 & G \end{pmatrix}.$$

Since G_2 and G_3 are compact and A is invertible, it follows that $\begin{pmatrix} G_1 & 0 \\ 0 & G \end{pmatrix}$ is Fredholm. Therefore G is Fredholm.

We now prove (2.5). It is easily seen from the properties of Fredholm operators that we can assume that $m_1 = 0$. Then G is lower triangular, and

$$\operatorname{Ker} G = \{0\}. \tag{6.1}$$

Let N be the orthogonal complement of $\operatorname{Im} G$ in ℓ^2_r. In order to complete the proof we only have to prove that

$$\dim N = p. \tag{6.2}$$

It is clear that $x = (x_n)_{n=0}^{\infty} \in N$, with $x_n \in \mathbb{C}^r$ for $n = 0, 1, \ldots$, if and only if

$$\sum_{n=0}^{\infty} \|x_n\|^2 < +\infty \tag{6.3}$$

and

$$x_n a_{n,n} + \cdots + x_{n+m_2} a_{n+m_2,n} = 0 \qquad (n = 0, 1, \ldots).$$

The last condition is equivalent to

$$(x_{n+1}, \ldots, x_{n+m_2}) = (x_n, \ldots, n_{n+m_2-1}) C_n \qquad (n = 0, 1, \ldots)$$

which, in turn, is equivalent to

$$(x_i, \ldots, x_{i+m_2-1}) = (x_0, \ldots, x_{m_2-1}) C_0 \cdots C_{i-1} \qquad (i = 0, 1, \ldots). \tag{6.4}$$

Now assume that $x = (x_n)_{n=0}^{\infty} \in N$. It follows from (6.3) and (6.4) that

$$\lim_{i \to +\infty} (x_0, \ldots, x_{m_2-1}) C_0, \ldots, C_{i-1} = 0.$$

Thus, if we denote by $(P_n)_{n=-\infty}^{\infty}$ the dichotomy of $(C_n)_{n=-\infty}^{\infty}$, it follows from Proposition 5.1 that

$$(x_0, \ldots, x_{m_2-1}) \in \operatorname{Im} P_0.$$

Moreover, if $(x_0, \ldots, x_{m_2-1}) = 0$, then $x = 0$ by (6.4). Consequently

$$\dim N \leq \dim(\operatorname{Im} P_0) = p. \tag{6.5}$$

On the other hand, let $(x_0, \ldots, x_{m_2-1}) \in \operatorname{Im}(P_0)$. Then for $i = 0, 1, \ldots$,

$$\|(x_0, \ldots, x_{m_2-1}) C_0 \cdots C_{i-1}\| \leq \|(x_0, \ldots, x_{m_2-1})\| M a^i,$$

where (a, M) is a dichotomy bound. Now define $x = (x_k)_{k=0}^{\infty}$ via the compatible system

$$(x_i, \ldots, x_{i+m_2-1}) = (x_0, \ldots, x_{m_2-1}) C_0 \cdots C_{i-1} \qquad (i = 1, 2, \ldots).$$

By the previous inequality we have $\sum_{n=0}^{\infty}\|x_n\|^2 < +\infty$, and therefore $x \in N$. Thus, $\dim N \geq \dim(\operatorname{Im} P_0) = p$. This inequality and (6.5) imply (6.2). □

PROOF OF THEOREM 1.1. Assume that $(C_n)_{n=-\infty}^{\infty}$ admits the dichotomy $(P_n)_{n=-\infty}^{\infty}$ with bound (a, M). Let K_1 and K_2 be defined as in (2.1) and (2.2). It follows from inequality (1.4) with $j = 0$, that $\|I - P_n\| \leq M$ $(n = 0, \pm 1, \ldots)$. Thus

$$K_1 \leq M\big(\sup_n \|a_{n-m_1,n}^{-1}\|\big). \tag{6.6}$$

Moreover, taking into account equalities (1.1) and (1.2), we have

$$\begin{aligned}(0, \ldots, a_{n+m_2,n}^{-1})P_{n+1} &= (-a_{n-m_1,n}^{-1}, 0, \ldots, 0)C_n P_{n+1} = \\ &= (-a_{n-m_1}^{-1}, 0, \ldots, 0)P_n C_n\end{aligned}$$

for $n = 0, \pm 1, \ldots$. On the other hand, inequality (1.3) with $j = 1$ implies that $\|P_n C_n\| \leq Ma$. Therefore

$$K_2 \leq Ma\big(\sup_n \|a_{n-m_1,n}^{-1}\|\big). \tag{6.7}$$

The theorem is a consequence of Theorems 2.1 and 2.2, using the inequalities (6.6) and (6.7) above. □

REFERENCES

[1] A. Ben-Artzi and I. Gohberg, Inertia Theorems for Nonstationary Discrete Systems and Dichotomy, to appear.

[2] A. Ben-Artzi and I. Gohberg, Extension of the Theorem of Krein on Orthogonal Polynomials for the Nonstationary Case, to appear.

[3] C. de Boor, Dichotomies for Band Matrices, SIAM J. Numer. Anal. 17(6) (1980), pp. 894–907.

Raymond and Beverly Sackler
Faculty of Exact Sciences
School of Mathematical Sciences
Tel-Aviv University
Ramat-Aviv, Israel

Operator Theory:
Advances and Applications, Vol. 32

A NOTE ON THE ALGEBRA GENERATED BY A SUBNORMAL OPERATOR [1)]

H. Bercovici and J.B. Conway

Dedicated to the memory of Constantin Apostol

Let H be a separable Hilbert space, and let $A \subset L(H)$ be a weak*-closed algebra. We say that A has property $(\mathbb{A}_1)$ if every weak*-continuous functional f on A can be represented as

(1) $f(T) = (Tx,y), \; T \in A,$

where x and y are vectors in H and $(\cdot,\cdot)$ denotes the scalar product in H. We say that A has property $(\mathbb{A}_1(r))$ if, in addition, given s>r and f, we can choose x and y satisfying (1) and

(2) $\|x\|\|y\| \leq s\|f\|$.

Clearly property $(\mathbb{A}_1(r))$ is only possible for $r \geq 1$, unless $A = \{0\}$.

Assume now that $A = A_S$ is the weak*-closed algebra generated by I and a given operator $S \in L(H)$. S. Brown [3] proved for certain subnormal operators S that A_S must have property $(\mathbb{A}_1(1))$ (and thereby deduced the existence of invariant subspaces for subnoemal operators). Later Olin and Thomson [7] proved that for all subnormal operators S, the algebra A_S has property $(\mathbb{A}_1(r))$ for some $r<882$. Another proof of this result was given by Thomson [10] who also improved the constant r to 144. Our purpose in this note is to show that in fact A_S has property $(\mathbb{A}_1(1))$ if S is sbnormal. The ingredients are a result in [2] about algebras isomorphic to H^∞ , and a refinement from [6] of Sarason's [9] decomposition theorem for $P^\infty(\mu)$. We state these two results for the reader's convenience. The following statement is a reformulation of the main result in [2] (see also [1], [4] and [7] for related earlier work).

---

1) The work in this paper was partially supported by grants from the National Science Foundation.

3. THEOREM. *Let* $A \subset L(H)$ *be a weak*-closed algebra which is isometrically isomorphic and weak*-homeomorphic to the algebra* H^∞ *of bounded analytic functions in the unit disc. Then* A *has property* $(\mathbb{A}_1(1))$.

The following result is well known to experts in subnormal operators. Since it is crucial to the proof of our main result and is not explicitly stated in the literature, a proof will be sketched. The notation will be that of [5], except that A_S will be used here rather than $P^\infty(S)$. Recall that A_S is said to be antisymmetric if the only selfadjoint operators in A_S are scalar multiples of the identity.

4. LEMMA. *If* S *is a subnormal operator and* A_S *is antisymmetric, then* A_S *is isometrically isomorphic and weak*-homeomorphic to* H^∞.

Proof. If μ is a scalar-valued spectral measure for the minimal normal extension of S, then A_S is isometrically isomorphic and weak*-homeomorphic to $P^\infty(\mu)$, the weak*-closure of the analytic polynomials in $L^\infty(\mu)$ (Corollary 12.11 on p.208 of [5]). If K denotes the spectrum of the identity polynomial $p(z) = z$ in the algebra $P^\infty(\mu)$, then $P^\infty(\mu)$ is isometrically isomorphic to $H^\infty(\text{int } K)$ ([9] and Theorem 4.5 on p. 398 of [5]). Moreover, under this isomorphism, a sequence in $P^\infty(\mu)$ converges weak* if and only if the corresponding sequence in $H^\infty(\text{int } K)$ converges boundedly pointwise (Lemma 4.3 on p. 441 of [5]).

Now A_S, and hence $P^\infty(\mu)$, is antisymmetric so int K must be connected. Moreover, int K must be simply connected (Theorem 4.5(a) on p. 398 and Theorem 9.21 on p. 356 of [5]). Let $g\colon \text{int } K \to \mathbf{D}$ be the Riemann map, $\mathbf{D} = \{z\colon |z|<1\}$. Thus $f \to f\circ g$ defines an isometric isomorphism of H^∞ onto $H^\infty(\text{int } K)$ and a sequence f_n in H^∞ converges boundedly pointwise if and only if $f_n\circ g$ converges boundedly pointwise in $H^\infty(\text{int } K)$. If this is combined with the results of the preceding paragraph, we see that there exists an isometric isomorphism $k\colon H^\infty \to A_S$ such that a sequence f_n in H^∞ converges boundedly pointwise to f if and only if $k(f_n)$ converges weak* to $k(f)$ in A_S. But a sequence in H^∞ converges weak* if and only if it converges boundedly pointwise. Hence k is weak* sequentially continuous. Since H^∞ and A_S are duals of separable Banach spaces, the Krein-Smulian theorem implies that k is a weak* homeomorphism (see p. 27 of [5]). The lemma is proved.

The statement of the next result involves the direct sum of weak*-closed algebras. We recall that, given weak*-closed algebras $A_n \subset L(H_n)$, $n=1,2,\ldots$, we can form the weak*-closed algebra $A = \oplus_{n=1}^\infty A_n$ on $\oplus_{n=1}^\infty H_n$. The algebra A simply consists of all operators of the form $\oplus_{n=1}^\infty T_n$, where $T_n \in A_n$ and

$\sup_n \|T_n\|$ is finite. In view of Lemma 4, the relevant result from [6] (Theorem 7.1; see also [5], Theorem 3.1 on p. 437) can now be formulated as follows.

5. THEOREM. *Let* S *be a subnormal operator on* H. *There exists a decomposition* $H = \oplus_{i=0}^{N} H_i$, $N \leq \infty$, *and weak*-closed subalgebras* A_i *of* $L(H_i)$ *with the following properties:*

(i) $A_S = \oplus_{i=0}^{N} A_i$;

(ii) A_0 *is isometrically isomorphic and weak*-homeomorphic to* $L^{\infty}(\mu)$ *for some compactly supported* μ *on* $\mathbb{C}$*; and*

(iii) A_i *is isometrically isomorphic and weak*-homeomorphic to* H^{∞} *for* $i>0$.

We are now ready for the main result of this paper.

6. THEOREM. *If* S *is a subnormal operator on* H *then* A_S *has property* $(\mathbb{A}_1(1))$.

Proof. Let A_i and H_i be as in Theorem 5 above. Theorem 3 implies that each A_i has property $(\mathbb{A}_1(1))$ for $i>0$. It is easy to see that A_0 also has property $(\mathbb{A}_1(1))$. Indeed, let $k: L^{\infty}(\mu) \to A_0$ be an isometric weak*-homeomorphism. Then elementary spectral theory implies that there exists an isometry $V: L^2(\mu) \to H_0$ such that $V^*k(u)Vh = uh$ for $u \in L^{\infty}(\mu)$ and $h \in L^2(\mu)$. Let now f be a weak*-continuous functional on A_0 so that $f \circ k$ is a weak*-continuous functional on $L^{\infty}(\mu)$. Thus there exists a function F in $L^1(\mu)$ such that $\|F\|_1 = \|f\|$ and $f(k(u)) = \int uF\, d\mu$ for u in $L^{\infty}(\mu)$. But now we can find h,g in $L^2(\mu)$ such that $F = h\bar{g}$ and $\|F\|_1 = \|h\|_2 \|g\|_2$. If we set $x = Vh$, $y = Vg$ it is immediate to verify that $f(T) = (Tx,y)$ for T in A_0, and $\|x\| \|y\| = \|f\|$.

Let us show now that A_S must also have property $(\mathbb{A}_1(1))$. Indeed, let f be a weak*-continuous functional on A_S, and $s>1$. It follows that there are weak*-continuous functionals f_i on A_i such that $\|f\| = \sum_{i=0}^{N} \|f_i\|$ and $f(T) = \sum_{i=0}^{N} f_i(T_i)$ if $T = \oplus_{i=0}^{N} T_i \in A_S$. Since each A_i has property $(\mathbb{A}_1(1))$, there exist vectors x_i, y_i in H such that $\|x_i\| = \|y_i\| \leq (s\|f_i\|)^{\frac{1}{2}}$, and $f_i(T_i) = (T_i x_i, y_i)$ for T_i in A_i. then the vectors $x = \oplus_{i=0}^{N} x_i$ and $y = \oplus_{i=0}^{N} y_i$ will clearly satisfy $\|x\| = \|y\| \leq (s\|f\|)^{\frac{1}{2}}$, and $f(T) = (Tx,y)$ for T in A. This completes the proof.

REFERENCES

1. Bercovici, H.: A contribution to the theory of operators in the class $(\mathbb{A})$, J. Funct. Anal., to appear.

2. Bercovici, H.: Factorization theorems and the structure of operators on Hilbert space, preprint.

3. Brown, S.: Some invariant subspaces for subnormal operators, Integral Equations Operator Theory 1(1978), 310-333.

4. Brown, S., Chvreau, B., Pearcy, C.: On the structure of contraction operators, II, J. Funct. Anal., to appear.

5. Conway, J.: Subnormal operators, Pitman, Boston, 1981.

6. Conway, J., Olin, R.: A functional calculus for subnormal operators, II, Memoirs Amr. Math. Soc. 10(1977), No. 184.

7. Chevreau, B., Pearcy, C.: On the structure of contraction operators, I, J. Funct. Anal., to appear.

8. Olin, R., Thomson, J.: Algebras of subnormal operators, J. Funct. Anal. 37(1980), 271-301.

9. Sarason, D.: Weak-star density of polynomials, J. Reine Angew. Math. 252(1972), 1-15.

10. Thomson, J.: Factorization over algebras of subnormal operators, preprint.

Department of Mathematics
Indiana University
Bloomington, IN 47405

Operator Theory:
Advances and Applications, Vol. 32

EXTREME POINTS IN THE SET OF CONTRACTIVE INTERTWINING DILATIONS

Zoia Ceaușescu and **I. Suciu**

Dedicated to the memory of Constantin Apostol

The following result is well known (cf. [5]): a function f in the unit ball of H^∞ is an extreme point if and only if

$$\int \log(1 - |f|)dt = -\infty.$$

Working in the context of the intertwining dilation theory (cf. [2]) we shall prove an operational version of this result: if A is a contraction which intertwines two contractions T_1, T_2 we give a necessary and sufficient condition for a contractive intertwining dilation B of A to be an extreme point of the set all contractive intertwining dilations of A (Theorem 1). When A is the null Hankel operator and f is a function in the unit ball of H^∞ considered as a symbol of A then our geometric condition of extremality is equivalent to the above nonintegrability condition.

In the case of a pair of two commuting contractions T, S, the extremality condition of a lifting V_o of the commutant S of T turns out to be the uniqueness condition for the Ando dilation of [T,S] which crosses through V_o (cf. [3]). Combining these results, a sufficient extremality condition is obtained for representing semi--spectral measures (on the bi-torus) of the pair [T,S].

1. PRELIMINARIES

We shall use the notation (H,T) for a contraction T acting on the Hilbert space H. $D_T = [I - T^*T]^{\frac{1}{2}}$ is the defect operator of T and $\mathcal{D}_T = \overline{D_T H}$ is the defect space of T.

Let $(\widetilde{K},\widetilde{U})$ be the minimal unitary dilation of (H,T). That is (cf. [6]) $\widetilde{U}$ is a unitary operator on $\widetilde{K}$, $H \subset \widetilde{K}$ and

$$T^n = P_H \widetilde{U}^n | H, \quad n \in \mathbf{Z},$$

$$\widetilde{K} = \bigvee_{n \in \mathbf{Z}} \widetilde{U}^n H.$$

Denoting

$$K = \bigvee_{n\geq 0} \tilde{U}^n H, \quad U = \tilde{U}\,|\,K$$

$$K_* = \bigvee_{n\geq 0} \tilde{U}^{*n} H, \quad U_* = \tilde{U}^*\,|\,K$$

then (K,U), (K_*,U_*) are the minimal isometric dilations of (H,T), (H,T^*) respectively. Recall also that

(1.1) $$P_H^K U = T P_H^K, \quad P_H^{K_*} U_* = T^* P_H^{K_*}$$

and

(1.2) $$\tilde{K} = [K_* \ominus H] \oplus H \oplus [K \ominus H].$$

Let now, for $j = 1,2$, (H_j,T_j) be a contraction and let A be a contraction from H_1 to H_2 such that $AT_1 = T_2A$. We shall denote by **CID**(A) the set of all intertwining dilations of A, i.e.

(1.3) $$\mathbf{CID}(A) = \{B \in L(K_1,K_2) : \|B\| \leq 1, BU_1 = U_2B, P_{H_2}^{K_2}B = AP_{H_1}^{K_1}\}.$$

By the Sz.-Nagy–Foiaş dilation theorem this set is nonempty. Clearly **CID**(A) is a convex subset in the unit ball of $L(K_1,K_2)$.

For $B \in$ **CID**(A) we shall denote by $\tilde{B}$ the Douglas extension of B (cf. [4]), i.e. $\tilde{B}$ is the unique contraction from $\tilde{K}_1$ to $\tilde{K}_2$ such that $\tilde{B}\tilde{U}_1 = \tilde{U}_2\tilde{B}$ and $\tilde{B}\,|\,K_1 = B$. Since $\tilde{B}(K_1 \ominus H) \subset K_2 \ominus H$ it results from (1.2) that $\tilde{B}^*K_{2*} \subset K_{1*}$. Let B_* be the contraction from K_{2*} to K_{1*} defined by $B_* = \tilde{B}^*\,|\,K_{2*}$. It is easy too see that

$$\mathbf{CID}(A^*) = \{B_* : B \in \mathbf{CID}(A)\}.$$

We shall use also the notations

$$\widetilde{\mathbf{CID}}(A) = \{\tilde{B} : B \in \mathbf{CID}(A)\}, \quad \widetilde{\mathbf{CID}}(A^*) = \{\tilde{B}^* : B \in \mathbf{CID}(A)\}.$$

All these sets are nonvoid convex sets of contractions. The maps $B \leftrightarrow \tilde{B} \leftrightarrow \tilde{B}^* \leftrightarrow B_*$ are one-to-one affine maps and consequently they preserve the extreme points.

2. THE MAIN RESULT

Let us fix now $B \in$ **CID**(A). Setting

(2.1) $$YD_B = D_B U_1, \quad Y_* D_{B_*} = D_{B_*} U_{2*}$$

we obtain the isometries Y, Y_* on $\mathcal{D}_B$, $\mathcal{D}_{B_*}$ respectively. Denote

(2.2) $$L = \mathcal{D}_B \ominus Y\mathcal{D}_B, \quad L_* = \mathcal{D}_{B_*} \ominus Y_* \mathcal{D}_{B_*}.$$

We have:

THEOREM 1. *The following assertions are equivalent:*

(i) B *is an extreme point in* **CID**(A).

(ii) *One of the subspace* L *or* L_* *is equal to* $\{0\}$.

PROOF. Let $\tilde{B}$ be the Douglas extension of B and denote

$$\tilde{Y}_1 = \tilde{U}_1 | \mathcal{D}_{\tilde{B}}, \quad \tilde{Y}_{2*} = \tilde{U}_2^* | \mathcal{D}_{\tilde{B}^*}.$$

The subspaces $\overline{D_{\tilde{B}} K_1}$ and $\overline{D_{\tilde{B}^*} K_{2*}}$ are invariant to $\tilde{Y}_1$, $\tilde{Y}_{2*}$ respectively. Using Wold decomposition theorem we can write

(2.3)
$$\overline{D_{\tilde{B}} K_1} = \bigoplus_{n \geq 0} \tilde{Y}_1^n \tilde{L}_1 \oplus \tilde{R}_1$$
$$D_{\tilde{B}^*} K_{2*} = \bigoplus_{n \geq 0} \tilde{Y}_{2*} \tilde{L}_{2*} \oplus \tilde{R}_{2*}$$

where

(2.4)
$$\tilde{L}_1 = \overline{D_{\tilde{B}} K_1} \ominus \tilde{Y}_1 \overline{D_{\tilde{B}} K_1}, \quad \tilde{R}_1 = \bigcap_{n \geq 0} \tilde{Y}_1^n \overline{D_{\tilde{B}} K_1}$$
$$\tilde{L}_{2*} = \overline{D_{\tilde{B}^*} K_{2*}} \ominus \tilde{Y}_{2*} \overline{D_{\tilde{B}^*} K_{2*}}, \quad \tilde{R}_{2*} = \bigcap_{n \geq 0} \tilde{Y}_{2*} \overline{D_{\tilde{B}^*} K_{2*}}.$$

Since

$$\mathcal{D}_{\tilde{B}} = \bigvee_{n \geq 0} \tilde{Y}_1^{*n} \overline{D_{\tilde{B}} K_1}, \quad \mathcal{D}_{\tilde{B}^*} = \bigvee_{n \geq 0} \tilde{Y}_{2*}^{*n} \overline{D_{\tilde{B}^*} K_{2*}}$$

we have

(2.5)
$$\mathcal{D}_{\tilde{B}} = \bigoplus_{n \in \mathbf{Z}} \tilde{Y}_1^n L_1 \oplus \tilde{\mathcal{R}}_1$$
$$\mathcal{D}_{\tilde{B}^*} = \bigoplus_{n \in \mathbf{Z}} \tilde{Y}_{2*}^n \tilde{L}_{2*} \oplus \tilde{\mathcal{R}}_{2*}.$$

Setting for $k_1 \in K_1$, $k_{2*} \in K_{2*}$

$$bD_B k_1 = D_{\tilde{B}} k_1, \quad b_* D_{B_*} k_{2*} = D_{\tilde{B}^*} k_{2*}$$

we obtain the isometries b from D_B onto $\overline{D_{\tilde{B}}K_1}$ and b_* from D_{B*} onto $\overline{D_{\tilde{B}^*}K_{2*}}$ such that

(2.7) $$\tilde{Y}_1 b = bY_1, \quad \tilde{Y}_{2*}b_* = b_*Y_{2*}.$$

Clearly we have

(2.8) $$bL = \tilde{L}_1, \quad b_*L_* = \tilde{L}_{2*}.$$

Suppose now that L and L_* are different from {0}. Form (2.8) it results $\tilde{L}_1 \neq \{0\}$, $\tilde{L}_{2*} \neq \{0\}$. Let $\tilde{Q}$ be a contraction from $\tilde{L}_1$ to $\tilde{L}_{2*}$. Using (2.5) we define the contraction $\tilde{R}$ from $D_{\tilde{B}}$ into $D_{\tilde{B}^*}$ by

$$\tilde{R} \sum_{n\in\mathbb{Z}} \tilde{Y}_1^n \ell_n = \sum_{n\in\mathbb{Z}} \tilde{Y}_{2*}^{*n+1}\tilde{Q}\ell_n, \quad \ell_n \in \tilde{L}_1, \quad \sum \|\ell_n\|^2 < \infty$$

$$\tilde{R}k = 0 \quad \text{for } k \in \tilde{R}_1.$$

We have $\|\tilde{R}\| = \|\tilde{Q}\|$, $R\tilde{Y}_1 = \tilde{Y}_{2*}^*\tilde{R}$ and from (2.3) and (2.5) we get

$$\tilde{R}\overline{D_{\tilde{B}}K_1} \subset \bigoplus_{n\geq 1} \tilde{Y}_{2*}^{*n}\tilde{L}_{2*} = D_{\tilde{B}^*} \ominus \overline{D_{\tilde{B}^*}K_{2*}}.$$

Hence $D_{\tilde{B}^*}\tilde{R}D_{\tilde{B}}K_1$ is orthogonal on K_{2*} and by (1.2) it results

(2.10) $$D_{\tilde{B}^*}\tilde{R}D_{\tilde{B}}K_1 \subset K_2 \ominus H.$$

Let us define the operator G from K_1 into K_2 by

(2.11) $$G = D_{\tilde{B}^*}\tilde{R}D_B | K_1.$$

It is clear that G is a contraction and

(2.12) $$GU_1 = U_2G, \quad P_{H_2}^{K_2}G = 0.$$

Let now $0 < \varepsilon < 1$ and suppose that $\tilde{Q}$ was chosen such that $0 < \|\tilde{Q}\| < \varepsilon$. Then $0 < \|\tilde{R}\| < \varepsilon$ and consequently

$$\|D_{\tilde{R}}D_{\tilde{B}}k_1\|^2 \geq (1 - \varepsilon^2)\|D_{\tilde{B}_1}k_1\|^2.$$

Since for any $k_1 \in K_1$ we have

$$\|Bk_1 \pm Gk_1\|^2 \leq \|\tilde{B}k_1\|^2 + \|D_{\tilde{B}^*}\tilde{R}D_{\tilde{B}}k_1\|^2 + 2|(\tilde{B}k_1, D_{\tilde{B}^*}\tilde{R}D_{\tilde{B}}k_1)| \leq$$

$$\leq \|k_1\|^2 - \|D_{\tilde{R}} D_{\tilde{B}} k_1\|^2 - \|\tilde{B}^* \tilde{R} D_{\tilde{B}} k_1\|^2 + 2\|R\| \, \|D_{\tilde{B}} k_1\|^2 \leq$$

$$\leq \|k_1\|^2 + (\varepsilon^2 + 2\varepsilon - 1)\|D_{\tilde{B}} k_1\|^2$$

and since ε can be chosen arbitrary small we deduce

(2.13) $$\|B \pm G\| \leq 1.$$

Letting now $B_1 = B + G$, $B_2 = B - G$ then $B_1, B_2 \in$ **CID**(A), $B_1 \neq B_2$ and $B = \frac{1}{2}(B_1 + B_2)$ and consequently B is not an extreme point in **CID**(A). The implication (i) $\Rightarrow$(ii) is proved.

Suppose now that B is not an extreme point in **CID**(A). By standard arguments we can suppose that there exist B_1, $B_2 \in$ **CID**(A), $B_1 \neq B_2$, such that

(2.14) $$B = \tfrac{1}{2}(B_1 + B_2).$$

Define

(5.2) $$\tilde{G} = \tilde{B} - \tilde{B}_1.$$

We have $\tilde{G} \neq 0$, $\tilde{G}\tilde{U}_1 = \tilde{U}_2\tilde{G}$ and $P_{H_2}^{K_2}\tilde{G} | K_1 = 0$. It results

(2.16) $$\tilde{G}K_1 \subset K_2 \ominus H_2.$$

Clearly $\|\tilde{B} \pm \tilde{G}\| \leq 1$ and consequently for any $\tilde{k}_1 \in \tilde{K}_1$ we have

$$\|B\tilde{k}_1\|^2 + \|Gk_1\|^2 = \tfrac{1}{2}[\|(B + G)k_1\|^2 + \|(B - G)k_1\|^2] \leq \|k_1\|^2.$$

Hence $\|\tilde{G}\tilde{k}_1\| \leq \|D_{\tilde{B}}\tilde{k}_1\|$ and we conclude that there exists a nonnull contraction $\tilde{R}$ from $D_{\tilde{B}}$ in $\tilde{K}_2$ such that

(2.17) $$\tilde{G} = \tilde{R}D_{\tilde{B}}.$$

Clearly $\tilde{R}\tilde{Y}_1 = \tilde{U}_2\tilde{R}$ and from (2.16) for any $k_1 \in K_1$ and $h_2 \in H_2$ we obtain

$$(\tilde{R}D_{\tilde{B}}k_1, \tilde{U}_2^{*n}h_2) = (U_2^n\tilde{R}D_{\tilde{B}}k_1, h_2) = (\tilde{R}D_{\tilde{B}}\tilde{U}_1^n k_1, h_2) = (\tilde{G}\tilde{U}_1^n k_1, h_2) = 0.$$

It results $\tilde{R}D_{\tilde{B}}K_1 \perp K_{*2}$ and using (1.2) we deduce

(2.18) $$\tilde{R}D_{\tilde{B}}K_1 \subset K_2 \ominus H_2.$$

Since $\tilde{U}_2 | K_2 \ominus H_2$ is unilateral shift we obtain

$$\tilde{R}\Big(\bigcap_{n\geq 0} \tilde{Y}_1^n \overline{D_{\tilde{B}}K_1}\Big) \subset \bigcap_{n\geq 0} \tilde{U}_1^n[K_2 \ominus H_2] = \{0\}$$

i.e. $\tilde{R}|\tilde{R}_1 = 0$.

Since $\tilde{R} \neq 0$ from (2.5) it results $\tilde{L}_1 \neq \{0\}$ and consequently $L \neq \{0\}$.

Working with B_* instead of B we obtain $L_* \neq \{0\}$ and the theorem is proved.

3. THE CASE OF HANKEL OPERATORS

If $\tilde{K} = L^2$ – the scalar Lebesgue space on the torus denote by U the unitary operator on $\tilde{K}$ given by the multiplication by coordinate function e^{it}. Let $H_1 = H^2$ – the Hardy subspace of L^2 and $H_2 = L^2 \ominus H^2$. Denote $T_1 = \tilde{U}|H_1$, $T_2^* = \tilde{U}^*|H_2$ and let A be a contraction from H_1 to H_2 such that $AT_1 = T_2A$. Then A is a Hankel operator. It is clear that $(\tilde{K},\tilde{U})$ is the minimal unitary dilation of (H_1,T_1) and (H_2,T_2). It results that the elements of $\widetilde{\mathbf{CID}}(A)$ are contractions on L^2 which commute with e^{it}. It results that $\widetilde{\mathbf{CID}}(A)$ can be identified by an affine isomorphism with the convex subset $S(A)$ of the unit ball of L^∞ consisting of all *contractive symbols* of A, i.e the set of all $\phi \in L^\infty$, $||\phi|| \leq 1$ and

$$Ah = P_{L^2 \ominus H^2}\phi h, \qquad h \in H^2. \tag{3.1}$$

If $\phi \in S(A)$ and $B = M_\phi \in \widetilde{\mathbf{CID}}(A)$ then

$$||D_{\tilde{B}}h||^2 = ||h||^2 - ||\tilde{B}h||^2 = (1/2\pi)\int(1 - |\phi|^2)|h|^2 dt = \int |h|^2 d\mu$$

where μ is the positive measure $d\mu = (1/2\pi)(1 - |\phi|^2)dt$.

It results that $D_{\tilde{B}} = D_{\tilde{B}^*}$ can be identified with $L^2(d\mu)$.

Using Szegö theorem we deduce from Theorem 1:

COROLLARY. *Let* A *be a Hankel contraction and denote by* $S(A)$ *the set of all contractive symbols of* A. *Then a function* ϕ *in* $S(A)$ *is an extreme point in* $S(A)$ *if and only if*

$$\int \log(1 - |\phi|)dt = -\infty.$$

If $A = 0$ – the null operator from H^2 into $L^2 \ominus H^2$ then $S(A) = H_1^\infty$ – the unit ball in H^∞, and we set the classical result in our context.

4. EXTREMAL ANDO DILATIONS AND REPRESENTING SEMI-SPECTRAL MEASURES

Let us denote by (H, [T,S]) a pair of commutating contractions T, S on the Hilbert space H. We shall denote by (K, [U,V]) an Ando dilation of (H, [T,S]). This means that K is Hilbert space containing H as a closed subspace, U, V are two commuting isometries on K such that

$$K = \bigvee_{n,m\geq 0} U^n V^m H$$

and

$$T^n S^m h = P_H U^n V^m h, \qquad n, m \geq 0, \ \ h \in H.$$

Let $K_0 = \bigvee_{n\geq 0} U^n H$, $U_0 = U|K_0$, $V_0 = P^K_{K_0} V|K_0$. Clearly (K_0,U_0) is (an identification for) the minimal isometric dilation of (H,T). Using the notations from Section 1 in case $T_1 = T_2 = T$ and A = S it is easy to see that $V_0 \in$ **CID**(S).

We say that the Ando dilation (K, [U,V]) *crosses through* $(K_0, [U_0,V_0])$ if $(K_0, [U_0,V_0])$ is associated to (K, [U,V]) as above.

Conversely if $V_0 \in$ **CID**(A) then taking (K,V) to be the minimal isometric dialtion of (K_0,V_0) then U_0 can be uniquely extended to an isometry U on K such that (K, [U,V]) is an Ando dilation of (H, [T,S]) which obviously crosses through $(K_0, [U_0,V_0])$. We call it the *distinguished Ando dilation* of (H, [T,S]) which crosses through $(K_0, [U_0, V_0])$.

Combining Theorem 1 with the uniqueness result proved in [3] we obtain:

THEOREM 2. *Let* (H, [T,S]) *be a pair of commuting contractions on* H *and* $V_0 \in$ **CID**(S). *The following assertions are equivalent:*

(i) V_0 *is an extreme point in* **CID**(S).

(ii) *The distinguished Ando dilation of* (H, [T,S]) *is the only Ando dilation crossing through* $(K_0, [U_0,V_0])$.

We shall see that this Ando dilation is extremal in the sense that it produces an extremal representing semi-spectral measure for (H, [T,S]).

For an Ando dilation (K, [U,V]) of (H, [T,S]) we shall denote by $(\tilde{K}, [\tilde{U},\tilde{V}])$ its minimal unitary extension. This means that $\tilde{K}$ is a Hilbert space containing K as a (closed) subspace, $\tilde{U}$, $\tilde{V}$ are two commuting unitary operators on $\tilde{K}$ which extend U and V respectively and

$$\tilde{K} = \bigvee_{n,m\leq 0} \tilde{U}^n \tilde{V}^m K.$$

Such an extension always exists and it is unique. The map $[n,m] \to \widetilde{U}^n\widetilde{V}^m$ is a unitary representation of the group $\mathbf{Z}^2$ on $\widetilde{K}$. It results that there exists a unique spectral measure E on the bi-torus $\mathbf{T}^2$ taking values orthogonal projections on $\widetilde{K}$ such that

$$\widetilde{U}^n\widetilde{V}^m = \int e^{-int}e^{-im\theta}dE(t,\theta). \tag{4.1}$$

If for a Borel set $\sigma \subset \mathbf{T}^2$ we set

$$F(\sigma) = P_H^{\widetilde{K}}\ E(\sigma)\,|\,H \tag{4.2}$$

we obtain a semi-spectral measure on $\mathbf{T}^2$ taking values positive operators on H which *represents* (H, [T,S]) in the sense that

$$T^nS^m = \int e^{-int}e^{-im\theta}dF(t,\theta). \tag{4.3}$$

Conversely if F is a semi-spectral measure on $\mathbf{T}^2$ which represents (H, [T,S]), taking $(\widetilde{K},E)$ its spectral (Naimark) dilation and $(\widetilde{K}, [\widetilde{U},\widetilde{V}])$ the corresponding unitary operators verifying (4.1) then setting $K = \bigvee_{n,m\geq 0} \widetilde{U}^n\widetilde{V}^m H$, $U = \widetilde{U}\,|\,K$, $V = \widetilde{V}\,|\,K$ we obatin an Ando dilation (K, [U,V]) of (H, [T,S]).

In this way we establish a one-to-one correspondence between the set of all Ando dilations of (H, [T,S]) and the convex set $F = F_{(H,\,[T,S])}$ of all representing semi-spectral measures for (H, [T,S]).

COROLLARY. *Let* $F \in F_{(H,\,[T,S])}$ *and let* (K, [U,V]) *be the Ando dilation of* (H, [T,S]) *corresponding to* F. *Suppose* (K, [U,V]) *crosses through* $(K_o, [U_o,V_o])$. *If* V_o *is an extreme point in* **CID**(S) *then* F *is an extreme point in* F.

PROOF. Suppose that $F = \alpha F' + (1-\alpha)F''$ with $0<\alpha<1$ and $F', F'' \in F$. Let (K', [U',V']) and (K", [U",V"]) be the Ando dilations corresponding to F', F" respectively. If (K', [U',V']) crosses through $(K_o, [U_o,V'_o])$ and (K", [U",V"]) crosses through $(K_o, [U_o,V''_o])$ with $V'_o, V''_o \in$ **CID**(S) it results $V_o = \alpha V'_o + (1-\alpha)V''_o$. Since V_o is an extreme point in **CID**(S) it results $V_o = V'_o = V''_o$. By Theorem 2 we obtain (K', [U',V']) and (K", [U",V"]) coincide with (K, [U,V]) and consequently F' = F" = F. The corollary is proved.

REMARK. For a pair (H, [T,S]) of commuting contractions on H let us consider $(K_o, [U_o,A_o])$ where (K_o,U_o) is the minimal isometric dilation of (H,T), $A_o \in$ **CID**(S) and

$(G_0, [V_0, B_0])$ where (G_0, V_0) is the minimal isometric dilation of (H,S) and $B_0 \in$ **CID**(T). A measure $F \in \mathcal{F} = \mathcal{F}_{[T,S]} = \mathcal{F}_{[S,T]}$ uniquely determines A_0, B_0 as above. If either A_0 is an extreme point in **CID**(S) or B_0 is an extreme point in **CID**(T), then F is an extreme point in $\mathcal{F}$. The two cases are generally disjoint so the converse assertion of the Corollary is not true.

REFERENCES

1. **Ando, T.,** On a pair of commutative contractions, *Acta Sci. Math. (Szeged)* **24** (1963), 88-90.
2. **Ceaușescu, Zoia ; Foiaș, C.,** On intertwining dilations. V, *Acta Sci. Math. (Szeged)* **40**(1978), 9-23.
3. **Ceaușescu, Zoia ; Suciu, I.,** Isometric dilations of commuting contractions. I, *J. Operator Theory* **12**(1984), 65-88.
4. **Douglas, R.G.,** On the operator equation S*XT = X and related topics, *Acta Sci. Math. (Szeged)* **30**(1969), 19-32.
5. **Hoffman, K.,** *Banach spaces of analytic functions*, Prentice-Hall Inc., Englewood Cliffs, New Jersey, 1962.
6. **Sz-Nagy, B. ; Foiaș, C.,** *Harmonic analysis of operators on Hilbert spaces*, Amsterdam – Budapest, 1970.

Department of Mathematics, INCREST
Bdul Păcii 220, 79622 Bucharest
Romania.

Operator Theory:
Advances and Applications, Vol. 32

Extreme Points in Quotients of Operator Algebras

Kenneth R. Davidson†, Timothy G. Feeman, and Allen L. Shields‡

Dedicated to the memory of Constantin Apostol,
a good friend and an inspiring mathematician.

Let $\mathcal{A}$ be a nest algebra and $\mathcal{K}$ the ideal of compact operators in $\mathcal{L}(\mathcal{H})$. We ask whether or not the closed unit ball of $\frac{\mathcal{L}(\mathcal{H})}{\mathcal{A}+\mathcal{K}}$ has any extreme points and find that the answer depends on the structure of the nest involved. For nests with order type of the extended integers and finite dimensional atoms, we completely characterize the extreme points and show that the closed convex hull of these is not all of $Ball(\frac{\mathcal{L}(\mathcal{H})}{\mathcal{A}+\mathcal{K}})$.

We prove some partial results concerning extreme points of $Ball(\frac{\mathcal{L}(\mathcal{H})}{\mathcal{A}})$ and obtain results on the existence of strong extreme points which closely parallel the situation in $\frac{\mathcal{L}(\mathcal{H})}{\mathcal{A}+\mathcal{K}}$.

1. Background and Preliminaries

In this paper we investigate some aspects of the geometric structure of spaces of the form $\frac{\mathcal{L}(\mathcal{H})}{\mathcal{A}}$ and $\frac{\mathcal{L}(\mathcal{H})}{\mathcal{A}+\mathcal{K}}$ where $\mathcal{L}(\mathcal{H})$ is the space of bounded linear operators on a Hilbert space $\mathcal{H}$, $\mathcal{K}$ is the ideal of compact operators, and $\mathcal{A}$ is a nest algebra of operators. Specifically, we are interested in obtaining information about the extreme points of the closed unit balls of these spaces.

The problem of characterizing the extreme points for the quotient spaces above has a parallel in the corresponding problem for the spaces $\frac{L^\infty}{H^\infty}$ and $\frac{L^\infty}{H^\infty+C}$ where L^∞

† *supported in part by a grant from NSERC*
‡ *supported in part by a grant from NSF*

and C are the spaces of bounded measureable and continuous functions, respectively, on the unit circle. H^∞, the space of bounded analytic functions in the open unit disc, is viewed as a subalgebra of L^∞ via identification with the boundary values. It is known that $\frac{L^\infty}{H^\infty}$ can be identified with the dual space of H^1 and hence the Krein–Milman Theorem implies that the closed unit ball of $\frac{L^\infty}{H^\infty}$ is the closed convex hull of its extreme points. These were completely characterized by Koosis in [8]. A key feature of his result is that for $f + H^\infty$ to be an extreme point there must be a unique $h \in H^\infty$ such that $||f-h||_\infty$ equals the distance from f to H^∞. In [2], Axler, Berg, Jewell, and Shields showed that for each $f \in L^\infty \backslash (H^\infty + C)$ there is more than one $h \in H^\infty + C$ with the property that $||f-h||_\infty$ equals the distance from f to $H^\infty + C$. Applying Koosis' reasoning, they were able to show that the closed unit ball of $\frac{L^\infty}{H^\infty+C}$ has no extreme points and, hence, that $\frac{L^\infty}{H^\infty+C}$ is not the dual space of any Banach space.

The situation in the operator setting is not so clear-cut. In the second section of this paper we illustrate some of the difficulties which arise in connection with the space $\frac{\mathcal{L}(\mathcal{H})}{\mathcal{A}}$. We present sufficient conditions and necessary conditions (though not conditions which are both necessary and sufficient) for $T + \mathcal{A}$ to be an extreme point for certain nest algebras and we give a characterization of those extreme points arising from Toeplitz operators. This result is analogous to that of Koosis.

In the third section we ask whether or not the closed unit ball of $\frac{\mathcal{L}(\mathcal{H})}{\mathcal{A}+\mathcal{K}}$ has any extreme points and find that the answer depends on the structure of the nest algebra involved. Interestingly, our analysis here does not make use of the fact (cf.[5],[7]) that best approximants from $\mathcal{A}+\mathcal{K}$ to operators not in $\mathcal{A}+\mathcal{K}$ are not unique. We do, however, make important use of the formula for the distance between an operator and the space $\mathcal{A}+\mathcal{K}$ (cf.[1],[5]). In case the unit ball of $\frac{\mathcal{L}(\mathcal{H})}{\mathcal{A}+\mathcal{K}}$ does contain extreme points, we show that for nests of a certain type the closed convex hull of these is not the entire closed unit ball. However, in general, this question remains open.

In the final section we return to the space $\frac{\mathcal{L}(\mathcal{H})}{\mathcal{A}}$ and obtain some facts concerning a special type of extreme points known as strong extreme points. The results suggest an intimate connection between strong extreme points in $\frac{\mathcal{L}(\mathcal{H})}{\mathcal{A}}$ and extreme points in $\frac{\mathcal{L}(\mathcal{H})}{\mathcal{A}+\mathcal{K}}$. The precise details of this connection are not yet fully known. We would like to thank J.Cima for bringing the notion of strong extreme points to our attention.

In what follows, $\mathcal{H}$ will be a separable infinite dimensional Hilbert space with $\mathcal{L}(\mathcal{H})$ denoting the algebra of bounded linear operators on $\mathcal{H}$ and $\mathcal{K}$ denoting the ideal of compact operators in $\mathcal{L}(\mathcal{H})$. By a subspace of $\mathcal{H}$ we mean a closed subspace and all projections are assumed self-adjoint. For a projection P let $P^{\perp} = 1 - P$. For two vectors $x, y \in \mathcal{H}$, the operator $x \otimes y^*$ is defined by $(x \otimes y^*)z = (z, y)x$ for $z \in \mathcal{H}$. Note that $||x \otimes y^*|| = ||x||\,||y||$.

If $\mathcal{S}$ is any subset of $\mathcal{L}(\mathcal{H})$ and $T \in \mathcal{L}(\mathcal{H})$ then the distance of T from $\mathcal{S}$ is given by $d(T, \mathcal{S}) = \inf\{||T - S|| : S \in \mathcal{S}\}$. An element S of $\mathcal{S}$ satisfying $||T - S|| = d(T, \mathcal{S})$ is called a closest approximant to T from $\mathcal{S}$. $Lat\mathcal{S}$ denotes the set of all projections P for which $PSP = SP$ whenever $S \in \mathcal{S}$. If $\mathcal{P}$ is a set of projections in $\mathcal{L}(\mathcal{H})$, then $Alg\mathcal{P}$ is the set of all operators T in $\mathcal{L}(\mathcal{H})$ for which $PTP = TP$ whenever $P \in \mathcal{P}$. A subalgebra $\mathcal{S}$ of $\mathcal{L}(\mathcal{H})$ is reflexive if and only $AlgLat\mathcal{S} = \mathcal{S}$.

A nest is a family of projections which is linearly ordered by range inclusion, contains 0 and 1, and is closed in the strong operator topology (SOT). A nest algebra is a subalgebra $\mathcal{A}$ of $\mathcal{L}(\mathcal{H})$ satisfying $\mathcal{A} = Alg\mathcal{P}$ for some nest $\mathcal{P}$. Equivalently, a nest algebra is a reflexive subalgebra $\mathcal{A}$ for which $Lat\mathcal{A}$ is linearly ordered. For an element P of a nest $\mathcal{P}$, define $P_- = \sup\{E \in \mathcal{P} : E < P\}$ and $P_+ = \inf\{E \in \mathcal{P} : E > P\}$. Thus, either $P_- = P$ or P_- is the immediate predecessor to P in the nest. Similar statements hold for P_+. Any non-zero projections $(P - P_-)$ are called atoms of the nest and $\mathcal{P}$ is said to be purely atomic if $\sum_{P\in\mathcal{P}}(P - P_-) = 1$ where the sum is taken in the strong operator topology In [1], Arveson established the following formula, valid for all nests $\mathcal{P}$.

$$d(T,\, Alg\mathcal{P}) = \sup\{||P^{\perp}TP|| \,:\, P \in \mathcal{P}\}\,,\ for\ T \in \mathcal{L}(\mathcal{H}). \tag{1.1}$$

A standard weak*-compactness argument shows that, given $T \in \mathcal{L}(\mathcal{H})$, there exists $A \in Alg\mathcal{P}$ such that $||T - A|| = d(T,\, Alg\mathcal{P})$. In [5] and [7], it is shown that $\frac{Alg\mathcal{P}+\mathcal{K}}{Alg\mathcal{P}}$ is an M-ideal in $\frac{\mathcal{L}(\mathcal{H})}{Alg\mathcal{P}}$. In particular, this implies that, given $T \in \mathcal{L}(\mathcal{H})$, there exists $A \in (Alg\mathcal{P} + \mathcal{K})$ satisfying $||T - A|| = d(T,\, Alg\mathcal{P} + \mathcal{K})$. If $T \notin (Alg\mathcal{P} + \mathcal{K})$, then the choice of A is not unique. The reader is referred to [4] for more information about nest algebras.

Throughout this paper, the symbol $\mathcal{E}$ will denote the following nest of projections. Taking $\{e_n : n \in \mathbf{N} \cup \{0\}\}$ to be an orthonormal basis for $\mathcal{H}$, define E_n, for $n \in \mathbf{N} \cup \{0\}$, to be the projection onto the subspace spanned by the set $\{e_j : j \leq n\}$. We set $\mathcal{E} = \{E_n : n \in \mathbf{N} \cup \{0\}\} \cup \{0, 1\}$. When more general nests are involved, they will be denoted by $\mathcal{P}$.

For a Banach space X, the closed unit ball of X is $Ball(X) = \{x \in X : ||x|| \leq 1\}$. The element x in X is an extreme point of $Ball(X)$ if and only if $||x|| = 1$ and there is no non-zero $y \in X$ satisfying $\max\{||x+y||, ||x-y||\} \leq 1$. This is equivalent to the condition that x is not a convex combination of other elements of $Ball(X)$. A unit vector $x \in X$ is said to be a strong extreme point of $Ball(X)$ if, and only if, for every $\epsilon > 0$ there exists $\delta > 0$ such that $||y|| \leq \epsilon$ whenever $\max\{||x+y||, ||x-y||\} \leq 1+\delta$. Every strong extreme point is an extreme point. Indeed, if x is not an extreme point then there exists $y \neq 0$ such that $\max\{||x \pm y||\} \leq 1$. Taking $\epsilon = ||y||/2$, we see that x is not a strong extreme point. In [3], Cima and Thomson studied the strong extreme points of the unit balls of H^p spaces.

2. Extreme Points of $Ball(\frac{\mathcal{L}(\mathcal{H})}{\mathcal{A}})$

In this section, we consider the problem of characterizing the extreme points of $Ball(\frac{\mathcal{L}(\mathcal{H})}{\mathcal{A}})$ where $\mathcal{A}$ is a nest algebra. If $T \in \mathcal{L}(\mathcal{H})$ satisfies $d(T, \mathcal{A}) = 1$, then $T + \mathcal{A}$ is an extreme point of $Ball(\frac{\mathcal{L}(\mathcal{H})}{\mathcal{A}})$ if and only if the condition $d(T \pm S, \mathcal{A}) \leq 1$ for some $S \in \mathcal{L}(\mathcal{H})$ implies $S \in \mathcal{A}$.

As a first step, suppose $d(T, \mathcal{A}) = 1$ where $\mathcal{A} = Alg\mathcal{P}$ and suppose $||P^\perp TP|| < 1$ for some $P \in \mathcal{P}$ satisfying $P_- < P < P_+$. Now, let $c > 0$ satisfy $||P^\perp TP|| + c \leq 1$ and let x and y be unit vectors in the ranges of $(P - P_-)$ and $(P_+ - P)$ respectively. Define $S = c(y \otimes x^*)$, that is $Sz = c(z,x)y \; for \; z \in \mathcal{H}$. Then $S \notin \mathcal{A}$ and, for $E \in \mathcal{P}$ satisfying $E \leq P_-$ or $E \geq P_+$, we have $E^\perp SE = 0$ and hence $||E^\perp(T \pm S)E|| = ||E^\perp TE|| \leq 1$. If $E = P$, then $||E^\perp(T \pm S)E|| \leq ||P^\perp TP|| + c \leq 1$. Thus, $d(T \pm S, \mathcal{A}) \leq 1$ and we see that $T + \mathcal{A}$ is not an extreme point of $Ball(\frac{\mathcal{L}(\mathcal{H})}{\mathcal{A}})$. We have proven the following.

Lemma 1: *Let $\mathcal{P}$ be a nest and let $\mathcal{A} = Alg\mathcal{P}$. If $T + \mathcal{A}$ is an extreme point* *of* $Ball(\frac{\mathcal{L}(\mathcal{H})}{\mathcal{A}})$, *then* $||P^{\perp}TP|| = 1$ *whenever* $P \in \mathcal{P}$ *satisfies* $P_- < P < P_+$.

In particular, if $T + Alg\mathcal{E}$ is an extreme point, then $||E_n{}^{\perp}TE_n|| = 1$ for all non-negative integers n. The converse of the lemma is not true, however, as the next example shows.

EXAMPLE 1: For $j \geq 0$, define

$$Te_j = \begin{cases} e_{j+2}, & \text{if } j \neq 1; \\ 0, & \text{if } j = 1. \end{cases}$$

Examination of the matrix representation for T shows that $||E_n{}^{\perp}TE_n|| = 1$ for all n. Yet, if we define $S = e_3 \otimes e_1{}^*$ then $S \notin Alg\mathcal{E}$ and $d(T \pm S,\ Alg\mathcal{E}) \leq 1$. Hence, $T + Alg\mathcal{E}$ is not an extreme point of $Ball(\frac{\mathcal{L}(\mathcal{H})}{Alg\mathcal{E}})$.

The next example shows it is possible to have $Te_k = 0$ for some values of k and still have $T + Alg\mathcal{E}$ as an extreme point.

EXAMPLE 2: Define

$$Te_j = \begin{cases} e_2, & \text{if } j = 0; \\ 0, & \text{if } j = 1; \\ e_{j+1}, & \text{if } j \geq 2. \end{cases}$$

Examination of the matrix for T again reveals that $||E_n{}^{\perp}TE_n|| = 1$ for all n and hence that $d(T,\ Alg\mathcal{E}) = 1$. Suppose now that $d(T \pm S,\ Alg\mathcal{E}) \leq 1$ for some $S \in \mathcal{L}(\mathcal{H})$. Since each column in the matrix of T, except the second, has norm 1 below the diagonal, it follows that the matrix of S may only have non-zero sub-diagonal entries in the second column. However, every row of T from the third row onward has norm 1 to the left of the diagonal which implies that S must have zero entries below the diagonal in the second column as well. In short, the condition $d(T \pm S,\ Alg\mathcal{E}) \leq 1$ implies that $S \in Alg\mathcal{E}$ and, hence, that $T + Alg\mathcal{E}$ is an extreme point.

It is clear from the last example that if $Te_k = 0$ for some k and if $T^*e_j = 0$ for some $j > k$ then $T + Alg\mathcal{E}$ is not an extreme point.

The situation in Example 2 can be generalized somewhat to yield a sufficient condition for $T + Alg\mathcal{P}$ to be an extreme point of $Ball(\frac{\mathcal{L}(\mathcal{H})}{Alg\mathcal{P}})$ in case the nest $\mathcal{P}$ is purely atomic. We first need the following lemma which will be used repeatedly throughout the remainder of the paper.

LEMMA 2: *Let S be an element of the C^* algebra $\mathcal{B}$. If either S or S^* is an isometry, then, for each $W \in \mathcal{B}$,* $\max\{||S+W||, ||S-W||\} \geq \sqrt{1+||W||^2}$.

PROOF: If S is an isometry then $S^*S = 1$ so that

$$\frac{1}{2}[(S+W)^*(S+W) + (S-W)^*(S-W)] = 1 + W^*W.$$

Hence,

$$\begin{aligned} 1 + ||W||^2 &= ||\frac{1}{2}[(S+W)^*(S+W) + (S-W)^*(S-W)]|| \\ &\leq \frac{1}{2}(||S+W||^2 + ||S-W||^2) \\ &\leq \max\{||S+W||^2, ||S-W||^2\}. \end{aligned}$$

Taking square roots yields the desired conclusion. Similar arguments apply in case $SS^* = 1$.

PROPOSITION 3: *Let $\mathcal{A} = Alg\mathcal{P}$ where $\mathcal{P}$ is a purely atomic nest. Let $T \in \mathcal{L}(\mathcal{H})$ satisfy $d(T, \mathcal{A}) = 1$ and suppose there is a projection $P_0 \in \mathcal{P}$ such that $P^\perp T(P-P_-)$ is isometric on the range of $(P-P_-)$ for each $P \leq P_0$ while $PT^*(P_+-P)$ is isometric on the range of $(P_+ - P)$ for all $P > P_0$. Then $T + \mathcal{A}$ is an extreme point of $Ball(\frac{\mathcal{L}(\mathcal{H})}{\mathcal{A}})$.*

PROOF: Let $S \in \mathcal{L}(\mathcal{H})$ satisfy $d(T \pm S, \mathcal{A}) \leq 1$. We will show that $S \in \mathcal{A}$ so that $T + \mathcal{A}$ is an extreme point. By the distance formula (1.1), the condition on S is that $||P^\perp(T \pm S)P|| \leq 1$ for all $P \in \mathcal{P}$.

Let $Q \in \mathcal{P}$. If $Q^\perp T(Q - Q_-)$ is isometric on range$(Q - Q_-)$ then, by Lemma 2, we have

$$\begin{aligned} 1 \geq \max\{||Q^\perp(T \pm S)Q||\} &\geq \max\{||Q^\perp(T \pm S)(Q - Q_-)||\} \\ &\geq (1 + ||Q^\perp S(Q - Q_-)||^2)^{\frac{1}{2}}. \end{aligned}$$

We conclude that $Q^\perp S(Q - Q_-) = 0$ whenever $Q^\perp T(Q - Q_-)$ is isometric.

Next, let $Q \in \mathcal{P}$ and suppose that $Q^\perp T(Q - Q_-)$ is not isometric. By hypothesis, $ET^*(E_+ - E)$ is isometric on the range of $(E_+ - E)$ for all $E \geq Q$. Using the Lemma again, we see that

$$\begin{aligned} 1 &\geq \max\{||E^\perp(T \pm S)E||\} \geq \max\{||(E_+ - E)(T \pm S)E||\} \\ &\geq (1 + ||(E_+ - E)SE||^2)^{\frac{1}{2}}. \end{aligned}$$

From this we conclude that $(E_+ - E)SE = 0$ for all $E \geq Q$. Since the nest is purely atomic, it follows that $Q^\perp = \sum_{E \geq Q}(E_+ - E)$. Hence,

$$\begin{aligned} Q^\perp SQ &= \sum_{E \geq Q} (E_+ - E)SQ \\ &= \sum_{E \geq Q} (E_+ - E)SEQ = 0. \end{aligned}$$

Now let $P \in \mathcal{P}$. Since $\mathcal{P}$ is atomic we have $P = \sum_{Q \leq P}(Q - Q_-)$ and, hence, $P^\perp SP = \sum_{Q \leq P} P^\perp S(Q - Q_-)$. For $Q \leq P$, either $Q^\perp T(Q - Q_-)$ is isometric in which case the above discussion implies that

$$P^\perp S(Q - Q_-) = P^\perp Q^\perp S(Q - Q_-) = 0$$

or $Q^\perp T(Q - Q_-)$ is co-isometric in which case we see that

$$P^\perp S(Q - Q_-) = P^\perp (Q^\perp SQ)(Q - Q_-) = 0.$$

In any case, $P^\perp S(Q - Q_-) = 0$ for all $Q \leq P$ and thus $P^\perp SP = 0$. Since P was arbitrary we conclude that $S \in \mathcal{A} = Alg\mathcal{P}$ which proves the proposition.

EXAMPLE 3: The converse of Proposition 3 is not true, however. For instance, let T have the following matrix with respect to the basis $\{e_n : n \geq 0\}$:

$$T = \begin{pmatrix} 0 & 0 & 0 & 0 & \dots \\ 1/\sqrt{2} & 0 & 0 & 0 & \dots \\ 1/2 & 1/2 & 0 & 0 & \dots \\ 1/2 & 1/2 & 0 & 0 & \dots \\ 0 & 0 & 1 & 0 & \dots \\ 0 & 0 & 0 & 1 & \ddots \\ \vdots & \vdots & \vdots & \ddots & \ddots \end{pmatrix}$$

It is clear that $d(T, Alg\mathcal{E}) = 1$ but T fails to satisfy the hypotheses of the Proposition since ${E_1}^\perp T(E_1 - E_0)$ is not isometric and $(E_2 - E_1)TE_1$ is not co-isometric. Nonetheless, $T + Alg\mathcal{E}$ is an extreme point. Indeed, suppose $d(T \pm S, Alg\mathcal{E}) \leq 1$. Since every column of T except the second has norm 1 below the diagonal, it follows that S may have non-zero sub-diagonal entries only in the second column. In addition, these non-zero entries may only occur in the third and fourth rows. Call them s_2 and s_3. We have,

$$\max\{\|E_1{}^\perp(T \pm S)E_1\|\} = \max\left\{\left\|\begin{pmatrix} 1/2 & 1/2 \pm s_2 \\ 1/2 & 1/2 \pm s_3 \end{pmatrix}\right\|\right\}.$$

Notice that $\binom{1/\sqrt{2}}{1/\sqrt{2}}$ is a unit vector and that

$$\begin{pmatrix} 1/2 & 1/2 \pm s_2 \\ 1/2 & 1/2 \pm s_3 \end{pmatrix}\begin{pmatrix} 1/\sqrt{2} \\ 1/\sqrt{2} \end{pmatrix} = \begin{pmatrix} 1/\sqrt{2} \\ 1/\sqrt{2} \end{pmatrix} \pm \begin{pmatrix} s_2/\sqrt{2} \\ s_3/\sqrt{2} \end{pmatrix}.$$

It follows that $\max\{\|E_1{}^\perp(T \pm S)E_1\|\} > 1$ unless $s_2 = s_3 = 0$. In other words, we must have $S \in Alg\mathcal{E}$ which implies that $T + Alg\mathcal{E}$ is an extreme point of $Ball(\frac{\mathcal{L}(\mathcal{H})}{Alg\mathcal{E}})$.

EXAMPLE 4: In contradistinction to Lemma 1, it is possible for $T + \mathcal{A}$ to be extreme, yet $P^\perp TP = 0$ for some proper element P in the nest. (By proper, we mean $0 < P < 1$.) Consider the following nest of projections acting on the space $\mathcal{H} \oplus \mathcal{H}$. Let $\{e_n : n \geq 1\}$ and $\{f_n : n \geq 1\}$ be two orthonormal bases for $\mathcal{H}$. For each $n \geq 1$, let E_n be the projection onto the subspace spanned by $\{e_k : 1 \leq k \leq n\}$ and take F_n to be projection onto $\mathcal{H} \oplus \vee\{f_k : 1 \leq k \leq n\}$. Define F_0 to be projection onto $\vee\{e_n : n \geq 1\}$ and set $\mathcal{P} = \{E_n, F_j, 0, 1 : n \geq 1,\ j \geq 0\}$. Now define the operator T by $Te_n = e_{n+1}$ and $Tf_n = f_{n+1}$ for all $n \geq 1$. Clearly, $F_0^\perp TF_0 = 0$. Nonetheless, $T + \mathcal{A}$ is extreme by Proposition 3. Note that $(F_0)_+ = F_1 > F_0 = (F_0)_-$. Similar examples can be constructed so that $P = P_+ = P_-$ or $P_+ = P > P_-$.

EXAMPLE 5: The set of extreme points of $Ball(\frac{\mathcal{L}(\mathcal{H})}{Alg\mathcal{E}})$ is not closed. Indeed, even if $\mathcal{H}$ is 6 dimensional this set is not closed. For $\epsilon > 0$, let $s = \sin\epsilon$, $c = \cos\epsilon$, and define

$$T_\epsilon = \begin{pmatrix} 0 & 0 & 0 & 0 & 0 & 0 \\ s & -c & 0 & 0 & 0 & 0 \\ 0 & 0 & 0 & 0 & 0 & 0 \\ c & s & 0 & 0 & 0 & 0 \\ 0 & 0 & s & 0 & -c & 0 \\ 0 & 0 & c & 0 & s & 0 \end{pmatrix}$$

To see that $T_\epsilon + Alg\mathcal{E}$ is extreme for $0 < \epsilon < \pi/2$, note that every subdiagonal column of T_ϵ has norm one except the second, fourth, and fifth, while every subdiagonal row has norm one except the second, third, and fifth. Thus, if S satisfies $d(T \pm S, Alg\mathcal{E}) \leq 1$, then the only possible non-zero subdiagonal entries of S are in the (3,2), (5,2), and

(5,4) positions. However, the condition on S also requires that the third and fifth rows of S be orthogonal to the fourth row of T_ϵ so that the (3,2) and (5,2) entries of S are 0. Similarly, the fourth column of S must be orthogonal to the third column of T_ϵ so that S has a 0 in the (5,4) entry as well. It follows that S is in $Alg\mathcal{E}$ and hence that $T_\epsilon + Alg\mathcal{E}$ is an extreme point.

As ϵ tends to 0, T_ϵ approaches a limit T_0 which does not represent an extreme point. For instance, if S has a 1 as its (3,2) entry and 0's elsewhere, then $d(T_0 \pm S, Alg\mathcal{E}) \leq 1$. Hence the set of extreme points is not closed. Note also that both T_ϵ and T_ϵ^* have non-trivial kernel and that T_ϵ does not satisfy the hypotheses of Proposition 3.

Let us now turn our attention to a special class of extreme points of $Ball(\frac{\mathcal{L}(\mathcal{H})}{Alg\mathcal{E}})$, namely those arising from the classical Toeplitz operators. For each function $f \in L^\infty$, the Toeplitz operator with symbol f, denoted by T_f, is the operator in $\mathcal{L}(\mathcal{H})$ defined by the equations $(T_f e_j, e_{j+n}) = \hat{f}(n)$ for all $j \geq 0$ and all integers n. Here $\hat{f}(n)$ denotes the nth Fourier coefficient of the function f. Note that $T_f \in Alg\mathcal{E}$ if and only if $\hat{f}(n) = 0$ for all $n > 0$, that is $\bar{f} \in H^\infty$. It is well known that, for $f \in L^\infty$, $||T_f|| = ||f||_\infty$ and $||f||_2 = (\sum_{n=-\infty}^{\infty} |\hat{f}(n)|^2)^{1/2}$. Moreover, $||f||_\infty = ||f||_2$ if and only if f has constant modulus almost everywhere on the unit circle. Keeping in mind the parallels which exist between nest algebras and the function space H^∞, our initial point of reference is the following theorem due to Koosis.

THEOREM 4 ([8]): *For $f \in L^\infty$, $f + H^\infty$ is an extreme point of $Ball(\frac{L^\infty}{H^\infty})$ if and only if there exists $u \in f + H^\infty$ such that $|u| = 1$ almost everywhere and $||u-h||_\infty > 1$ for all non-zero $h \in H^\infty$.*

The last condition of the Theorem says, in particular, that f must have a unique closest approximant from H^∞ in order for $f + H^\infty$ to be an extreme point. That this is not the case for $Ball(\frac{\mathcal{L}(\mathcal{H})}{Alg\mathcal{E}})$ is illustrated by Example 2 above. For instance, both $A = 0$ and $A = e_1 \otimes e_1{}^*$ are best approximants from $Alg\mathcal{E}$ to the operator T in that example. However, the Toeplitz operators are better behaved as the following result shows.

THEOREM 5: *Let $f \in L^\infty$. Then $T_f + Alg\mathcal{E}$ is an extreme point of*

$Ball(\frac{\mathcal{L}(\mathcal{H})}{Alg\mathcal{E}})$ if and only if there is an inner function $u \in H^\infty$ such that $u(0) = 0$ and $T_u \in T_f + Alg\mathcal{E}$.

PROOF: Suppose first that $u \in H^\infty$ is an inner function (so that $|u| = 1$ a.e.) satisfying $u(0) = 0$. We have $||T_u|| = ||u||_\infty = 1$ and also

$$||E_0{}^\perp T_u E_0||^2 = \sum_{n\geq 1} |\hat{u}(n)|^2 = ||u||_2{}^2 = 1.$$

Arveson's distance formula (1.1) implies that $1 \leq d(T_u, Alg\mathcal{E}) \leq ||T_u|| = 1$ so equality holds throughout. Since $||E_k{}^\perp T_u e_k||^2 = ||u||_2{}^2 = 1$ for all $k \geq 0$, Proposition 3 implies that $T_u + Alg\mathcal{E}$ is an extreme point.

For the converse, suppose that $T_f + Alg\mathcal{E}$ is an extreme point of $Ball(\frac{\mathcal{L}(\mathcal{H})}{Alg\mathcal{E}})$. Lemma 1 implies that $1 = ||E_0{}^\perp T_f E_0||^2 = \sum_{n\geq 1} |\hat{f}(n)|^2$ which in turn implies that $||E_k{}^\perp T_f e_k|| = 1$ for all $k \geq 0$. If $A \in Alg\mathcal{E}$ is chosen so that $||T_f - A|| = d(T_f, Alg\mathcal{E}) = 1$, then, for each $k \geq 1$, we have

$$1 = ||E_k^\perp T_f e_k|| = ||E_k^\perp (T_f - A)e_k|| \leq ||(T_f - A)e_k|| \leq 1.$$

It follows that $(T_f - A)e_k = E_k^\perp T_f e_k$ for all k so that $T_f - A$ is strictly lower triangular. It is immediate that the choice of A is unique. Moreover, $T_f - A$ is a Toeplitz operator T_u satisfying

$$||u||_\infty = ||T_u|| = 1 = ||E_0^\perp T_u e_0|| = ||u||_2.$$

Thus, $u \in H^\infty$, $u(0) = 0$, and $|u| = 1$ almost everywhere (so that u is an inner function). This proves the Theorem.

Comparison of Theorem 5 with Koosis' result leads to the following corollary.

COROLLARY 6: *$T_f + Alg\mathcal{E}$ is an extreme point of $Ball(\frac{\mathcal{L}(\mathcal{H})}{Alg\mathcal{E}})$ if and only if the following conditions hold:*

(i) $\bar{f} + H^\infty$ is an extreme point of $Ball(\frac{L^\infty}{H^\infty})$;

(ii) $\sum_{n\geq 1} |\hat{f}(n)|^2 = 1$.

PROOF: The proof of Theorem 5 shows that if $T_f + Alg\mathcal{E}$ is an extreme point then, in fact, $f \in H_0^\infty + \overline{H^\infty}$ and moreover $\sum_{n\geq 1} |\hat{f}(n)|^2 = 1$. This together with Theorem 4 then implies that $\bar{f} + H^\infty$ is an extreme point in $Ball(\frac{L^\infty}{H^\infty})$.

Conversely, suppose f in L^∞ satisfies the two conditions given above. From Theorem 4, we may assume without loss of generality that $|f| = 1$ a.e. and, hence, $1 = ||f||_2^2 = \sum_{n=-\infty}^{\infty} |\hat{f}(n)|^2$. This and condition (ii) now imply that $\hat{f}(n) = 0$ for all $n \leq 0$. Thus, f is in H_0^∞ and is inner. Theorem 5 now yields the desired conclusion.

The problem of completely classifying the extreme points of $Ball(\frac{\mathcal{L}(\mathcal{H})}{\mathcal{A}})$ for an arbitrary nest algebra $\mathcal{A}$ seems fairly intractable. We will see in the last section that some general information can be obtained by focussing attention on a special type of extreme points known as strong extreme points.

3. Extreme Points in $Ball(\frac{\mathcal{L}(\mathcal{H})}{\mathcal{A}+\mathcal{K}})$

Let $\mathcal{P}$ be a nest and $\mathcal{A} = Alg\mathcal{P}$. It is known (cf.[6], [7]) that $\frac{\mathcal{L}(\mathcal{H})}{\mathcal{A}}$ can be identified with the dual space of $(tc) \cap \mathcal{A}_0$ where (tc) denotes the trace class in $\mathcal{L}(\mathcal{H})$ and $\mathcal{A}_0 = \{T \in \mathcal{L}(\mathcal{H}) : TP = P_-TP, \forall P \in \mathcal{P}\}$. We wish to determine whether or not $\frac{\mathcal{L}(\mathcal{H})}{\mathcal{A}+\mathcal{K}}$ is also a dual space and, to this end, we ask whether or not $Ball(\frac{\mathcal{L}(\mathcal{H})}{\mathcal{A}+\mathcal{K}})$ has any extreme points. As we shall see, the answer is "no" in case $\mathcal{P}$ is a sequence of finite rank projections increasing strongly to the identity. Thus, $\frac{\mathcal{L}(\mathcal{H})}{Alg\mathcal{P}+\mathcal{K}}$ is not a dual space for such nests. However, extreme points do exist for more general nests. In case the nest in question is indexed by the integers with finite dimensional atoms, we give a complete characterization of the extreme points of $Ball(\frac{\mathcal{L}(\mathcal{H})}{\mathcal{A}+\mathcal{K}})$ from which it follows that the closed convex hull of these extreme points is not the entire closed unit ball of $\frac{\mathcal{L}(\mathcal{H})}{\mathcal{A}+\mathcal{K}}$. We conclude that $\frac{\mathcal{L}(\mathcal{H})}{\mathcal{A}+\mathcal{K}}$ is not a dual space in this case either. For other nests, determining the closed convex hull of the extreme points of $Ball(\frac{\mathcal{L}(\mathcal{H})}{\mathcal{A}+\mathcal{K}})$ remains an open problem.

In the discussion below, we shall make use of Lemma 2 from the previous section. Also, for later use we observe now that if A is an $n \times n$ matrix each entry of which is the number 1 then the operator norm of A is n. Indeed, $||A|| = ||(e_1 + \cdots + e_n) \otimes (e_1 + \cdots + e_n)^*|| = ||(e_1 + \cdots + e_n)||^2 = n$. Consequently, an $n \times n$ matrix each entry of which is the number $1/n$ has operator norm equal to 1. The norm of each row and column of such a matrix is $1/\sqrt{n}$.

We shall make use of the formula for $d(T, \mathcal{A}+\mathcal{K})$ established by Davidson and Power in [5]. This formula is obtained as follows. Given a nest $\mathcal{P}$, let $C_{s^*}(\mathcal{P}, \mathcal{L}(\mathcal{H}))$ and $C_n(\mathcal{P}, \mathcal{K})$ respectively denote the C^*-algebra of all *- strongly continuous functions from$\mathcal{P}$ into $\mathcal{L}(\mathcal{H})$ and the norm-closed two-sided ideal of norm continuous functions from $\mathcal{P}$ into $\mathcal{K}$. Let $q : C_{s^*}(\mathcal{P}, \mathcal{L}(\mathcal{H})) \longrightarrow \frac{C_{s^*}(\mathcal{P}, \mathcal{L}(\mathcal{H}))}{C_n(\mathcal{P}, \mathcal{K})}$ be the quotient map and, for $F \in C_{s^*}(\mathcal{P}, \mathcal{L}(\mathcal{H}))$, define $||F||_e = ||qF||$. Next, define the map $\Phi : \mathcal{L}(\mathcal{H}) \longrightarrow C_{s^*}(\mathcal{P}, \mathcal{L}(\mathcal{H}))$ by

$$\Phi(A)(P) = P^{\perp}AP, \quad \forall A \in \mathcal{L}(\mathcal{H}), \forall P \in \mathcal{P}.$$

Thus, Φ is a contractive linear map whose kernel is $Alg\mathcal{P}$. Moreover, the distance formula (1.1) for nest algebras asserts that

$$d(A, Alg\mathcal{P}) = ||\Phi(A)||, \quad \forall A \in \mathcal{L}(\mathcal{H}).$$

Fall, Arveson, and Muhly proved in [6] that $A \in (Alg\mathcal{P}+\mathcal{K})$ if and only if $q(\Phi(A)) = 0$. The distance formula proved by Davidson and Power asserts that

$$d(A, Alg\mathcal{P}+\mathcal{K}) = ||\Phi(A)||_e, \quad \forall A \in \mathcal{L}(\mathcal{H}). \tag{3.1}$$

In case $\mathcal{P} = \{P_n\}$ is a sequence of finite rank projections increasing strongly to the identity, formula (3.1) simplifies to give

$$d(A, Alg\mathcal{P}+\mathcal{K}) = \limsup_{n\longrightarrow\infty} ||P_n{}^{\perp}AP_n||. \tag{3.2}$$

This was first established by Arveson in [1].

THEOREM 7: *If $\mathcal{P} = \{P_n\}$ is a sequence of finite rank projections increasing strongly to the identity and if $\mathcal{A} = Alg\mathcal{P}$, then $Ball(\frac{\mathcal{L}(\mathcal{H})}{\mathcal{A}+\mathcal{K}})$ has no extreme points.*

PROOF: For simplicity, we will work with the nest $\mathcal{E}$ which has one dimensional atoms. Let $\mathcal{A} = Alg\mathcal{E}$ and let $T \in \mathcal{L}(\mathcal{H})$ satisfy $d(T, \mathcal{A}+\mathcal{K}) = 1$. Since best approximants to T from $\mathcal{A}+\mathcal{K}$ exist [cf. 4, 6] we may assume that $||T|| = 1$ as well. Given $\delta > 0$ there is an increasing sequence of non-negative integers $n_0 < n_1 < n_2 < \cdots$ such that $||E_{n_k}^{\perp}T(E_k - E_{k-1})|| < \delta/2^{k+1}$ for each $k \geq 0$. (Set $E_{-1} = 0$.) Define $T_1 = \sum_{k=0}^{\infty} E_{n_k}{}^{\perp}T(E_k - E_{k-1})$ and set $T_2 = T - T_1$. Note that T_1 is compact and that

$||T_1|| < \delta/2$. Now choose integers $m_0 < m_1 < m_2 < \cdots$ so that, for each $k \geq 0$, there is at most one m_i satisfying $k \leq m_i \leq n_k$. For each natural number l, define

$$W_l = \frac{1}{2^l} \sum_{i=1}^{2^l} \sum_{j=1}^{2^l} (e_{m_{2^{l+1}+i}} \otimes e_{m_{2^l+j}}{}^*)$$

and let

$$W = \sum_{l=1}^{\infty} W_l.$$

The non-zero entries in the matrix of W_l form a $2^l \times 2^l$ block each entry of which is the number $1/2^l$. By our earlier remarks, it follows that $||W_l|| = 1$ for all l. Also, the sets of rows and columns containing the non-zero entries of W_l and W_{l+1} are disjoint for all values of l. This implies that W is actually a direct sum and hence $||W|| = 1$. Also, since $E^{\perp}_{m_{2^{l+1}}} W E_{m_{2^{l+1}}} = W_l$, for each $l \geq 1$, formula (3.2) shows that $d(W, \mathcal{A} + \mathcal{K}) = 1$.

For each $n \geq 0$, there exists $k \geq 0$ such that $n_{k-1} < n \leq n_k$. Thus, $E_n{}^{\perp} T_2 E_n = (E_{n_n} - E_n) T_2 (E_n - E_{k-1})$. Moreover, by construction, there is at most one i for which $k \leq m_i \leq n_k$ and at most one j for which $n \leq m_j \leq n_n$. There is at most one l satisfying both $2^{l+1} + 1 \leq j \leq 3(2^l)$ and $2^l + 1 \leq i \leq 2^{l+1}$ in which case we have

$$\begin{aligned}
||E_n{}^{\perp}(T_2 \pm W)E_n|| &= ||(E_{n_n} - E_n)T_2(E_n - E_{k-1}) \pm E_n{}^{\perp} W_l E_n|| \\
&\leq \max\{||(E_{n_n} - E_n)(T_2 \pm W_l)(E_n - E_{k-1})||, ||E^{\perp}_{n_n} W_l E_{k-1}||\} \\
&\quad + \max\{||(E_{n_n} - E_n) W_l E_{k-1}||, ||E^{\perp}_{n_n} W_l (E_n - E_{k-1})||\} \\
&\leq \max\{||(E_{n_n} - E_n)T_2(E_n - E_{k-1})|| + ||1/2^l(e_{m_j} \otimes e_{m_i}{}^*)||, 1\} \\
&\quad + \max\{||1/2^l(e_{m_j} \otimes (\sum_{r=2^l+1}^{i-1} e_{m_r})^*)||, \\
&\qquad\qquad ||1/2^l((\sum_{r=j+1}^{3(2^l)} e_{m_r}) \otimes e_{m_i}{}^*)||\} \\
&\leq \max\{||T|| + ||E_n{}^{\perp} T_1 E_n|| + 2^{-l}, 1\} + \max\{2^{-l/2}, 2^{-l/2}\} \\
&= 1 + 2^{-l/2} + 2^{-l} + ||E_n{}^{\perp} T_1 E_n||.
\end{aligned}$$

Thus,

$$\begin{aligned}
||E_n{}^{\perp}(T \pm W)E_n|| &\leq ||E_n{}^{\perp}(T_2 \pm W)E_n|| + ||E_n{}^{\perp} T_1 E_n|| \\
&\leq 1 + 2^{-l} + 2^{-l/2} + 2||E_n{}^{\perp} T_1 E_n||.
\end{aligned}$$

As $n \to \infty$, the value of l also approaches ∞. Moreover, since $T_1 \in \mathcal{K}$, it follows that $\lim_{n\to\infty} ||E_n{}^{\perp} T_1 E_n|| = 0$. Hence,

$$d(T \pm W, \mathcal{A} + \mathcal{K}) = \limsup_{n\to\infty} ||E_n{}^{\perp}(T \pm W)E_n|| \leq 1.$$

Since $W \notin (\mathcal{A} + \mathcal{K})$, we see that $T + (\mathcal{A} + \mathcal{K})$ is not an extreme point of $Ball(\frac{\mathcal{L}(\mathcal{H})}{\mathcal{A}+\mathcal{K}})$.

This construction can clearly be modified to yield the desired conclusion in case the atoms of the nest are of arbitrary finite dimension. The proof of the theorem is now completed.

For each coset $T + \mathcal{A} + \mathcal{K}$ with quotient norm equal to 1 (that is, $d(T, \mathcal{A} + \mathcal{K}) = 1$), we have constructed $W + \mathcal{A} + \mathcal{K}$ of (quotient) norm 1 such that $d(T \pm W, \mathcal{A} + \mathcal{K}) = 1$. Repetition of this argument shows that, for any $n \geq 1$, there exist $W_1, W_2, \ldots, W_n$ such that $d(W_k, \mathcal{A} + \mathcal{K}) = 1$ for all k and $d(T \pm W_1 \pm W_2 \pm \ldots \pm W_n, \mathcal{A} + \mathcal{K}) = 1$. Thus, the unit sphere of $\frac{\mathcal{L}(\mathcal{H})}{\mathcal{A}+\mathcal{K}}$ has many flat spots.

We remark that the conclusion of the previous Theorem also holds, by taking adjoints, when the nest consists of a sequence of projections of finite co-rank decreasing strongly to 0. The key to the construction, as we shall see below, lies in our ability to construct the small compact operator T_1. This relies on the assumption that each projection in the nest has finite rank (or co-rank). When this condition on the nest fails, then $Ball(\frac{\mathcal{L}(\mathcal{H})}{Alg\mathcal{P}+\mathcal{K}})$ does admit extreme points.

PROPOSITION 8: *Let $\mathcal{P}$ be a nest and set $\mathcal{A} = Alg\mathcal{P}$. If $T \in \mathcal{L}(\mathcal{H})$ has the property that, for all $P \in \mathcal{P}$, either $P^{\perp}TP$ is isometric on the range of P or $PT^*P^{\perp}$ is isometric on the range of $P^{\perp}$ then $T + (\mathcal{A} + \mathcal{K})$ is an extreme point of $Ball(\frac{\mathcal{L}(\mathcal{H})}{\mathcal{A}+\mathcal{K}})$.*

PROOF: With T as hypothesized, suppose that $W \in \mathcal{L}(\mathcal{H})$ satisfies $d(T \pm W, \mathcal{A} + \mathcal{K}) \leq 1$. We will show that $W \in \mathcal{A} + \mathcal{K}$ which amounts to showing that the map $\Phi(W)$ is norm continuous from $\mathcal{P}$ into $\mathcal{K}$.

Denote $\Phi_+ = \Phi(T + W)$ and $\Phi_- = \Phi(T - W)$. The condition $d(T \pm W, \mathcal{A} + \mathcal{K}) \leq 1$ is equivalent to $||\Phi_\pm||_e \leq 1$. Also, choose compact operators K_1 and K_2 so that $\max\{||\Phi(T + W - K_1)||, ||\Phi(T - W - K_2)||\} \leq 1$. (Such operators exist by the results of [5] and [7].)

If $\Phi(W)(P) = P^{\perp}WP \notin \mathcal{K}$ for some $P \in \mathcal{P}$, then $||P^{\perp}WP||_{ess} = \delta > 0$. By hypothesis, $P^{\perp}TP$ is either isometric or co-isometric and it will retain this property when we project into the Calkin algebra, $\frac{\mathcal{L}(\mathcal{H})}{\mathcal{K}}$. Applying Lemma 2 in the Calkin algebra yields

$$\max\{||P^{\perp}(T+W)P||_{ess}, ||P^{\perp}(T-W)P||_{ess}\} \geq (1+\delta^2)^{1/2} > 1.$$

This implies that either $||\Phi_+||_e > 1$ or $||\Phi_-||_e > 1$ which is a contradiction. We conclude that $\Phi(W)$ maps $\mathcal{P}$ into $\mathcal{K}$.

Now suppose that $\Phi(W)$ is discontinuous at $P_0 \in \mathcal{P}$. Without loss of generality, we suppose that there is an increasing sequence $\{P_n\} \subset \mathcal{P}$ such that $P_n \to P_0$ (SOT) but $||\Phi(W)(P_n) - \Phi(W)(P_0)|| > 2\delta > 0$ for all n. Note that

$$\begin{aligned}\Phi(W)(P_n) - \Phi(W)(P_0) &= P_n^{\perp}WP_n - P_0^{\perp}WP_0 \\ &= (P_0 - P_n)WP_n - P_0^{\perp}W(P_0 - P_n) \\ &= (P_0 - P_n)WP_n - P_0^{\perp}WP_0P_n^{\perp}.\end{aligned}$$

Since $\Phi(W)(P_0)$ is compact, it follows that $\lim_{n\to\infty} ||(P_0^{\perp}WP_0)P_n^{\perp}|| = 0$. Thus, for sufficiently large n, $||(P_0 - P_n)WP_n|| > \delta$. Without loss of generality, we assume that this inequality holds for all n.

By restricting to a subsequence if necessary, we may assume that $P_n^{\perp}TP_n$ is either isometric on $P_n\mathcal{H}$ for all n or co-isometric onto $P_n^{\perp}\mathcal{H}$ for all n. In the first case, for each n choose a unit vector $e_n = P_ne_n$ satisfying

$$||(P_0 - P_n)We_n|| > \max\{(1 - \frac{1}{n})||(P_0 - P_n)WP_n||, \delta\}.$$

Since $\lim_{n\to\infty} ||(P_0 - P_n)Wx|| = 0$ for each fixed vector $x \in \mathcal{H}$, it follows that the sequence $\{e_n\}$ converges weakly to 0. Indeed, suppose to the contrary that there is a subsequence $\{e_{n_i}\}$ converging weakly to $f \neq 0$. (Since $\mathcal{H}$ is separable, it suffices to look at sequences.) Choose an integer n_0 and a positive constant c so that $||P_{n_0}f|| > c > 0$ and set $x = P_{n_0}f/||P_{n_0}f||$. Note that $\lim_{i\to\infty}(e_{n_i}, x) = ||P_{n_0}f|| > c$. Thus, there is an integer $n_1 \geq n_0$ satisfying $||(P_0 - P_{n_i})Wx|| < c\delta/4$ and $|(e_{n_i}, x)| > c$ for all $n_i \geq n_1$. Write $e_{n_i} = a_ix + (1 - |a_i|^2)^{1/2}y_i$ where y_i is a unit vector orthogonal to x. For each

$n_i \geq n_1$, we then have

$$\begin{aligned}||(P_0 - P_{n_i})We_{n_i}|| &\leq |a_i|c\delta/4 + (1-|a_i|^2)^{1/2}||(P_0 - P_{n_i})WP_{n_i}|| \\ &\leq (|a_i|c/4 + 1 - |a_i|^2/2)||(P_0 - P_{n_i})WP_{n_i}|| \\ &\leq (1 - c^2/4)||(P_0 - P_{n_i})WP_{n_i}||.\end{aligned}$$

For n_i sufficiently large, this contradicts the choice of the vectors $\{e_n\}$. Thus, $\{e_n\}$ converges weakly to 0 as claimed. With K_1 and K_2 chosen as above, we have $\lim_{n\to\infty} ||P_n^\perp K_i P_n e_n|| \leq \lim_{n\to\infty} ||K_i e_n|| = 0$ so that $||P_n^\perp K_i P_n e_n|| < \delta^2/8$ for $i = 1, 2$, and n sufficiently large. Hence, for large n, it follows from Lemma 2 that

$$\begin{aligned}&\max\{||P_n^\perp(T + W - K_1)P_n||, ||P_n^\perp(T - W - K_2)P_n||\} \\ &\geq \max\{||P_n^\perp(T + W - K_1)P_n e_n||, ||P_n^\perp(T - W - K_2)P_n e_n||\} \\ &\geq \max\{||P_n^\perp(T \pm W)P_n e_n||\} - \delta^2/8 \\ &\geq (1+\delta^2)^{1/2} - \delta^2/8 > 1.\end{aligned}$$

This implies that either $||\Phi(T + W - K_1)|| > 1$ or $||\Phi(T - W - K_2)|| > 1$, a contradiction.

In case each $P_n^\perp T P_n$ is co-isometric, we choose unit vectors $f_n = (P_0 - P_n)f_n$ so that $||P_n W^*(P_0 - P_n)f_n|| > \max\{\delta, (1 - \frac{1}{n})||P_n W^*(P_0 - P_n)||\}$. The sequence $\{f_n\}$ converges weakly to 0 so we may proceed as above to reach a contradiction. Thus $\Phi(W)$ is a continuous map from $\mathcal{P}$ into $\mathcal{K}$ and W is in $\mathcal{A} + \mathcal{K}$ as claimed. We conclude that $T + \mathcal{A} + \mathcal{K}$ is an extreme point in $Ball(\frac{\mathcal{L}(\mathcal{H})}{\mathcal{A}+\mathcal{K}})$.

THEOREM 9: *Suppose the nest $\mathcal{P}$ contains a projection having both infinite rank and infinite co-rank. Then $Ball(\frac{\mathcal{L}(\mathcal{H})}{Alg\mathcal{P}+\mathcal{K}})$ has extreme points.*

PROOF: Let $P_0 \in \mathcal{P}$ have both infinite rank and infinite co-rank. Let V be a partial isometry with initial space (range P_0) and final space (range $P_0{}^\perp$). Thus, $V^*V = P_0$ and $VV^* = P_0{}^\perp$.

For a projection $P \in \mathcal{P}$ there are two possibilities. If $P \leq P_0$, then

$$\begin{aligned}(PV^*P^\perp)(P^\perp VP) &= PV^*P_0{}^\perp VP = PV^*(VV^*)VP \\ &= PV^*VP = PP_0P = P.\end{aligned}$$

Therefore, $P^\perp VP$ is isometric on the range of P.

If $P \geq P_0$, then $(P^\perp VP)(PV^*P^\perp) = P^\perp$ by similar reasoning. This says that $P^\perp VP$ is co-isometric onto the range of $P^\perp$. The preceding proposition now implies that $V + (Alg\mathcal{P} + \mathcal{K})$ is an extreme point of $Ball(\frac{\mathcal{L}(\mathcal{H})}{Alg\mathcal{P}+\mathcal{K}})$.

COROLLARY 10: *If $\mathcal{P}$ is a nest, then $Ball(\frac{\mathcal{L}(\mathcal{H})}{Alg\mathcal{P}+\mathcal{K}})$ has extreme points if and only if there is a projection P in $\mathcal{P}$ having both infinite rank and infinite co-rank.*

If the nest $\mathcal{N}$ is indexed by the extended integers and has finite dimensional atoms then the previous theorem implies that $Ball(\frac{\mathcal{L}(\mathcal{H})}{Alg\mathcal{N}+\mathcal{K}})$ does have extreme points. But more can be said. Indeed, for such nests we can completely characterize the extreme points of $Ball(\frac{\mathcal{L}(\mathcal{H})}{Alg\mathcal{N}+\mathcal{K}})$ and , on this basis, determine that the closed convex hull of the extreme points is not all of $Ball(\frac{\mathcal{L}(\mathcal{H})}{Alg\mathcal{N}+\mathcal{K}})$. These results are the contents of the following theorem and its corollary.

THEOREM 11: *Let $\mathcal{N} = \{E_n : n \in \mathbf{Z}\} \cup \{0,\ 1\}$ be a nest indexed by the extended integers and suppose that $(E_{n+1} - E_n)$ is finite dimensional for all $n \in \mathbf{Z}$. Set $\mathcal{A} = Alg\mathcal{N}$ and let $T \in \mathcal{L}(\mathcal{H})$ satisfy $||T|| = d(T,\ Alg\mathcal{N} + \mathcal{K}) = 1$. The following are equivalent:*

1) $T + (Alg\mathcal{N} + \mathcal{K})$ is an extreme point of $Ball(\frac{\mathcal{L}(\mathcal{H})}{Alg\mathcal{N}+\mathcal{K}})$;

2) $E_0^\perp T E_0$ is an essential unitary in $\mathcal{L}(E_0\mathcal{H}, E_0^\perp\mathcal{H})$;

3)
$$\lim_{n\to\infty} ||E_n^\perp - E_n^\perp T E_n T^* E_n^\perp|| =$$
$$\lim_{n\to-\infty} ||E_n - E_n T^* E_n^\perp T E_n|| = 0.$$

Before providing a proof, we establish the following corollary result and an elementary lemma.

COROLLARY 12: *With $\mathcal{N}$ as in the Theorem, the closed convex hull of the set of extreme points of $Ball(\frac{\mathcal{L}(\mathcal{H})}{Alg\mathcal{N}+\mathcal{K}})$ is equal to $E_0^\perp(Ball(\mathcal{L}(\mathcal{H}))E_0 + Alg\mathcal{N} + \mathcal{K}$. This is not all of $Ball(\frac{\mathcal{L}(\mathcal{H})}{Alg\mathcal{N}+\mathcal{K}})$.*

PROOF: The first part follows immediately from the Theorem. For the second statement, let S be any partial isometry of the form $S = \sum_{n\in\mathbf{Z}} x_n \otimes x_{n-1}^*$, where $x_n = (E_{k_{n+1}} - E_{k_n})x_n$ are unit vectors and $k_n < k_{n+1}$ for all $n \in \mathbf{Z}$. Then $||S|| =$

$d(S, Alg\mathcal{N} + \mathcal{K}) = 1$. But $E_0^{\perp} S E_0$ is rank one so S is not in the closed convex hull of the extreme points of $Ball(\frac{\mathcal{L}(\mathcal{H})}{Alg\mathcal{N}+\mathcal{K}})$.

We conjecture that the closed convex hull of the extreme points of $Ball(\frac{\mathcal{L}(\mathcal{H})}{Alg\mathcal{P}+\mathcal{K}})$ does not coincide with $Ball(\frac{\mathcal{L}(\mathcal{H})}{Alg\mathcal{P}+\mathcal{K}})$ for any nest $\mathcal{P}$ provided $Alg\mathcal{P} + \mathcal{K} \neq \mathcal{L}(\mathcal{H})$.

LEMMA 13: *Suppose that $(E_0^{\perp} T E_0)^*$ is not essentially isometric and let $(E_0^{\perp} T E_0)^* = VA$ be the polar decomposition. Then $A - A^2$ is compact if and only if there is a projection $P \leq E_0^{\perp}$ such that P is in the double commutant of A and such that $A - P$ is compact.*

PROOF: With P as stated, we have $A - A^2 = (A - P)(1 - A - P)$ which is compact. Conversely, suppose $A - A^2$ is compact. Then the essential spectrum of A is contained in the set $\{0, 1\}$ and it follows that there is a spectral projection P for A such that $A - P$ is compact. P is in the double commutant of A since it is a spectral projection. In particular, P commutes with V^*V.

We remark that the conclusion of the preceding lemma remains true if we replace E_0 by any fixed E_n. In addition, arguments analogous to those given in the lemma apply in case $E_0^{\perp} T E_0$ is not essentially isometric and has polar decomposition WB.

Let us now turn to the proof of Theorem 11.

PROOF OF THEOREM 11: **2)⇔3):** For this, notice that, for each $n \in \mathbf{Z}$,

$$E_n - E_n T^* E_n^{\perp} T E_n = E_n(1 - T^*T)E_n + E_n T^* E_n T E_n \geq 0.$$

For $n \leq 0$, it follows that

$$E_n(E_0 - E_0 T^* E_0^{\perp} T E_0)E_n - (E_n - E_n T^* E_n^{\perp} T E_n) = E_n T^*(E_0 - E_n) T E_n \geq 0.$$

Thus, $||E_n - E_n T^* E_n^{\perp} T E_n||$ is decreasing as n tends to $-\infty$. If condition 2) holds, then $E_0^{\perp} T E_0$ is an essential isometry so that $E_0 - E_0 T^* E_0^{\perp} T E_0$ is compact, whence $E_n(E_0 - E_0 T^* E_0^{\perp} T E_0)E_n$ tends to zero in norm. It follows that $\lim_{n \to -\infty} ||E_n - E_n T^* E_n^{\perp} T E_n|| = 0$ as desired.

Conversely, since $E_0 - E_0T^*E_0^\perp TE_0$ differs from $E_n - E_nT^*E_n^\perp TE_n$ by a compact operator (here we use the hypothesis that the atoms are finite rank), the second limit in condition 3) implies that $E_0 - E_0T^*E_0^\perp TE_0$ is compact. Thus, $E_0^\perp TE_0$ is an essential isometry.

The equivalence between the first limit of 3) and the property that $E_0^\perp TE_0$ is an essential co-isometry is proven analogously. So 2) is equivalent to 3).

2)⇒1): If 2) holds then, since $||T|| = 1$, it follows that $E_0TE_0 + E_0^\perp TE_0^\perp$ is compact. Therefore, $T + (Alg\mathcal{N} + \mathcal{K})$ contains a representative $T_0 = E_0^\perp T_0E_0$ for which $E_0^\perp T_0E_0$ is either isometric or co-isometric (as an element of $\mathcal{L}(E_0\mathcal{H}, E_0^\perp\mathcal{H})$) with finite index (index(A)=nullity(A)−nullity(A^*)). We will treat only the isometric case with index $-n$. The other possibility is handled analogously. We may and do assume that T_0 has been further modified so that $range(T_0) = E_n^\perp\mathcal{H}$. Let J be any partial isometry taking $(E_n - E_0)\mathcal{H}$ onto itself such that $(E_n - E_k)J(E_k - E_0)$ is either isometric or co-isometric on its range for each $0 \le k \le n$. (For example, with one dimensional atoms $\mathbf{C}e_j$, set $Je_j = e_{n+1-j}$.) Then, $T_0 + J$ is in $T + (Alg\mathcal{N} + \mathcal{K})$ and satisfies the hypotheses of Proposition 8. We conclude that $T + (Alg\mathcal{N} + \mathcal{K})$ is an extreme point. This proves that 2) implies 1).

1)⇒2): Suppose 2) fails. Without loss of generality, we may suppose that $E_0^\perp TE_0$ is not an essential co-isometry, that is that $E_0^\perp - (E_0^\perp TE_0)(E_0T^*E_0^\perp)$ is not a compact operator. We shall construct an operator W not in $Alg\mathcal{N} + \mathcal{K}$ satisfying $d(T \pm W, Alg\mathcal{N} + \mathcal{K}) \le 1$. W will satisfy $W = E_0^\perp WE_0^\perp$ and will be much like the operator constructed in the proof of Theorem 7.

Let $(E_0^\perp TE_0)^* = E_0T^*E_0^\perp = VA$ be the polar decomposition. We break our construction into two separate cases according to whether or not $A - A^2$ is compact.

Let us assume first that $A - A^2$ is not compact. Since VA is not an essential isometry and $A - A^2$ is not compact, there exists $0 < \delta < 1/2$ such that the spectral projection F for A corresponding to the interval $[\delta, 1-\delta]$ is infinite dimensional. We can then choose unit vectors $\{x_k : k = 1, 2, \ldots\}$, scalars $\{\lambda_k : k = 1, 2, \ldots\}$, and an increasing sequence of integers $0 = n_0 < n_1 < n_2 < \ldots$ such that, for all $k \ge 1$, the following hold:

(i) $||E_{n_k}^{\perp}T(E_{n_{k-1}} - E_0)|| < 2^{-k-1}$;

(ii) $x_k = (E_{n_k} - E_{n_{k-1}})x_k$;

(iii) $||F^{\perp}x_k|| < 2^{-k}$;

(iv) $\delta \leq \lambda_k \leq 1 - \delta$;

(v) $||Ax_k - \lambda_k x_k|| < 2^{-k-1}$.

(Condition (i) is obtained as in the proof of Theorem 7 while the fact that F has infinite rank enables us to find approximate eigenvectors x_k for A satisfying conditions (ii) through (v).)

One immediate consequence of our choices is that, for each $k \geq 1$, if $n_{k-1} < n < n_k$, then

$$||E_n^{\perp}T(E_n - E_0) - E_{n_{k+1}}E_n^{\perp}TE_nE_{n_{k-2}}^{\perp}|| < 2^{-k}.$$

Also, for later use, note that $A = AV^*V = V^*VA$ and that $||Vx_k|| \geq ||AV^*Vx_k|| = ||Ax_k|| \geq ||\lambda_k x_k|| - 2^{-k-1} \geq \delta - 2^{-k-1}$. Thus, for k sufficiently large, $||Vx_k|| \geq \delta/2$.

Next, for each integer $r \geq 0$, define

$$W_r = 2^{-r} \sum_{i=2^{r+1}+1}^{3(2^r)} \sum_{j=3(2^r)+1}^{2^{r+2}} x_j \otimes x_i{}^*,$$

and set $W = \sum_{r \geq 0} W_r$. It is clear that $W = E_0^{\perp}WE_0^{\perp}$ and, as in the proof of Theorem 7, one can check that W is in fact a direct sum and a partial isometry satisfying $||W|| = d(W, Alg\mathcal{N} + \mathcal{K}) = 1$. By the distance formula (3.1), we see that

$$d(T \pm \delta W, Alg\mathcal{N} + \mathcal{K}) = \limsup_{|n| \to \infty} ||E_n^{\perp}(T \pm \delta W)E_n||.$$

We will show that this is no greater than 1 and, hence, that $T + (Alg\mathcal{N} + \mathcal{K})$ is not an extreme point.

For $n \leq 0$, $E_n^{\perp}WE_n = 0$ so $||E_n^{\perp}(T \pm \delta W)E_n|| = ||E_n^{\perp}TE_n|| \leq 1$. Now take $n > n_1$ and let k and s be the integers satisfying $n_{k-1} < n \leq n_k$ and $2^{s+1} < k \leq 2^{s+2}$. Since $E_n^{\perp}x_j = 0$ if $j \leq k-1$ and $E_nx_i = 0$ whenever $i \geq k+1$, it follows that $E_n^{\perp}W_r = 0$ whenever $r \leq s-1$ while $W_rE_n = 0$ if $r \geq s+1$. Thus, $E_n^{\perp}WE_n = E_n^{\perp}W_sE_n$. Moreover, $||E_n^{\perp}W_sE_n - E_{n_{k+1}}^{\perp}W_sE_{n_{k-2}}|| \leq 2(2^{-s/2})$. So, with an error in norm of at most $2 \cdot 2^{-s/2} + 2^{-k}$, we can consider the operator $(E_n^{\perp}TE_0 + E_{n_{k+1}}E_n^{\perp}TE_nE_{n_{k-2}}^{\perp} \pm \delta E_{n_{k+1}}^{\perp}W_sE_{n_{k-2}})$ in place of $E_n^{\perp}(T \pm \delta W)E_n$.

Letting P_s denote the projection onto the span of the vectors $\{x_i : 2^{s+1} + 1 \leq i \leq 3(2^s),\ E_{n_{k-2}} x_i \neq 0\}$ and Q_s the projection onto the span of the vectors $\{x_j : 3(2^s) + 1 \leq j \leq 2^{s+2},\ E_{n_{k+1}}^{\perp} x_j \neq 0\}$ we have $E_{n_{k+1}}^{\perp} W_s E_{n_{k-2}} = Q_s W_s P_s$ and, moreover, $E_{n_{k+1}} E_n^{\perp} T E_n E_{n_{k-2}}^{\perp} = Q_s^{\perp} E_{n_{k+1}} E_n^{\perp} T E_n E_{n_{k-2}}^{\perp} P_s^{\perp}$. Also

$$
\begin{aligned}
||Q_s^{\perp} A Q_s|| &= || \sum_{j=k+2}^{2^{s+2}} Q_s^{\perp} A x_j \otimes x_j^* || \\
&\leq \sum_{j=k+2}^{2^{s+2}} ||Q_s^{\perp} A x_j|| \\
&\leq \sum_{j \geq k+2} ||Q_s^{\perp} \lambda_j x_j|| + ||Q_s^{\perp}(A x_j - \lambda_j x_j)|| \\
&\leq \sum_{j \geq k+2} 2^{-j-1} \\
&= 2^{-k-1}.
\end{aligned}
$$

It follows that $||Q_s^{\perp} A Q_s||$, $||Q_s A Q_s^{\perp}||$, and $||AQ_s - Q_s A||$ all tend to 0 as n (and therefore s) tends to infinity. Moreover, from the definitions of the projection F and of the vectors $\{x_j\}$, it follows that $||F^{\perp} Q_s|| < 2^{-k-1}$ and that $||FAV^*|| \leq 1 - \delta$.

Let R_s be the projection onto the closure of the range of the operator VQ_s and set $R_s' = VQ_s V^*$. We claim that $\lim_{s \to \infty} ||AV^*(R_s' - R_s)|| = 0$. For this, suppose that x is orthogonal to the range of R_s. We then have $R_s' x = V(Q_s V^* x) = 0$. On the other hand, the range of R_s is spanned by the (finite) set $\{V x_j : x_j = Q_s x_j\}$ and we have seen that, for s and j sufficiently large, $||V x_j|| \geq \delta/2$. Thus for large s and $x_j = Q_s x_j$, we have

$$
\begin{aligned}
\frac{||AV^*(R_s - R_s')V x_j||}{||V x_j||} &\leq 2/\delta ||AV^* V Q_s x_j - AV^* V Q_s V^* V x_j|| \\
&\leq 2/\delta ||AQ_s(1 - V^* V) x_j|| \\
&\leq 2/\delta ||AQ_s (V^* V)^{\perp}|| \\
&= 2/\delta ||AQ_s (V^* V)^{\perp} - Q_s A (V^* V)^{\perp}|| \\
&\leq 2/\delta ||AQ_s - Q_s A||
\end{aligned}
$$

which tends to 0 as $s \to \infty$. This establishes the claim.

Using this last claim, it now follows that

$$\begin{aligned}||Q_s(E_n^\perp TE_0)R_s^\perp|| &= ||E_n^\perp Q_s AV^*(1-R_s)|| \\ &\le ||E_n^\perp Q_s AV^*(1-R'_s)|| + ||E_n^\perp Q_s AV^*(R'_s - R_s)|| \\ &\le ||E_n^\perp (Q_s A - Q_s A Q_s)V^*|| + ||AV^*(R'_s - R_s)|| \\ &\le ||Q_s A Q_s^\perp|| + ||AV^*(R'_s - R_s)||\end{aligned}$$

which tends to 0 as $s \to \infty$. We also have

$$\begin{aligned}||Q_s^\perp(E_n^\perp TE_0)R_s|| &\le ||E_n^\perp Q_s^\perp AV^* R'_s|| + ||E_n^\perp Q_s^\perp AV^*(R'_s - R_s)|| \\ &\le ||E_n^\perp (Q_s^\perp A Q_s)V^*|| + ||AV^*(R'_s - R_s)|| \\ &\le ||Q_s^\perp A Q_s|| + ||AV^*(R'_s - R_s)||\end{aligned}$$

which tends to 0 as $s \to \infty$. If we define

$$\xi_s \equiv ||E_n^\perp(E_0^\perp TE_0)E_n - \{Q_s(E_n^\perp TE_0)R_s \oplus Q_s^\perp(E_n^\perp TE_0)R_s^\perp\}||,$$

we then have $\lim_{s\to\infty} \xi_s = 0$.

Putting all of this information together, we see that, for n sufficiently large,

$$\begin{aligned}||E_n^\perp(T \pm \delta W)E_n|| &\le ||E_n^\perp TE_0 + E_{n_{k+1}} E_n^\perp TE_n E_{n_{k-2}}^\perp \pm \delta E_{n_{k+1}}^\perp W_s E_{n_{k-2}}|| \\ &\qquad + 2\cdot 2^{-s/2} + 2^{-k} \\ &\le ||Q_s(E_n^\perp TE_0)R_s + Q_s^\perp(E_n^\perp TE_0)R_s^\perp \\ &\qquad + Q_s^\perp(E_{n_{k+1}} E_n^\perp TE_n E_{n_{k-2}}^\perp)P_s^\perp \pm \delta(Q_s W_s P_s)|| \\ &\qquad\quad + \xi_s + 2\cdot 2^{-s/2} + 2^{-k} \\ &\le \max\{||Q_s(E_n^\perp TE_0)R_s \pm \delta(Q_s W_s P_s)||, \\ &\qquad\quad ||Q_s^\perp(E_n^\perp TE_0)R_s^\perp + Q_s^\perp(E_{n_{k+1}} E_n^\perp TE_n E_{n_{k-2}}^\perp)P_s^\perp||\} \\ &\qquad\quad + \xi_s + 2\cdot 2^{-s/2} + 2^{-k} \\ &\le \max\{||E_n^\perp Q_s FAV^* R_s|| + ||E_n^\perp Q_s F^\perp AV^* R_s|| + \delta,\ 1 + 2^{-k-2}\} \\ &\qquad + \xi_s + 2\cdot 2^{-s/2} + 2^{-k} \\ &\le \max\{(1-\delta)+\delta,\ 1\} + \xi_s + 2\cdot 2^{-s/2} + 2^{-k} + 2^{-k-2} + ||F^\perp Q_s|| \\ &= 1 + \xi_s + 2\cdot 2^{-s/2} + 2^{-k} + 2^{-k-2} + 2^{-k-1}\end{aligned}$$

which tends to 1 as n tends to infinity. It follows that

$$d(T \pm \delta W,\ Alg\mathcal{N} + \mathcal{K}) = \limsup_{|n|\to\infty} ||E_n^{\perp}(T \pm \delta W)E_n|| \leq 1$$

as desired.

The preceding work proves that 2) implies 1) in case $A - A^2$ is not compact. In the case where $A - A^2$ is compact then, as we saw in Lemma 13, there is a spectral projection F for A such that $F \leq E_0^{\perp}$ and $A - F$ is compact. Let $T_0 = T - (A - F)V^*$. Then $T_0 + Alg\mathcal{N} + \mathcal{K} = T + Alg\mathcal{N} + \mathcal{K}$ and $(E_0^{\perp}T_0E_0)^*$ has polar decompositon $V(V^*VF)$. Since $(E_0^{\perp}T_0E_0)^*$ is not essentially isometric, it follows that $E_0^{\perp} - V^*VF$ is infinite dimensional and that $(E_0^{\perp} - V^*VF)T_0E_0 = 0$. Now choose unit vectors $\{x_k : k \geq 1\}$ and an increasing sequence of integers $0 \leq n_0 < n_1 < n_2 < \ldots$ so that $x_k = (E_{n_k} - E_{n_{k-1}})x_k$, $||(E_0^{\perp} - V^*VF)x_k|| > 1 - 2^{-k-1}$, and $||E_{n_k}^{\perp}T(E_{n_{k-1}} - E_0)|| < 2^{-k-1}$ for all $k \geq 1$. With W as before, it can be shown that $d(T_0 \pm W,\ Alg\mathcal{N} + \mathcal{K}) \leq 1$ and hence that $T + Alg\mathcal{N} + \mathcal{K}$ is not an extreme point of $Ball(\frac{\mathcal{L}(\mathcal{H})}{Alg\mathcal{N}+\mathcal{K}})$. This completes the proof of the theorem.

4. Strong Extreme Points in $Ball(\frac{\mathcal{L}(\mathcal{H})}{Alg\mathcal{P}})$

In this section we apply the ideas and results of the previous section to obtain information about strong extreme points in $Ball(\frac{\mathcal{L}(\mathcal{H})}{\mathcal{A}})$, where $\mathcal{A}$ is a nest algebra. In this setting, given $T \in \mathcal{L}(\mathcal{H})$ with $d(T, \mathcal{A}) = 1$, $T + \mathcal{A}$ is a strong extreme point of $Ball(\frac{\mathcal{L}(\mathcal{H})}{\mathcal{A}})$ provided that for every $\epsilon > 0$ there exists $\delta > 0$ such that $d(S, \mathcal{A}) \leq \epsilon$ whenever $d(T \pm S, \mathcal{A}) \leq 1 + \delta$. As in the previous section, the structure of the underlying nest will prove decisive in our analysis.

Proposition 14: *If $\mathcal{P} = \{P_n\}$ is a sequence of finite rank projections increasing strongly to the identity, then $Ball(\frac{\mathcal{L}(\mathcal{H})}{Alg\mathcal{P}})$ has no strong extreme points.*

Proof: For simplicity we will provide a proof for the case $\mathcal{P} = \mathcal{E}$. Set $\mathcal{A} = Alg\mathcal{E}$.

Set $\epsilon_0 = 1/2$ and let $\delta > 0$. Let $T \in \mathcal{L}(\mathcal{H})$ satisfy $||T|| = d(T, \mathcal{A}) = 1$ and define operators T_1 and T_2 , as in the proof of Theorem 7, so that $T_1 \in \mathcal{K}$, $||T_1|| \leq$

$\delta/4$, and $T = T_1 + T_2$. For each l, define W_l as in Theorem 7 and note that

$$\begin{aligned}||E_n{}^{\perp}(T \pm W_l)E_n|| &\leq ||E_n{}^{\perp}(T_2 \pm W_l)E_n|| + ||E_n{}^{\perp}T_1E_n|| \\ &\leq 1 + 2^{-l} + 2^{-l/2} + \delta/2, \quad \forall n \geq 0.\end{aligned}$$

Given $\delta > 0$, choose l so that $2^{-l} + 2^{-l/2} \leq \delta/2$. This yields

$$d(T \pm W_l, \mathcal{A}) = \sup_{n \geq 0} \{||E_n{}^{\perp}(T \pm W_l)E_n||\} \leq 1 + \delta.$$

Examination of W_l shows that $W_l = E^{\perp}_{m_{2^l+1}} W_l E_{m_{2^l+1}}$ so that $d(W_l, \mathcal{A}) = ||W_l|| = 1 > \epsilon_0$. We conclude that $T + Alg\mathcal{E}$ is not a strong extreme point of $Ball(\frac{\mathcal{L}(\mathcal{H})}{Alg\mathcal{E}})$.

As with Theorem 7, the above argument can be modified to yield the same result in case the nest consists of a sequence of projections of finite co-rank decreasing to 0. For nests other than these types, however, strong extreme points do exist, as the next proposition and its corollary show.

PROPOSITION 15: *Let $\mathcal{P}$ be a nest and let $\mathcal{A} = Alg\mathcal{P}$. If $T \in \mathcal{L}(\mathcal{H})$ satisfies the property that, for each $P \in \mathcal{P}$, either $P^{\perp}TP$ is isometric on the range of P or $PT^*P^{\perp}$ is isometric on the range of $P^{\perp}$, then $T + Alg\mathcal{P}$ is a strong extreme point of $Ball(\frac{\mathcal{L}(\mathcal{H})}{Alg\mathcal{P}})$.*

PROOF: Let $\delta > 0$ and suppose that $d(T \pm S, \mathcal{A}) \leq 1 + \delta$. Then $||P^{\perp}(T \pm S)P|| \leq 1 + \delta$ for all $P \in \mathcal{P}$. Since $P^{\perp}TP$ is a partial isometry by hypothesis, it follows from Lemma 2 that $\max\{||P^{\perp}(T \pm S)P||\} \geq (1 + ||P^{\perp}SP||^2)^{1/2}$. Hence, $(1 + ||P^{\perp}SP||^2)^{1/2} \leq 1 + \delta$ which implies that $||P^{\perp}SP||^2 \leq \delta(2 + \delta)$ for all $P \in \mathcal{P}$.

Thus, given $\epsilon > 0$, let $\delta = \min\{1, \epsilon^2/3\}$. Then the condition $d(T \pm S, \mathcal{A}) \leq 1 + \delta$ implies that

$$||P^{\perp}SP||^2 \leq \delta(2 + \delta) \leq 3(\frac{\epsilon^2}{3}) = \epsilon^2, \ \forall P \in \mathcal{P}.$$

Hence, $d(S, \mathcal{A}) \leq \epsilon$. We conclude that $T + Alg\mathcal{P}$ is a strong extreme point of $Ball(\frac{\mathcal{L}(\mathcal{H})}{Alg\mathcal{P}})$.

COROLLARY 16: *If $\mathcal{P}$ is any nest other than one which is indexed by $\mathcal{N}$ or $-\mathcal{N}$ and which has finite dimensional atoms then $Ball(\frac{\mathcal{L}(\mathcal{H})}{Alg\mathcal{P}})$ has strong extreme points.*

PROOF: As in the proof of Theorem 9, the hypotheses imply the existence of an operator V satisfying the hypotheses of Proposition 15.

REFERENCES

[1] William Arveson, *Interpolation problems in nest algebras* , J.Functional Analysis, **20** (1975), 208–233.

[2] Sheldon Axler, I.David Berg, Nicholas Jewell, and Allen Shields , *Approximation by compact operators and the space* $H^{\infty}+C$, Ann. of Math., **109** (1979), 601–612.

[3] J.A.Cima and James Thomson, *On strong extreme points in* H^p, Duke Math.J., **40** (1973), 529–532.

[4] Kenneth R.Davidson, Nest Algebras, Research Notes in Mathematics, Pitman-Longman Pub., Boston-London-Melbourne, to appear.

[5] Kenneth R.Davidson and Stephen C.Power, *Best approximation in* C^**-algebras*, J.für die reine und angew.Math, **368** (1986), 43–62.

[6] T.Fall, W.Arveson, and P.Muhly, *Perturbations of nest algebras* , J.Operator Theory, **1** (1979), 137–150.

[7] Timothy G.Feeman, *M-ideals and quasi-triangular algebras* , Illinois J.Math., **31** (1987), 89–98.

[8] Paul Koosis, *Weighted quadratic means of Hilbert transforms* , Duke Math.J., **38** (1971), 609–634.

Kenneth R.Davidson
Univ. of Waterloo
Waterloo, Ontario
N2L 3G1 CANADA

Timothy G.Feeman
Villanova Univ.
Villanova, PA
19085 USA

Allen L.Shields
Univ. of Michigan
Ann Arbor, MI
48109 USA

Operator Theory:
Advances and Applications, Vol. 32

ON THE FOUR BLOCK PROBLEM, I

Ciprian Foias   Allen Tannenbaum

Dedicated to the memory of Constantin Apostol

This paper is concerned with the study of the singular values of a "four block operator" which naturally appears in control engineering and which possesses a number of interesting mathematical properties. The study of this operator will be shown to be reducible to a skew Toeplitz operator problem of the kind studied in [1]. The main theoretical fact proven here is an explicit closed form formula for the essential norm of the four block operator (see Proposition 1.1). We dedicate this paper to the memory of Constantin Apostol who was a great master of the essential properties of operators in Hilbert space.

1. INTRODUCTION

In this paper we will study the singular values of the *the four block operator* (to be defined below), which is closely connected to certain operators that appear in many engineering H^∞ control problems. Indeed following the framework of the monograph of Francis [7], almost all such robust design problems can be formulated in terms of the spectral properties of such operators. This includes the problems of sensitivity and mixed sensitivity minimization, model-matching and certain tracking problems, and the μ-synthesis procedure of Doyle. We refer the reader to [7], the lecture notes of Doyle [2], and the references therein for more details about this area, and the excellent work done on the part of both the engineering and mathematical communities in its connection. We should also add that very recently Dym and Gohberg [3] have made some important contributions to the subject using the theory of banded matrices.

In order to state a precise mathematical problem, we will first need to set up some notation. Accordingly, let $w, f, g, h, m \in H^\infty$, where w, f, g, h are rational and m is nonconstant inner. (All of our Hardy spaces will be defined on the unit disc D in the standard way.)

Set

$$A := \begin{bmatrix} Pw(S) & f(S) \\ g(S) & h(S) \end{bmatrix}$$

where $S : H^2 \to H^2$ denotes the unilateral shift, and P denotes orthogonal projection from H^2 to $H(m) := H^2 \ominus mH^2$. Note that if we let T denote the compression of S to $H(m)$, then $w(T)P = Pw(S)$. The problem we study here is the calculation of the norm and singular values of A. The point of this paper is to give a determinantal formula (see Theorem 1) for explicitly carrying out this computation.

We should note that the techniques given here were heavily influenced by (and are based on) the previous work in [1], [4], [5], [6], [12]. In [11], these ideas were applied to the mixed sensitivity ("two block") problem. (See also the closely related work in [8].) The present treatment essentially follows that of [1], [4], and [6].

It is very important to add that the skew Toeplitz framework developed in [1] also leads to the complete determination of the singular values of operators of the form A when the four blocks are taken to be **matrix-valued.** (See [1] for the precise definition and a more complete discussion on skew Toeplitz operators.) In this case the first block would correspond to a block Hankel operator, while the other blocks would correspond to matrix-valued Toeplitz operators. Since it is this kind of problem that one encounters in the H^∞ design of multiple input/multiple output systems, results on the norm of the matrix version of the four block operator should also be of interest even for finite dimensional systems. (At present certain iterative methods are used for such problems [2], [7].) This will be the topic of the sequel to this paper in which we will apply the block determinantal formula of [1] (see Theorem (7.5)) to the specific four block structure. We moreover plan to have an applied version of the present work in which the connections to engineering problems are more explicitly discussed.

We would like to thank John Doyle and Bruce Francis for exciting our interest in this problem. Finally, we feel that the results given here should complement some of the beautiful results announced by John Doyle and Keith Glover on the parametrization of the sub-optimal controllers for the four block problem.

This research was supported in part by grants from the Research Fund of Indiana University, NSF (ECS-8704047), and the AFOSR-88-0020.

2. PRELIMINARY RESULTS AND NOTATION

In this section we make some remarks, and prove a result which will allow us to give the determinantal formula for the singular values and vectors of A in Section 3. We are basically following the line of argument given in [1], [4], and [6]. Thus we must first identify the essential norm of A (denoted by $\|A\|_e$), and then give an algorithmic procedure for determining a singular value of A, $\rho > \|A\|_e$. We are using the standard notation from operator theory as, for example, given in [9], [10]. In particular σ_e will denote the essential spectrum, and $A(\overline{D})$ will stand for the set of analytic functions on D which are continuous on the closed disc $\overline{D}$. We begin with:

PROPOSITION 1. *Notation as in the Introduction. Set*

$$\alpha := \max \{\| \begin{bmatrix} w(\zeta) & f(\zeta) \\ g(\zeta) & h(\zeta) \end{bmatrix} \| : \zeta \in \sigma_e(T)\} \tag{1}$$

$$\beta := \max \{\| \begin{bmatrix} 0 & f(\zeta) \\ g(\zeta) & h(\zeta) \end{bmatrix} \| : \zeta \in \partial D\} \tag{2}$$

where the norms in (1) and (2) are those in $L(\mathbf{C}^2)$. Then

$$\|A\|_e = \max(\alpha, \beta). \tag{3}$$

REMARK. We will actually prove Proposition 1 in the more general case when $w, f, g, h \in A(\overline{D})$.

PROOF OF PROPOSITION 1. Let $\zeta \in \sigma_e(T)$. Then there exits a sequence $x_n \in H := H(m)$, $\|x_n\| = 1$, $x_n \to 0$ weakly, such that $\|(T - \zeta)x_n\| \to 0$. Clearly, we have that $\|(S - \zeta)x_n\| \to 0$ as well. Consequently, $\|(p(T) - p(\zeta))x_n\| \to 0$ and $\|(p(S) - p(\zeta))x_n\| \to 0$ for any (analytic) polynomial $p(z)$. It follows easily that $\|(\phi(T) - \phi(\zeta))x_n\| \to 0$ for any $\phi \in A(\overline{D})$.

Now let ξ, and η be such that $|\xi|^2 + |\eta|^2 = 1$. Thus from the above, we have that

$$\|A\|_e \geq \overline{\lim_{n\to\infty}} \left\| \begin{bmatrix} w(T)P & f(S) \\ g(S) & h(S) \end{bmatrix} \begin{bmatrix} \xi x_n \\ \eta x_n \end{bmatrix} \right\| \tag{4}$$

$$= \overline{\lim_{n\to\infty}} \left\| \begin{bmatrix} w(\zeta) & f(\zeta) \\ g(\zeta) & h(\zeta) \end{bmatrix} \begin{bmatrix} \xi x_n \\ \eta x_n \end{bmatrix} \right\| = \left\| \begin{bmatrix} w(\zeta) & f(\zeta) \\ g(\zeta) & h(\zeta) \end{bmatrix} \begin{bmatrix} \xi \\ \eta \end{bmatrix} \right\|$$

where the last norm is in $\mathbf{C}^2$. Hence, we have that

$$\|A\|_e \geq \alpha. \tag{5}$$

Let now $\zeta \in \partial D$, and $y_n \in H^2$ be chosen such that $y_n \to 0$ weakly, $\|y_n\| = 1$, and $\|(S - \zeta)y_n\| \to 0$. Then certainly the sequence $z_n := my_n = m(S)y_n$ also satisfies $\|(S - \zeta)z_n\| \to 0$, $z_n \to 0$ weakly, and $\|z_n\| = 1$. Then exactly as above we have that $\|(\phi(S) - \phi(\zeta))z_n\| \to 0$ for all $\phi \in A(\overline{D})$, and

$$\|A\|_e \geq \overline{\lim_{n\to\infty}} \left\| \begin{bmatrix} w(T)P & f(S) \\ g(S) & h(S) \end{bmatrix} \begin{bmatrix} \xi z_n \\ \eta z_n \end{bmatrix} \right\|$$

$$= \overline{\lim_{n\to\infty}} \left\| \begin{bmatrix} 0 & f(\zeta) \\ g(\zeta) & h(\zeta) \end{bmatrix} \begin{bmatrix} \xi z_n \\ \eta z_n \end{bmatrix} \right\| = \left\| \begin{bmatrix} 0 & f(\zeta) \\ g(\zeta) & h(\zeta) \end{bmatrix} \begin{bmatrix} \xi \\ \eta \end{bmatrix} \right\|$$

from which we see immediately that

$$\|A\|_e \geq \beta. \tag{6}$$

The other direction is a bit more intricate. Let

$$z_n := \begin{bmatrix} x_n \\ y_n \end{bmatrix}$$

be such that $\|z_n\| = 1$, $z_n \to 0$ weakly, and

$$\|Az_n\| \to \|A\|_e . \tag{7}$$

We now use a few facts from [6]. (See in particular lemma (2.1) and the proof of (2.2).) So first of all, we may write $T = V + F$ where V is unitary on H, and F has rank ≤ 1. Thus as in [6], we see that $p(T) = p(V) +$ (finite rank operator) for any polynomial p, and $\phi(T) = \phi(V) +$ (compact operator), for any $\phi \in A(\overline{D})$. Thus $\|(\phi(T) - \phi(V))w_n\| \to 0$, for any weakly converging sequence $w_n \to 0$, in H. On the other hand, $\|T^m w_n - V^m w_n\| \to 0$ implies that $\|T^m w_n - S^m w_n\| \to 0$, and so it follows as above that

$$\|(\phi(S) - \phi(V))w_n\| \to 0 \tag{8}$$

for every $w_n \to 0$ in H, and for all $\phi \in A(\overline{D})$. Setting $p_n := Px_n$, $q_n := Py_n$, $p'_n := x_n - p_n$, $q'_n := y_n - q_n$, we have by virtue of (8) that

$$\|f(S)y_n - (f(V)q_n + f(S)q'_n)\| \to 0$$

and similarly,

$$\|g(S)x_n - (g(V)p_n + g(S)p'_n)\| \to 0$$

$$\|(h(S)y_n - (h(V)q_n + h(S)p'_n)\| \to 0$$

and

$$\|w(T)p_n - w(V)p_n\| \to 0.$$

Consequently,

$$\|A\|_e = \lim_{n\to\infty} \| \begin{bmatrix} w(V) & f(V) \\ g(V) & h(V) \end{bmatrix} \begin{bmatrix} p_n \\ q_n \end{bmatrix} + \begin{bmatrix} 0 & f(S) \\ g(S) & h(S) \end{bmatrix} \begin{bmatrix} p'_n \\ q'_n \end{bmatrix} \|.$$

Thus using the fact that

$$\begin{bmatrix} w(V) & f(V) \\ g(V) & h(V) \end{bmatrix} \begin{bmatrix} p_n \\ q_n \end{bmatrix} \in H(m) \oplus H(m)$$

and

$$\begin{bmatrix} 0 & f(S) \\ g(S) & h(S) \end{bmatrix} \begin{bmatrix} p'_n \\ q'_n \end{bmatrix} \in mH^2 \oplus mH^2$$

we see that

$$\|A\|_e^2 = \lim_{n\to\infty} \left\{ \| \begin{bmatrix} w(V) & f(V) \\ g(V) & h(V) \end{bmatrix} \begin{bmatrix} p_n \\ q_n \end{bmatrix} \|^2 + \| \begin{bmatrix} 0 & f(S) \\ g(S) & h(S) \end{bmatrix} \begin{bmatrix} p'_n \\ q'_n \end{bmatrix} \|^2 \right\}.$$

Now

$$\| \begin{bmatrix} 0 & f(S) \\ g(S) & h(S) \end{bmatrix} \begin{bmatrix} p'_n \\ q'_n \end{bmatrix} \|^2$$

$$= \frac{1}{2\pi} \int_0^{2\pi} \| \begin{bmatrix} 0 & f(e^{it}) \\ g(e^{it}) & h(e^{it}) \end{bmatrix} \begin{bmatrix} p'_n(e^{it}) \\ q'_n(e^{it}) \end{bmatrix} \|^2 \, dt \le \frac{\beta^2}{2\pi} \int_0^{2\pi} \| \begin{bmatrix} p'_n(e^{it}) \\ q'_n(e^{it}) \end{bmatrix} \|^2 \, dt = \beta^2 \| \begin{bmatrix} p'_n \\ q'_n \end{bmatrix} \|^2 . \tag{9}$$

On the other hand $\sigma_e(V) = \sigma_e(T)$, and if E denotes the spectral projection of V associated to $\sigma_e(V)$ ([9]), then

$$\|\phi(V)(I - E)w_n\| \to 0 \tag{10}$$

for every weakly convergent $w_n \to 0$ in H, and for every $\phi \in A(\overline{D})$. It follows that

$$\|A\|_e^2 \le \overline{\lim_{n\to\infty}} \left\{ \| \begin{bmatrix} w(V) & f(V) \\ g(V) & h(V) \end{bmatrix} \begin{bmatrix} Ep_n \\ Eq_n \end{bmatrix} \|^2 + \beta^2 \| \begin{bmatrix} p'_n \\ q'_n \end{bmatrix} \|^2 \right\}. \tag{11}$$

Next $V \mid (EH)$ can be viewed as the restriction of the multiplication by ζ on some $L_K^2(\mu)$, where K is a separable Hilbert space, and $\mu \ge 0$ is a measure whose support is $\sigma_e(V)$. Thus if $\hat{p}$ denotes the representation of $p \in EH$ in $L_K^2(\mu)$, we have

$$\| \begin{bmatrix} w(V) & f(V) \\ g(V) & h(V) \end{bmatrix} \begin{bmatrix} Ep_n \\ Eq_n \end{bmatrix} \|^2 = \int_{\sigma_e(V)} \| \begin{bmatrix} w(\zeta) & f(\zeta) \\ g(\zeta) & h(\zeta) \end{bmatrix} \begin{bmatrix} \widehat{Ep_n}(\zeta) \\ \widehat{Eq_n}(\zeta) \end{bmatrix} \|^2 \, d\mu(\zeta) \le$$

$$\alpha^2 \int_{\sigma_e(V)} \| \begin{bmatrix} \widehat{Ep_n}(\zeta) \\ \widehat{Eq_n}(\zeta) \end{bmatrix} \|^2 \, d\mu(\zeta) \le \alpha^2 \| \begin{bmatrix} p_n \\ q_n \end{bmatrix} \|^2 .$$

Thus since we have that

$$\| \begin{bmatrix} p_n \\ q_n \end{bmatrix} \|^2 + \| \begin{bmatrix} p'_n \\ q'_n \end{bmatrix} \|^2 = \|z_n\|^2 = 1$$

we see using (11) that

$$\|A\|_e^2 \le \overline{\lim_{n\to\infty}} \left\{ \alpha^2 \left\| \begin{bmatrix} p_n \\ q_n \end{bmatrix} \right\|^2 + \beta^2 \left\| \begin{bmatrix} p'_n \\ q'_n \end{bmatrix} \right\|^2 \right\} \le$$

$$\overline{\lim_{n\to\infty}} \{\max(\alpha, \beta)\}^2 \left\{ \left\| \begin{bmatrix} p_n \\ q_n \end{bmatrix} \right\|^2 + \left\| \begin{bmatrix} p'_n \\ q'_n \end{bmatrix} \right\|^2 \right\} = \{\max(\alpha,\beta)\}^2,$$

which completes the proof. □

In the next section we will see how Proposition 1 leads to an algorithm for finding $\|A\|$.

3. ALGORITHM FOR NORM OF FOUR BLOCK OPERATOR

In this section we will discuss an algorithm for finding the singular values of A, which we will implement via a determinantal formula in Section 4. Again the line of argument we use here follows very closely that of [1], [4], and [6]. Using the notation of Section 2, we let $\rho > \max(\alpha, \beta)$. Notice that if $\|A\| > \|A\|_e$, then $\|A\|^2$ is an eigenvalue of A^*A.

So we begin by writing $w = a/k$, $f = b/k$, $g = c/k$, $h = d/k$, where a, b, c, d, k are polynomials of degree $\le n$. Then ρ^2 is an eigenvalue of A^*A if and only if

$$\begin{bmatrix} a(S)^*P & c(S)^* \\ b(S)^* & d(S)^* \end{bmatrix} \begin{bmatrix} Pa(S) & b(S) \\ c(S) & d(S) \end{bmatrix} \begin{bmatrix} x \\ y \end{bmatrix} - \begin{bmatrix} \rho^2 k(S)^* k(S) & 0 \\ 0 & \rho^2 k(S)^* k(S) \end{bmatrix} \begin{bmatrix} x \\ y \end{bmatrix} = 0 \qquad (12)$$

for some non-zero

$$\begin{bmatrix} x \\ y \end{bmatrix} \in H^2 \oplus H^2.$$

Next for any polynomial p of degree $\le n$, we set

$$\hat{p}(z) := z^n \overline{p(\frac{1}{\bar{z}})}$$

for $z \in \mathbf{C}$, $z \ne 0$. With this notation if we multiply (12) by ζ^n, $\zeta \in \partial D$, we see that

$$\begin{bmatrix} \hat{a}(S)P & \hat{c}(S) \\ \hat{b}(S) & \hat{d}(S) \end{bmatrix} \begin{bmatrix} Pa(S) & b(S) \\ c(S) & d(S) \end{bmatrix} \begin{bmatrix} x \\ y \end{bmatrix} - \begin{bmatrix} \rho^2 \hat{k}(S) k(S) & 0 \\ 0 & \rho^2 \hat{k}(S) k(S) \end{bmatrix} \begin{bmatrix} x \\ y \end{bmatrix} \qquad (13)$$

$$= \begin{bmatrix} u_1 \\ v_1 \end{bmatrix} \zeta^{n-1} + \begin{bmatrix} u_2 \\ v_2 \end{bmatrix} \zeta^{n-2} + \cdots + \begin{bmatrix} u_n \\ v_n \end{bmatrix}$$

for some

$$\begin{bmatrix} u_j \\ v_j \end{bmatrix} \in \mathbf{C}^2$$

$j=1, \cdots, n$. Then applying P to (13), we obtain

$$\left\{ \begin{bmatrix} \hat{a}(T) & \hat{c}(T) \\ \hat{b}(T) & \hat{d}(T) \end{bmatrix} \begin{bmatrix} a(T) & b(T) \\ c(T) & d(T) \end{bmatrix} - \begin{bmatrix} \rho^2 \hat{k}(T) k(T) & 0 \\ 0 & \rho^2 \hat{k}(T) k(T) \end{bmatrix} \right\} \begin{bmatrix} p \\ q \end{bmatrix} \tag{14}$$

$$= \begin{bmatrix} u_1 \\ v_1 \end{bmatrix} T^{n-1} P 1 + \cdots + \begin{bmatrix} u_n \\ v_n \end{bmatrix} P 1$$

where

$$\begin{bmatrix} p \\ q \end{bmatrix} := \begin{bmatrix} Px \\ Py \end{bmatrix}.$$

Next, we note that if $Px = Py = 0$, then $x = mx'$, $y = my'$ with $x', y' \in H^2$, and (13) becomes

$$\begin{bmatrix} 0 & \hat{c}(S) \\ \hat{b}(S) & \hat{d}(S) \end{bmatrix} \begin{bmatrix} 0 & b(S) \\ c(S) & d(S) \end{bmatrix} \begin{bmatrix} x' \\ y' \end{bmatrix} - \begin{bmatrix} \rho^2 \hat{k}(S) k(S) & 0 \\ 0 & \rho^2 \hat{k}(S) k(S) \end{bmatrix} \begin{bmatrix} x' \\ y' \end{bmatrix} \tag{15}$$

$$= \begin{bmatrix} u_1 \\ v_1 \end{bmatrix} \overline{m(\zeta)}\, \zeta^{n-1} + \cdots + \begin{bmatrix} u_n \\ v_n \end{bmatrix} \overline{m(\zeta)}.$$

Multiplying (15) by ζ^{-n}, and applying the orthogonal projection onto $H^2 \oplus H^2$, it is easy to see that

$$\left\| \begin{bmatrix} 0 & f(S) \\ g(S) & h(S) \end{bmatrix} \begin{bmatrix} x' \\ y' \end{bmatrix} \right\|^2 = \rho^2 \left\| \begin{bmatrix} x' \\ y' \end{bmatrix} \right\|^2$$

and hence,

$$\left\| \begin{bmatrix} 0 & f(S) \\ g(S) & h(S) \end{bmatrix} \begin{bmatrix} x' \\ y' \end{bmatrix} \right\| = \rho \left\| \begin{bmatrix} x' \\ y' \end{bmatrix} \right\|.$$

Since

$$\left\| \begin{bmatrix} 0 & f(S) \\ g(S) & h(S) \end{bmatrix} \right\| \leq \beta < \rho.$$

we deduce that $x' = y' = 0$, and thus $x = y = 0$.

Now let us assume that

$$\begin{bmatrix} p \\ q \end{bmatrix} \in H(m) \oplus H(m)$$

satisfies (14) and that

$$\begin{bmatrix} p \\ q \end{bmatrix} \neq 0.$$

Then

$$\left\{ \begin{bmatrix} \hat{a}(S)P & \hat{c}(S) \\ \hat{b}(S) & \hat{d}(S) \end{bmatrix} \begin{bmatrix} Pa(S) & b(S) \\ c(S) & d(S) \end{bmatrix} - \begin{bmatrix} \rho^2 \hat{k}(S)k(S) & 0 \\ 0 & \rho^2 \hat{k}(S)k(S) \end{bmatrix} \right\} \begin{bmatrix} p \\ q \end{bmatrix} = \tag{16}$$

$$\begin{bmatrix} u_1 \\ v_1 \end{bmatrix} \zeta^{n-1} + \cdots + \begin{bmatrix} u_n \\ v_n \end{bmatrix} + m \begin{bmatrix} \xi \\ \eta \end{bmatrix}$$

for some $\xi, \eta \in H^2$.

But (13) and (16) can be re-written as

$$\begin{bmatrix} \hat{a}(S)a(T) & \hat{a}(S)b(T) \\ \hat{b}(S)a(T) & 0 \end{bmatrix} \begin{bmatrix} Px \\ Py \end{bmatrix} + B(S) \begin{bmatrix} x \\ y \end{bmatrix} = \begin{bmatrix} u_1 \\ v_1 \end{bmatrix} \zeta^{n-1} + \cdots + \begin{bmatrix} u_n \\ v_n \end{bmatrix}, \tag{13a}$$

respectively,

$$\begin{bmatrix} \hat{a}(S)a(T) & \hat{a}(S)b(T) \\ \hat{b}(S)a(T) & 0 \end{bmatrix} \begin{bmatrix} p \\ q \end{bmatrix} + B(S) \begin{bmatrix} p \\ q \end{bmatrix} = \begin{bmatrix} u_1 \\ v_1 \end{bmatrix} \zeta^{n-1} + \cdots + \begin{bmatrix} u_n \\ v_n \end{bmatrix} + m \begin{bmatrix} \xi \\ \eta \end{bmatrix} \tag{16a}$$

where

$$B(\zeta) := \begin{bmatrix} 0 & \hat{c}(\zeta) \\ \hat{b}(\zeta) & \hat{d}(\zeta) \end{bmatrix} \begin{bmatrix} 0 & b(\zeta) \\ c(\zeta) & d(\zeta) \end{bmatrix} - \begin{bmatrix} \rho^2(\hat{k}\ k)(\zeta) & 0 \\ 0 & \rho^2(\hat{k}\ k)(\zeta) \end{bmatrix} \tag{17}$$

for $\zeta \in \mathbf{C}$. Obviously, (13a) and (16a) coincide provided

$$\begin{bmatrix} \xi \\ \eta \end{bmatrix} = B(S) \begin{bmatrix} \xi' \\ \eta' \end{bmatrix} \tag{18}$$

for some $\xi', \eta' \in H^2$.

We can now summarize the above discussion with the following:

PROPOSITION 2. *There exists an eigenvector*

$$z = \begin{bmatrix} x \\ y \end{bmatrix} \in H^2 \oplus H^2 \, , \, z \neq 0$$

satisfying (12) (i.e., ρ *is a singular value of* A*) if and only if there exists a non-zero*

$$\begin{bmatrix} p \\ q \end{bmatrix} \in H(m) \oplus H(m)$$

satisfying (16a) and (18).

In the next section we will construct the determinant of the linear system associated to (16a) and (18).

4. DETERMINANTAL FORMULA

In this section, we will give the explicit determinantal formula for the singular values of the four block operator A. The line of reasoning we are using here, closely parallels that of [1] and [6] which was applied to the one block Hankel case. We use the previous notation here. Moreover, by slight abuse of notation, ζ will denote a complex variable as well as an element of ∂D (the unit circle). The context will always make the meaning clear. Of course, when $\zeta \in \partial D$, then $\bar{\zeta} = 1/\zeta$.

As in Section 3, we let $\rho > \max(\alpha, \beta)$, and as we have seen from Proposition 2, we are looking for a non-zero

$$\begin{bmatrix} p \\ q \end{bmatrix} \in H(m) \oplus H(m)$$

and vectors

$$\begin{bmatrix} u_j \\ v_j \end{bmatrix} \in \mathbf{C}^2,$$

$$\begin{bmatrix} \xi' \\ \eta' \end{bmatrix} \in H^2 \oplus H^2$$

satisfying

$$\begin{bmatrix} \hat{a}(S)a(T) & \hat{a}(S)b(T) \\ \hat{b}(S)a(T) & 0 \end{bmatrix} \begin{bmatrix} p \\ q \end{bmatrix} + B \begin{bmatrix} p \\ q \end{bmatrix} = \begin{bmatrix} u_1 \\ v_1 \end{bmatrix} \zeta^{n-1} + \cdots + \begin{bmatrix} u_n \\ v_n \end{bmatrix} + mB \begin{bmatrix} \xi' \\ \eta' \end{bmatrix}. \tag{19}$$

We now introduce the matrix-valued polynomial operator

$$C := \begin{bmatrix} \hat{a}a & \hat{a}b \\ \hat{b}a & 0 \end{bmatrix} + B \tag{20}$$

Noting that $\overline{m}p$, $\overline{m}q \in L^2 \ominus H^2$, we can express

$$\begin{aligned} m &= m_0 + m_1\zeta + \ldots.. \\ \overline{m}p &= p_{-1}\zeta^{-1} + p_{-2}\zeta^{-2} + \cdots \\ \overline{m}q &= q_{-1}\zeta^{-1} + q_{-2}\zeta^{-2} + \cdots \end{aligned} \tag{21}$$

and write

$$C = \sum_{l=0}^{2n} C_l \zeta^l, \quad B = \sum_{l=0}^{2n} B_l \zeta^l, \quad a = \sum_{l=0}^{n} a_l \zeta^l, \quad b = \sum_{l=0}^{n} b_l \zeta^l. \tag{22}$$

Next we let P_{H^2} denote orthogonal projection from L^2 to H^2. Then

$$\begin{aligned} a(T)p &= ap - m\ P_{H^2}(a\overline{m}p) \\ b(T)q &= bq - m\ P_{H^2}(b\overline{m}q) \end{aligned}$$

and thus (19) is equivalent to

$$C \begin{bmatrix} p \\ q \end{bmatrix} - \sum_{j=1}^{n} \zeta^{n-j} \begin{bmatrix} u_j \\ v_j \end{bmatrix} = m \left\{ \begin{bmatrix} \hat{a} & \hat{a} \\ \hat{b} & 0 \end{bmatrix} \begin{bmatrix} P_{H^2}(a\overline{m}p) \\ P_{H^2}(b\overline{m}q) \end{bmatrix} + B \begin{bmatrix} \xi' \\ \eta' \end{bmatrix} \right\} =: m \begin{bmatrix} \xi'' \\ \eta'' \end{bmatrix}. \tag{23}$$

Now it is well-known (and easy to compute) that

$$T^j p = \zeta^j p - m \sum_{i=1}^{j} \zeta^{j-i} p_{-i}$$

and similarly, of course, for q and $P1 = 1 - m\ \overline{m(0)}$. (Recall that $P : H^2 \to H(m)$ denotes orthogonal projection.) By applying P to (23), we get that

$$C(T)\begin{bmatrix} p \\ q \end{bmatrix} - \sum_{j=1}^{n} T^{n-j} P 1 \begin{bmatrix} u_j \\ v_j \end{bmatrix} = 0. \tag{24}$$

Notice that this means

$$C\begin{bmatrix} p \\ q \end{bmatrix} - \sum_{j=1}^{n} \zeta^{n-j} \begin{bmatrix} u_j \\ v_j \end{bmatrix} = mP_{H^2}\left\{ C\begin{bmatrix} \overline{m}p \\ \overline{m}q \end{bmatrix} - \sum_{j=1}^{n} \zeta^{n-j}\overline{m} \begin{bmatrix} u_j \\ v_j \end{bmatrix} \right\}. \tag{24a}$$

Consequently, we see that (19) is equivalent to (24a) and

$$P_{H^2}C\begin{bmatrix} \overline{m}p \\ \overline{m}q \end{bmatrix} - \sum_{j=1}^{n} P_{H^2}(\zeta^{n-j}\overline{m}) \begin{bmatrix} u_j \\ v_j \end{bmatrix} = \begin{bmatrix} \xi'' \\ \eta'' \end{bmatrix}. \tag{24b}$$

Now

$$P_{H^2}C\begin{bmatrix} \overline{m}p \\ \overline{m}q \end{bmatrix} = P_{H^2}\left\{ \sum_{l=0}^{2n} C_l \zeta^l \sum_{j=1}^{\infty} \zeta^{-j} \begin{bmatrix} p_{-j} \\ q_{-j} \end{bmatrix} \right\} = \sum_{s=0}^{2n-1} \zeta^s \sum_{j=1}^{2n-s} C_{s+j} \begin{bmatrix} p_{-j} \\ q_{-j} \end{bmatrix}. \tag{25a}$$

Similarly, we have

$$P_{H^2}\begin{bmatrix} a & 0 \\ 0 & b \end{bmatrix}\begin{bmatrix} \overline{m}p \\ \overline{m}q \end{bmatrix} = \sum_{s=0}^{n-1} \zeta^s \sum_{j=1}^{n-s} \begin{bmatrix} a_{s+j} & 0 \\ 0 & b_{s+j} \end{bmatrix}\begin{bmatrix} p_{-j} \\ q_{-j} \end{bmatrix}, \tag{25b}$$

and

$$P_{H^2}\sum_{j=1}^{n}(\zeta^{n-j}\overline{m}) \begin{bmatrix} u_j \\ v_j \end{bmatrix} = \sum_{j=0}^{n-1} \zeta^s \sum_{j=1}^{n-s} \overline{m}_{n-s-j} \begin{bmatrix} u_j \\ v_j \end{bmatrix}. \tag{25c}$$

Thus (24a) and (24b) are equivalent to

$$C\begin{bmatrix} p \\ q \end{bmatrix} - \sum_{j=1}^{n} \zeta^{n-j} \begin{bmatrix} u_j \\ v_j \end{bmatrix} = m\left\{ \sum_{s=0}^{2n-1} \zeta^s \sum_{j=1}^{2n-s} C_{s+j} \begin{bmatrix} p_{-j} \\ q_{-j} \end{bmatrix} - \sum_{s=0}^{n-1} \zeta^s \sum_{j=1}^{n-s} \overline{m}_{n-s-j} \begin{bmatrix} u_j \\ v_j \end{bmatrix} \right\} \tag{24c}$$

and

$$
B\begin{bmatrix}\xi' \\ \eta'\end{bmatrix}
$$

$$
= \sum_{s=0}^{2n-1} \zeta^s \sum_{j=1}^{2n-s} C_{s+j} \begin{bmatrix} p_{-j} \\ q_{-j} \end{bmatrix} - \sum_{l=0}^{n} \sum_{s=0}^{n-1} \zeta^{s+l} \sum_{j=1}^{n-s} \left\{ \begin{bmatrix} \overline{a}_{n-l}\, a_{s+j} & \overline{a}_{n-l}\, b_{s+j} \\ \overline{b}_{n-l}\, a_{s+j} & 0 \end{bmatrix} \begin{bmatrix} p_{-j} \\ q_{-j} \end{bmatrix} + \overline{m}_{n-s-j} \begin{bmatrix} u_j \\ v_j \end{bmatrix} \right\}. \tag{24d}
$$

Taking a respite (!) from all of these computations, let us summarize the above discussion with:

PROPOSITION 3. *Equality (19) is equivalent to the two equalities (24c) and (24d). The eigenvalue equation (12) with*

$$
\begin{bmatrix} x \\ y \end{bmatrix} \neq 0
$$

is thus equivalent (see Proposition 2) to (24c), (24d), and

$$
\begin{bmatrix} p \\ q \end{bmatrix} \neq 0. \tag{26}
$$

Next set

$$
d_c(\zeta) := \det C(\zeta) \tag{27}
$$

and

$$
C^+(\zeta) = \sum_{j=0}^{2n} C_j^+ \zeta^j
$$

where

$$
C_j = \begin{bmatrix} c_{j,11} & c_{j,12} \\ c_{j,21} & c_{j,22} \end{bmatrix}
$$

$$
C_j^+ = \begin{bmatrix} c_{j,22} & -c_{j,12} \\ -c_{j,21} & c_{j,11} \end{bmatrix}
$$

for $0 \leq j \leq 2n$. Note that $C^+(\zeta)C(\zeta) = d_c(\zeta)I$.

We will now make a technical assumption in order to simplify our exposition. Below we will discuss how to remove this *assumption of genericity*. Explicitly, we assume that

$$d_c \text{ has distinct roots all of which are non-zero.} \tag{28}$$

We can now prove the following:

LEMMA 1. *Under assumption (28), $d_c(\zeta)$ has r zeros $\alpha_1, \ldots, \alpha_r \in D$, r zeros $1/\overline{\alpha}_1, \cdots, 1/\overline{\alpha}_r \notin \overline{D}$, and $4n-2r$ zeros $\alpha_{2r+1}, \cdots, \alpha_{4n} \in \partial D \backslash \sigma(T)$.*

PROOF.

Let $\lambda \in \sigma_e(T) = \sigma(T) \cap \partial D$. Then since

$$\lambda^{-2n} C(\lambda) \geq (\rho^2 - \alpha^2) I$$

and $\lambda^{-4n} d_c(\lambda)$ is the determinant of $\lambda^{-2n} C(\lambda)$, the zeros of d_c are outside of $\sigma_e(T)$. Next it is easy to compute that

$$\zeta^{2n} \overline{C(1/\overline{\zeta})} = C(\zeta)$$

and hence

$$\zeta^{4n} \overline{d_c(1/\overline{\zeta})} = d_c(\zeta). \tag{29}$$

The assertion of the lemma now immediately follows from (29). □

REMARK. Lemma 1 is the precise analogue of (3.5) of [6] for the computation of the singular values of the Hankel. See also [4], pages 153-154.

Continuing our computation, we note that by multiplying by $C^+(\zeta)$, we can express (24c) equivalently as

$$d_c \begin{bmatrix} p \\ q \end{bmatrix} = \sum_{j=1}^{n} (E_j + mF_j) \begin{bmatrix} u_j \\ v_j \end{bmatrix} + m \sum_{j=1}^{2n} G_j \begin{bmatrix} p_{-j} \\ q_{-j} \end{bmatrix} \tag{30}$$

where the matrix-valued polynomials E_j, F_j, G_j are given by

$$E_j := \sum_{l=0}^{2n} \zeta^{n-j+l} C_l^+,$$
$$F_j := -\sum_{s=0}^{n-j} \sum_{l=0}^{2n} \zeta^{s+l} \overline{m}_{n-s-j} C_l^+ \tag{30a}$$
$$G_j := \sum_{s=0}^{2n-j} \sum_{l=0}^{2n} \zeta^{s+l} C_l^+ C_{s+j}$$

for $1 \leq j \leq n$.

We are almost done now! Indeed arguing precisely as in [6], we note that since $p, q \in H(m)$, they must be analytic in a neighborhood of $\partial D \backslash \sigma(T)$ as well as in D. Hence using Lemma 1, from (30) we see that

$$\sum_{j=1}^{n} (E_j(\alpha_i) + m(\alpha_i) F_j(\alpha_i)) \begin{bmatrix} u_j \\ v_j \end{bmatrix} + m(\alpha_i) \sum_{j=1}^{2n} G_j(\alpha_i) \begin{bmatrix} p_{-j} \\ q_{-j} \end{bmatrix} = 0 \tag{31a}$$

for $1 \leq i \leq r$, and for $2r+1 \leq i \leq 4n$ For $|\zeta|=1$, if we multiply (30) by $\overline{\zeta}^{4n} \overline{m}$, we get

$$\overline{\zeta}^{4n} d_c \begin{bmatrix} \overline{m}p \\ \overline{m}q \end{bmatrix} = \sum_{j=1}^{n} (\overline{m} \overline{\zeta}^{4n} E_j + \overline{\zeta}^{4n} F_j) \begin{bmatrix} u_j \\ v_j \end{bmatrix} + \sum_{j=1}^{2n} \overline{\zeta}^{4n} G_j \begin{bmatrix} p_{-j} \\ q_{-j} \end{bmatrix}. \tag{32}$$

But this last equation admits an analytic extension to the complement of the disc, i.e. all the functions are analytic in $1/\zeta$. Set

$$E_j^o(1/\zeta) := (1/\zeta)^{4n} E_j(\zeta)$$

and similarly for d_c^o, F_j^o, and G_j^o. Hence we can express (32) equivalently as

$$d_c^o(1/\zeta) \begin{bmatrix} (\overline{m}p)(1/\zeta) \\ (\overline{m}q)(1/\zeta) \end{bmatrix} = \sum_{j=1}^{n} (\overline{m(1/\overline{\zeta})} E_j^o(1/\zeta) + F_j^o(1/\zeta)) \begin{bmatrix} u_j \\ v_j \end{bmatrix} + \sum_{j=1}^{2n} G_j^o(1/\zeta) \begin{bmatrix} p_{-j} \\ q_{-j} \end{bmatrix}. \tag{32a}$$

Then from (32a) and Lemma 1, we get that

$$\sum_{j=1}^{n} \overline{(m(\alpha_i)} E_j^o(\overline{\alpha_i}) + F_j^o(\overline{\alpha_i})) \begin{bmatrix} u_j \\ v_j \end{bmatrix} + \sum_{j=1}^{2n} G_j^o(\overline{\alpha_i}) \begin{bmatrix} p_{-j} \\ q_{-j} \end{bmatrix} = 0 \tag{32b}$$

for $1 \leq i \leq r$.

We now play the analogous game with the B operator that we just did with the C operator. Indeed, setting

$$d_b(\zeta) := \det B(\zeta)$$

we have the analogy of (29), and since $\rho > \beta$, we have that $d_b(\zeta) \neq 0$ for $\zeta \in \partial D$. Again we make the assumption of genericity that

$$d_b \text{ has distinct roots all of which are non-zero.} \tag{33}$$

Just like for lemma 1 above, we see that d_b has $2n$ zeros $\beta_1, \ldots, \beta_{2n} \in D$, and $2n$ zeros $1/\overline{\beta}_1, \cdots, 1/\overline{\beta}_{2n}$ in the complement of $\overline{D}$.

We now set (as above)

$$B^+(\zeta) := \sum_{j=0}^{2n} B_j^+ \zeta^j$$

where

$$B_j = \begin{bmatrix} b_{j,11} & b_{j,12} \\ b_{j,21} & b_{j,22} \end{bmatrix}$$

$$B_j^+ = \begin{bmatrix} b_{j,22} & -b_{j,12} \\ -b_{j,21} & b_{j,11} \end{bmatrix}$$

for $0 \leq j \leq 2n$. Note that $B^+(\zeta)B(\zeta) = d_b(\zeta)$. Finally, since

$$\begin{bmatrix} \xi' \\ \eta' \end{bmatrix} \in H^2 \oplus H^2$$

(24d) implies for $1 \leq i \leq 2n$, the equations

$$\sum_{j=1}^{2n} H_j(\beta_i) \begin{bmatrix} p_{-j} \\ q_{-j} \end{bmatrix} + \sum_{j=1}^{n} K_j(\beta_i) \begin{bmatrix} u_j \\ v_j \end{bmatrix} = 0 \tag{34}$$

where

$$\begin{aligned} H_j(\zeta) &:= \sum_{s=0}^{2n-j} \sum_{l=0}^{2n} \zeta^{s+l} B_l^+ C_{s+j} - \sum_{i=0}^{n} \sum_{s=0}^{n-j} \sum_{l=0}^{2n} \zeta^{s+l+i} B_l^+ \begin{bmatrix} \overline{a}_{n-i} a_{s+j} & \overline{a}_{n-i} b_{s+j} \\ \overline{b}_{n-i} a_{s+j} & 0 \end{bmatrix} \\ K_j(\zeta) &:= -\sum_{s=0}^{n-j} \sum_{l=0}^{2n} \begin{bmatrix} \zeta^{s+l} & 0 \\ 0 & \zeta^{s+l} \end{bmatrix} B_l^+ \overline{m}_{n-s-j}. \end{aligned} \tag{34a}$$

We are at long last ready to state our main result:

THEOREM 1. *Assume the genericity conditions (28) and (33) hold. Then* ρ *is a singular value of the four block operator A if and only if the* $12n \times 6n$ *matrix*

$$M := \begin{bmatrix} M_1 & M_2 \\ N_2 & N_1 \end{bmatrix}$$

has rank $< 6n$ *where*

$$M_1 := \begin{bmatrix} E_1(\alpha_1) + m(\alpha_1)F_1(\alpha_1) & \cdot & E_n(\alpha_1) + m(\alpha_1)F_n(\alpha_1) \\ \cdot & \cdot & \cdot \\ \cdot & \cdot & \cdot \\ \cdot & \cdot & \cdot \\ E_1(\alpha_r) + m(\alpha_r)F_1(\alpha_r) & \cdot & E_n(\alpha_r) + m(\alpha_r)F_n(\alpha_r) \\ E_1(\alpha_{2r+1}) + m(\alpha_{2r+1})F_1(\alpha_{2r+1}) & \cdot & E_n(\alpha_{2r+1}) + m(\alpha_{2r+1})F_n(\alpha_{2r+1}) \\ \cdot & \cdot & \cdot \\ \cdot & \cdot & \cdot \\ \cdot & \cdot & \cdot \\ E_1(\alpha_{4n}) + m(\alpha_{4n})F_1(\alpha_{4n}) & \cdot & E_n(\alpha_{4n}) + m(\alpha_{4n})F_n(\alpha_{4n}) \\ \overline{m(\alpha_1)}E_1^o(\overline{\alpha}_1) + F_1^o(\overline{\alpha}_1) & \cdot & \overline{m(\alpha_1)}E_n^o(\overline{\alpha}_1) + F_n^o(\overline{\alpha}_1) \\ \cdot & \cdot & \cdot \\ \cdot & \cdot & \cdot \\ \cdot & \cdot & \cdot \\ \overline{m(\alpha_r)}E_1^o(\overline{\alpha}_r) + F_1^o(\overline{\alpha}_r) & \cdot & \overline{m(\alpha_r)}E_n^o(\overline{\alpha}_r) + F_n^o(\overline{\alpha}_r) \end{bmatrix}$$

$$N_1 := \begin{bmatrix} K_1(\beta_1) & \cdot & K_n(\beta_1) \\ \cdot & \cdot & \cdot \\ \cdot & \cdot & \cdot \\ \cdot & \cdot & \cdot \\ K_1(\beta_{2n}) & \cdot & K_n(\beta_{2n}) \end{bmatrix}$$

$$M_2 := \begin{bmatrix} m(\alpha_1)G_1(\alpha_1) & . & m(\alpha_1)G_{2n}(\alpha_1) \\ . & . & . \\ . & . & . \\ . & . & . \\ m(\alpha_r)G_1(\alpha_r) & . & m(\alpha_r)G_{2n}(\alpha_r) \\ m(\alpha_{2r+1})G_1(\alpha_{2r+1}) & . & m(\alpha_{2r+1})G_{2n}(\alpha_{2r+1}) \\ . & . & . \\ . & . & . \\ . & . & . \\ m(\alpha_{4n})G_1(\alpha_{4n}) & . & m(\alpha_{4n})G_{2n}(\alpha_{4n}) \\ G_1^o(\overline{\alpha}_1) & . & G_{2n}^o(\overline{\alpha}_1) \\ . & . & . \\ . & . & . \\ . & . & . \\ G_1^o(\overline{\alpha}_r) & . & G_{2n}^o(\overline{\alpha}_r) \end{bmatrix}$$

$$N_2 := \begin{bmatrix} H_1(\beta_1) & . & H_{2n}(\beta_1) \\ . & . & . \\ . & . & . \\ . & . & . \\ H_1(\beta_{2n}) & . & H_{2n}(\beta_{2n}) \end{bmatrix}.$$

PROOF. Indeed, we have shown that (12) (the eigenvalue equation) with

$$\begin{bmatrix} x \\ y \end{bmatrix} \neq 0$$

implies the existence of a non-zero solution of the linear system of equations given by (31a), (32b), and (34). Since all of the matrices E_j, F_j, G_j, H_j, K_j, E_j^o, F_j^o, and G_j^o are polynomially valued, it is easy to check that from the above argument, we can define vectors

$$\begin{bmatrix} p \\ q \end{bmatrix} \in H(m) \oplus H(m), \quad \begin{bmatrix} \xi' \\ \eta' \end{bmatrix} \in H^2 \oplus H^2$$

which satisfy (24c) and (24d), respectively. Hence (19) will also be satisfied. By Proposition 2,

$$\begin{bmatrix} p \\ q \end{bmatrix} \neq 0$$

which completes the proof of the theorem. □

REMARKS.

(i) Note that in order to find the singular values of A, we are required by the theorem to determine the non-injectivity of a $6n \times 3n$ matrix with block 2×2 entries given by the above explicit formulae .

(ii) The matrix $[\ M_1 \quad M_2\ ]$ above is the analogue of the matrix given in Theorem (3.6) of [6].

(iii) One can easily remove the genericity assumptions (28) and (33), and get a determinantal formula for the singular values when the determinants d_c and d_b have multiple roots. Basically one uses a limiting argument, precisely as in [6]. We leave this as an exercise for the interested reader.

5. CONCLUSIONS

In this note, we have derived an explicit determinantal formula for the singular values of the four block operator in the numerical case. The general case of matrix-valued entries will be handled in a forthcoming paper. Basically, we will show that the computation of the norm of the general four block operator reduces to a skew Toeplitz operator problem, and our plan is to apply the determinantal formula of [1] Theorem (7.5) to this setting.

Finally, from our computational experience, it appears that the digital implementation of determinantal formulae like that given in Theorem 1 is not very difficult to carry out. Indeed, some systems engineers at Honeywell (Blaise Morton and James Krause) have already developed software for the determinantal expression of [6] which has performed quite well on several non-trivial examples. At present, we are also writing the software for the algorithms of [1], which as we have noted in conjunction with the recently announced results of Doyle and Glover should lead to a reliable and direct way of solving the four block problem.

REFERENCES

[1] H. BERCOVICI, C. FOIAS, and A. TANNENBAUM, On skew Toeplitz operators, I, to appear in *Integral Equations and Operator Theory*.

[2] J. DOYLE, Lecture Notes, ONR/Honeywell Workshop on Advances in Multivariable Control, Minneapolis, Minnesota, 1984.

[3] H. DYM and I. GOHBERG, A new class of contractive interpolants and maximal entropy principles, to appear in *Integral Equations and Operator Theory*.

[4] C. FOIAS and A. TANNENBAUM, On the Nehari problem for a certain class of L^∞ functions appearing in control theory, *J. Functional Analysis* **74** (1987), pp. 146-159.

[5] C. FOIAS, A. TANNENBAUM, and G. ZAMES, On the H^∞-optimal sensitivity problem for systems with delays, *SIAM J. Control and Optimization* **25** (1987), 686-706.

[6] C. FOIAS, A. TANNENBAUM, and G. ZAMES, Some explicit formulae for the singular values of certain Hankel operators with factorizable symbol, Technical Report, Department of Electrical Engineering, Univ. of Minnesota, March 1987. Submitted for publication.

[7] B.A. FRANCIS, "A Course in H^∞ Control Theory," Lecture Notes in Control and Information Science, Springer, New York, 1987.

[8] E. JONCKHEERE and M. VERMA, A spectral characterization of H^∞ optimal performance and its efficient computation, *Systems and Control Letters* **8** (1986), pp. 13-22.

[9] N. K. NIKOLSKII, "Treatise on the Shift Operator," Springer, New York, 1986.

[10] B. SZ.-NAGY and C. FOIAS, "Harmonic Analysis of Operators on Hilbert Space," North-Holland, Amsterdam, 1970.

[11] G. ZAMES and S. MITTER, On Hankel + Toeplitz operators, MTNS Conference, Phoenix, Arizona, 1987.

[12] G. ZAMES, A. TANNENBAUM, and C. FOIAS, Optimal H^∞ interpolation: a new approach, Proceedings of the CDC, Athens, Greece, 1986, pp. 350-355.

Ciprian Foias
Department of Mathematics
Indiana University
Bloomington, Indiana 47405

Allen Tannenbaum
Department of Electrical Engineering
University of Minnesota
123 Church Street SE
Minneapolis, Minnesota 55455

Operator Theory:
Advances and Applications, Vol. 32

VARIATION OF THE POINT SPECTRUM UNDER COMPACT PERTURBATIONS

Domingo A. Herrero,[1] Thomas J. Taylor and Zong Y. Wang

To the memory of our friends
and colleagues Constantin
Apostol and Douglas McMahon

This article provides a complete answer to the following problem: given a Hilbert space operator T, find all possible pairs of subsets Σ,Ξ of the complex plane such that for some compact operator K, Σ is the point spectrum of $T-K$ and Ξ is the point spectrum of its adjoint. The result is applied to study the continuity properties of the function $T \to \sigma_p(T)^-$, and several related spectral functions.

1. INTRODUCTION

Let $L(H)$ denote the algebra of all (bounded linear) operators acting on a complex, separable, infinite dimensional Hilbert space H. The *point spectrum* of $T \in L(H)$,

$$\sigma_p(T) = \{\lambda \in \mathbb{C}: \text{ there exists a unit vector x in } H \text{ such that } Tx = \lambda x\},$$

is a (possibly empty) subset of the *spectrum* of T, $\sigma(T) = \{\lambda \in \mathbb{C}: T-\lambda \text{ is not invertible in } L(H)\}$, and therefore a bounded set.

J. Dixmier and C. Foiaş have shown that $\sigma_p(T)$ is always an F_σ set; moreover, given a bounded F_σ set Σ, it is possible to find $T \in L(H)$ such that $\sigma_p(T) = \Sigma$ and $\sigma(T) = \Sigma^-$, the closure of Σ [14]. (See also the examples provided by G. K. Kalish in [28], [29].)

Recall that $A \in L(H)$ is a *semi-Fredholm operator* if ran $A := AH$ is closed (that is, ran A is a *subspace* of H) and ei-

---

1) The research of the first author was partially supported by a Grant of the National Science Foundation.

ther the kernel, ker A, or the cokernel, ker A*, of A is a finite dimensional subspace of H. (Here A* denotes the adjoint of the operator A.) In this case, the index of A is the cardinal number defined by

$$\text{ind } A = \text{nul } A - \text{nul } A^*,$$

where nul A = dim ker A. (The reader is referred to [18], [30] for the properties of the semi-Fredholm operators.)

Let $\rho_{s-F}(T) = \{\lambda \in \mathbb{C}: \lambda - T \text{ is semi-Fredholm}\}$ (= the semi-Fredholm domain of T), and let $\rho^+_{s-F}(T)$ $(\rho^-_{s-F}(T)) = \{\lambda \in \mathbb{C}: \lambda - T$ is semi-Fredholm and $\text{ind}(\lambda - T) > 0$ $(\text{ind}(\lambda - T) < 0$, resp.)$\}$. If $K(H)$ denotes the ideal of all compact operators acting on H, and $\tilde{T} = T + K(H)$ is the canonical projection of $T \in K(H)$ in the quotient Calkin algebra, then the *essential spectrum* of T, $\sigma_e(T) = \sigma(\tilde{T})$, coincides with the set

$$\sigma_{\ell re}(T) \cup \{\lambda \in \rho_{s-F}(T): \text{ind}(\lambda - T) = \pm\infty\},$$

where $\sigma_{\ell re}(T) = \mathbb{C} \setminus \rho_{s-F}(T)$. (If $\sigma_{\ell e}(T)$ and $\sigma_{re}(T)$ denote the left and, respectively, the right spectrum of $\tilde{T}$, then $\sigma_e(T) = \sigma_{\ell e}(T) \cup \sigma_{re}(T)$ and $\sigma_{\ell re}(T) = \sigma_{\ell e}(T) \cap \sigma_{re}(T)$.)

The point spectrum of an operator T always includes $\rho^+_{s-F}(T)$, as well as the set $\sigma_o(T)$ of all *normal eigenvalues* of T. ($\sigma_o(T) = \{\lambda \in \mathbb{C}: \lambda$ is an isolated point of $\sigma(T) \setminus \sigma_e(T)\}$.) By taking adjoints, we deduce that $\sigma_p(T^*)^* \supset \rho^-_{s-F}(T) \cup \sigma_o(T)$, where $\Gamma^* = \{\bar{\lambda}: \lambda \in \Gamma\}$ $(\Gamma \subset \mathbb{C})$.

The main result of this article says that, given T in $L(H)$ and two bounded F_σ sets, Σ and Ξ, satisfying the obvious necessary conditions (derived from the above observations), it is possible to find a compact operator K such that

$$\sigma_p(T - K) = \Sigma \text{ and } \sigma_p((T - K)^*) = \Xi.$$

The results include an upper estimate on the size of a compact operator K necessary to attain these results. By using this estimate, the following (rather strange) corollary is obtained:

Let C be the family of all compact subsets of $\mathbb{C}$ with the modified Hausdorff distance $d_H(\Gamma,\emptyset) = 1$ for all $\Gamma \neq \emptyset$, and $d_H(\Gamma,\Delta) = \min[1,\min\{\varepsilon \geq 0: \Gamma \subset \Delta_\varepsilon, \Delta \subset \Gamma_\varepsilon\}]$ (if Γ and Δ are nonempty sets), where $\Gamma_\varepsilon = \{\lambda \in \mathbb{C}: \text{dist}[\lambda,\Gamma] \leq \varepsilon\}$.

Consider the spectral function $\bar{\sigma}_p : L(H) \to C$, defined by

$$\bar{\sigma}_p(T) = \sigma_p(T)^-.$$

If $\bar{\sigma}_p$ is discontinuous at T, then there exist $\delta > 0$ and a sequence of *compact* operators, $\{K_n\}_{n=1}^{\infty}$, such that $\|K_n\| \to 0$ $(n \to \infty)$ and

$$d_H[\bar{\sigma}_p(T), \bar{\sigma}_p(T - K_n)] \geq \delta$$

for all n = 1,2,... . (The points of continuity of $\bar{\sigma}_p$ were determined by J. B. Conway and B. B. Morrel in [12]; see also [4,Chapter 4] and [24].)

Indeed, the article includes analogous results for several spectral functions related to $\bar{\sigma}_p$.

Sections 2, 3, 4 and 5 deal with the "tools" necessary to attack the solution of the main problem, and are extremely technical. The core of the article is Theorem 6.1 of Section 6. Finally, Section 7 is devoted to the analysis of $\bar{\sigma}_p$ and related functions.

The authors wish to thank Professors J. Agler, L. Avon and J. Mc Donald for many stimulating discussions.

2. CLEANSING. FIRST PART

The process to reach the main result consists of two steps. In the first step, we "cleanse" whatever can be eliminated from the point spectra of T and T*. In the second one, we put some "make up" to add the points that must be in the new point spectra.

The first cleansing step is based on the results of [27] for biquasitriangular operators: if $A \in L(H)$ is biquasitriangular (that is, $\rho^+_{s-F}(A) = \rho^-_{s-F}(A) = \emptyset$; see [4], [21]), then there exists a compact operator K such that both, A - K and (A - K)*, are triangular operators; moreover, if

$$A - K = \begin{pmatrix} \lambda_1 & & & & * \\ & \lambda_2 & & & \\ & & \lambda_3 & & \\ & & & \cdot & \\ & 0 & & & \cdot \\ & & & & & \cdot \end{pmatrix} \begin{matrix} e_1 \\ e_2 \\ e_3 \\ \cdot \\ \cdot \\ \cdot \end{matrix}$$

with respect to some orthonormal basis $\{e_j\}_{j=1}^{\infty}$, then

$$(A-K)^* = \begin{pmatrix} \bar{\lambda}_1 & & & & \\ & \bar{\lambda}_2 & & * & \\ & & \bar{\lambda}_3 & & \\ & & & \cdot & \\ & 0 & & & \cdot \\ & & & & & \cdot \end{pmatrix} \begin{matrix} f_1 \\ f_2 \\ f_3 \\ \cdot \\ \cdot \\ \cdot \end{matrix}$$

with respect to some other orthonormal basis $\{f_j\}_{j=1}^{\infty}$, and

$$\sigma_p(A) = \sigma_p(A^*)^* = \{\lambda_j\}_{j=1}^{\infty}.$$

Furthermore, if $\sigma_e(A)$ is connected, then the λ_j's can be chosen so that $\lambda_j = \lambda_0$ for all $j = 1,2,\ldots$, for some prescribed point λ_0 of $\sigma_e(A)$. By using this result, C. Apostol proved that if A is biquasitriangular and $\sigma_e(A)$ is connected, then there exists $K \in K(H)$ such that $\sigma_p(A-K) = \sigma_p((A-K)^*) = \emptyset$ (personal communication). Lemma 2.7 below is just the extension of this result to arbitrary biquasitriangular operators.

LEMMA 2.1. *Let*

$$T = \begin{pmatrix} A & C \\ 0 & B \end{pmatrix} \begin{matrix} H_1 \\ H_2 \end{matrix} \in L(H) \quad (H = H_1 \oplus H_2);$$

then $\sigma_p(T)$ *is the union of* $\sigma_p(A)$ *and* $\{\lambda \in \mathbb{C}:$ *there exist a unit vector* $y \in H_2$ *and a vector* $x \in H_1$ *such that* $(B-\lambda)y = 0$ *and* $Cy = (A-\lambda)x\}$.

In particular, $\sigma_p(A) \subset \sigma_p(T) \subset \sigma_p(A) \cup \sigma_p(B)$.

PROOF. It is completely apparent that $(A-\lambda)x = 0$ ($x \in H_1$) implies $(T-\lambda)x = 0$. Therefore, $\sigma_p(A) \subset \sigma_p(T)$.

Suppose $(T-\lambda)\begin{pmatrix} x \\ y \end{pmatrix} = 0$ for some vector $\begin{pmatrix} x \\ y \end{pmatrix}$ with $y \neq 0$; then

$$\begin{pmatrix} A-\lambda & C \\ 0 & B-\lambda \end{pmatrix} \begin{pmatrix} x \\ y \end{pmatrix} = \begin{pmatrix} (A-\lambda)x+Cy \\ (B-\lambda)y \end{pmatrix} = \begin{pmatrix} 0 \\ 0 \end{pmatrix}$$

implies $(B-\lambda)y = 0$ (so that $\lambda \in \sigma_p(B)$) and $Cy = (A-\lambda)(-x)$, whence the results follow. ∎

REMARK 2.2. The proof shows that, if $\lambda \in \sigma_p(B)$ and $(B-\lambda)y = 0$, then $Cy \in \operatorname{ran}(\lambda - A)$. Thus, $\lambda \in \sigma_p(T)$ if and only if either $\lambda \in \sigma_p(A)$, or $(B-\lambda)y = 0$ and $Cy = 0$ for some unit vector $y \in H_2$, or

$$C[\ker(B-\lambda)] \cap \operatorname{ran}(\lambda - A) \neq \{0\}.$$

The following is a folklore result. For completeness, we include a proof here.

LEMMA 2.3. *Let* $V \in L(L^2([0,1],dt))$ *be the Volterra operator, defined by*

$$(Vf)(t) = \int_{[0,t]} f(s)\, ds, \quad t \in [0,1].$$

There exists a family $\{U_r\}_{0 \le r < 1}$ *of unitary operators such that*

$$\operatorname{ran} U_r V U_r^* \cap \operatorname{ran} U_s V U_s^* = \operatorname{ran} V \cap \operatorname{ran} U_r V U_r^* = \{0\}$$

for $0 \le r < s < 1$.

PROOF. Let χ_n denote the characteristic function of $[q_n,\infty)$, where $\{q_n\}_{n=1}^{\infty}$ is an enumeration of the rational numbers in the interval $[0,1)$. Define

$$u(t) = \exp\{i\textstyle\sum_{n=1}^{\infty} 2^{-n} \chi_n(t)\},$$

for $0 \le t < 1$, and extend it to a periodic function of period 1. For $0 \le r < 1$, define

$$(U_r f)(t) = u(t-r) f(t), \quad t \in [0,1].$$

Suppose $0 \le r < s < 1$ and $(U_r V U_r^*)g = (U_s V U_s^*)h$; that is,

$$g_1 = V(U_r^* g) = (U_r^* U_s)[V(U_s^* h)] = u(t+r)u(t-s)h_1,$$

where $g_1 = V(U_r^* g)$ and $h_1 = V(U_s^* h)$ are absolutely continuous functions with the same modulus everywhere on $[0,1]$.

If $|g_1(t)| \ge \delta > 0$ on an interval $(c,d) \subset [0,1]$, then

$$0 = \|(U_r V U_r^*)g - (U_s V U_s^*)h\|^2$$

$$\ge \delta^2 \int_{(c,d)} |[g_1(t)/h_1(t)] - u(t+r)u(t-s)|^2\, ds > 0,$$

a contradiction. (If r-s is irrational, then it is obvious that $u(t+r)u(t-s)$ cannot coincide a.e. with the continuous function g_1/h_1 on any interval; if r-s is rational, we arrive to the same conclusion by observing that all the jumps are distinct. The details are left to the reader.)

It follows that $g_1(t) \equiv 0$ on $[0,1]$, and therefore

$$\operatorname{ran} U_r V U_r^* \cap \operatorname{ran} U_s V U_s^* = \{0\}.$$

That $\operatorname{ran} V \cap \operatorname{ran} U_r V U_r^* = \{0\}$ for all $r \in [0,1]$ follows by the same argument. ∎

The sequence of s-numbers of the Volterra operator is equal to $\{[(n+\frac{1}{2})\pi]^{-1}\}_{n=0}^{\infty}$ (see [18] for details). It is not difficult to see that if K is compact and $s_n(K) = O(1/n)$, then we can find U unitary such that

$$\text{ran } K \subset \text{ran } UVU^* = U(\text{ran } V).$$

By combining this observation with the above lemma, we obtain the following.

COROLLARY 2.4. *If* $K \in K(H)$ *and its sequence of* s-*numbers satisfies* $s_n(K) = O(\frac{1}{n})$, *then there exists a unitary operator* U_0 *such that*

$$\text{ran } K \cap \text{ran } U_0 V U_0^* = \{0\}.$$

Let $\{U_r\}_{0 \le r < 1}$ be defined as in Lemma 2.3, and let 1 denote the identity operator on H (H a complex, separable, infinite dimensional Hilbert space); then $\{U_r V U_r^* \otimes 1\}_{0 \le r < 1}$ is a family of *non-compact* operators with the property that $\text{ran}(U_r V U_r^* \otimes 1)$ is dense for all r, $0 \le r < 1$, and

$$\text{ran } (U_r V U_r^* \otimes 1) \cap \text{ran } (U_s V U_s^* \otimes 1) = \{0\} \text{ for } 0 \le r < s < 1.$$

By using this result, it is not difficult to show that if $A \in L(H)$ and ran A is not a (closed!) subspace of finite codimension, then there exists $B \simeq V \otimes 1$ in $L(H)$ such that $\text{ran } A \cap \text{ran } B = \{0\}$. The interested reader can consult [13], [15].

LEMMA 2.5. *If* $A \in L(H)$ *is a bitriangular operator with diagonal sequence* $\{\lambda_j\}_{j=1}^{\infty}$ *and* N *is a normal operator of uniform infinite multiplicity and purely point spectrum such that* $\sigma_p(N) = \{\lambda_j\}_{j=1}^{\infty}$, *then given* $\varepsilon > 0$ *there exists* $K_\varepsilon \in K(H)$, *with* $\|K_\varepsilon\| < \varepsilon$, *such that*

$$\sigma_p(A \oplus N - K_\varepsilon) = \sigma_p([A \oplus N - K_\varepsilon]^*) = \emptyset.$$

PROOF. By hypothesis, $N = (\text{diag}\{\nu_1, \nu_2, \ldots\})^{(\infty)}$, where $\nu_m \ne \nu_n$ if $m \ne n$, $\{\nu_n\} = \{\lambda_j\}_{j=1}^{\infty}$ (as a set) and $T^{(\alpha)}$ denotes the direct sum of α copies of $T \in L(H)$ acting on the orthogonal direct sum $H^{(\alpha)}$ of α copies of H. Clearly, $N \simeq N \oplus N \oplus N$, so we can replace (if necessary) A by $A \oplus N$ and assume directly that $\text{nul}(A - \lambda_j) = \text{nul}(A - \lambda_j)^* = \infty$ for all $j = 1, 2, \ldots$, and proceed to prove that some small compact perturbation of $N \oplus A \oplus N$ has the desired properties.

In what follows, we shall assume that $\{\nu_n\}$ is an infinite (denumerable) set. (The case when this set is finite follows by the same argument.) Recall that $\sigma_p(A) = \sigma_p(A^*)^* = \{\lambda_j\}_{j=1}^{\infty} = \{\nu_n\}_{n=1}^{\infty}$ [27].

We have

$$N \oplus A \oplus N = \left(\begin{array}{cccc|c|cccc}
\nu_1 & & & & & & & & \\
& \nu_2 & 0 & & & & & & \\
& & \cdot & & 0 & & 0 & & \\
& 0 & & \cdot & & & & & \\
& & & & \cdot & & & & \\
\hline
& 0 & & & A & & 0 & & \\
\hline
& & & & & \nu_1 & & & \\
& & & & & & \nu_2 & 0 & \\
& 0 & & & 0 & & & \cdot & \\
& & & & & & 0 & & \cdot \\
& & & & & & & & \cdot
\end{array}\right)\begin{array}{l} H_1 \\ H_2 \\ \cdot \\ \cdot \\ \cdot \\ H \\ H_1^* \\ H_2^* \\ \cdot \\ \cdot \\ \cdot \end{array},$$

where $H_1, H_2, \dots, H_1^*, H_2^*, \dots$ are copies of H.

Let $\{e_i^n\}_{i=1}^{\infty}$ be an orthonormal basis of $\ker(A-\nu_n)$ and let $\{f_i^n\}_{i=1}^{\infty}$ be an orthonormal basis of $\ker(N-\nu_n)$ $(n=1,2,\dots)$. Define $Q_{2,n}: H \to H_n$ by $Q_{2,n} = \sum_{i=1}^{\infty} (\varepsilon/2^{n+i})\, f_i^n \otimes e_i^n$. (As usual, $f \otimes g \in L(H)$ is the rank one operator defined by $(f\otimes g)x = \langle x,e\rangle f$, $x \in H$.) By Corollary 2.4, there exists an operator $Q_{1,n}' \simeq V$ such that $\operatorname{ran} Q_{2,n} \cap \operatorname{ran} Q_{1,n}' = \{0\}$. Define $Q_{1,n} = (\varepsilon/2^n)\, Q_{1,n}'$ $(n = 1,2,\dots)$ and $Q_1 = \oplus_{n=1}^{\infty} Q_{1,n}$.

CLAIM. $\sigma_p(N-Q_1)$, $\sigma_p([N-Q_1]^*)$ and the point spectrum of

$$B = \left(\begin{array}{cccc|c}
\nu_1 - Q_{1,1} & & & & -Q_{2,1} \\
& \nu_2 - Q_{1,2} & 0 & & -Q_{2,2} \\
& & \cdot & & \cdot \\
& 0 & & \cdot & \cdot \\
& & & & \cdot \\
\hline
& 0 & & & A
\end{array}\right) = \begin{pmatrix} N - Q_1 & -Q_2 \\ 0 & A \end{pmatrix}$$

are empty sets. (Here Q_2 is the operator corresponding to the column of operators $Q_{2,1}, Q_{2,2}, \dots$.)

Since $Q_{1,n}$ is unitarily equivalent to a positive multiple of the Volterra operator, $\sigma_p(V) = \sigma_p(V^*) = \emptyset$, $N - Q_1 = \oplus_{n=1}^{\infty} (\nu_n - Q_{1,n})$ and $\sigma_p(\oplus_{n=1}^{\infty} M_n) = \cup_{n=1}^{\infty} \sigma_p(M_n)$ (for every direct sum), it is not difficult to infer that $\sigma_p(N-Q_1) = \sigma_p([N-Q_1]^*) = \emptyset$.

Assume that $\begin{pmatrix} x \\ y \end{pmatrix} \in \ker(B-\lambda)$ for some $\lambda \in \mathbb{C}$, where $x = (x_1, x_2, \dots)$ $(x_n \in H_n,\ n = 1,2,\dots)$ and $y \in H$. Since $\sigma_p(N-Q_1) = \emptyset$, it follows from Lemma 2.1 and Remark 2.2 that $\lambda = \nu_m$ for some

m, $y \in \ker(A - \nu_m)$ and

$$Q_2 y \in \operatorname{ran}(N - Q_1 - \nu_m).$$

Let P_m denote the orthogonal projection onto H_m; then

$$P_m Q_2 y = Q_{2,m} y = (\nu_m - Q_{1,m} - \nu_m) z = -Q_{1,m} z$$

for some $z \in H_m$. Since $\operatorname{ran} Q_{2,m} \cap \operatorname{ran} Q_{1,m} = \{0\}$, we infer that $Q_{2,m} y = 0$. Since $Q_{2,m}$ is an injective mapping from $\ker(A - \nu_m)$ into $\ker(N - \nu_m)$, we deduce that $y = 0$ and $(B - \lambda)\begin{pmatrix} x \\ 0 \end{pmatrix} = (N - Q_1 - \lambda)x = 0$. Since $\sigma_p(N - Q_1) = \emptyset$, we conclude that $x = 0$. Therefore, $\sigma_p(B) = \emptyset$.

Similarly, we can construct compact operators

$$Q_3 = (Q_{3,1}, Q_{3,2}, \ldots) \quad \text{and} \quad Q_4 = \oplus_{n=1}^{\infty} Q_{4,n},$$

where $Q_{3,n}^* : H \to H_n$ maps $\ker(A - \nu_n)^*$ injectively into $\ker(N - \nu_n)$, $\ker Q_{3,n}^* = [\ker(A - \nu_n)^*]^\perp$, and $Q_{4,n}$ is unitarily equivalent to a positive multiple of the Volterra operator, in such a way that

$$\sigma_p(N - Q_4) = \sigma_p([N - Q_4]^*)$$

$$= \sigma_p\left(\begin{pmatrix} A & -Q_3 \\ 0 & N-Q_4 \end{pmatrix}^*\right) = \emptyset.$$

Clearly, the compact operators Q_1 and Q_2 satisfy

$$\|(Q_1 \ \ Q_2)\| \le \|Q_1\| + \|Q_2\| < \varepsilon/\pi + \varepsilon/2 < \varepsilon,$$

and Q_3, Q_4 can also be constructed so that

$$\left\| \begin{pmatrix} Q_3 \\ Q_4 \end{pmatrix} \right\| \le \|Q_3\| + \|Q_4\| < \varepsilon.$$

Therefore,

$$K_\varepsilon = \begin{pmatrix} Q_1 & Q_2 & 0 \\ 0 & 0 & Q_3 \\ 0 & 0 & Q_4 \end{pmatrix}$$

is compact, and

$$\|K_\varepsilon\| \le \max\left\{\|(Q_1 \ \ Q_2)\|, \left\| \begin{pmatrix} Q_3 \\ Q_4 \end{pmatrix} \right\|\right\} < \varepsilon.$$

Moreover, by construction and Lemma 2.1,

$$\sigma_p(N \oplus A \oplus N - K_\varepsilon) = \sigma_p\left(\begin{pmatrix} N-Q_1 & -Q_2 & 0 \\ 0 & A & -Q_3 \\ 0 & 0 & N-Q_4 \end{pmatrix}\right)$$

$$\subset \sigma_p\left(\begin{pmatrix} N-Q_1 & -Q_2 \\ 0 & A \end{pmatrix}\right) \cup \sigma_p(N - Q_4) = \emptyset.$$

On the other hand, by Lemma 2.1,

$$\sigma_p([N\oplus A\oplus N - K_\varepsilon]^*) = \sigma_p\left(\begin{pmatrix} (N-Q_1)^* & 0 & 0 \\ -Q_2^* & A^* & 0 \\ 0 & -Q_3^* & (N-Q_4)^* \end{pmatrix}\right)$$

$$= \sigma_p\left(\begin{pmatrix} (N-Q_4)^* & -Q_3^* & 0 \\ 0 & A^* & -Q_2^* \\ 0 & 0 & (N-Q_1)^* \end{pmatrix}\right)$$

$$\subset \sigma_p\left(\begin{pmatrix} (N-Q_4)^* & -Q_3^* \\ 0 & A^* \end{pmatrix}\right) \cup \sigma_p([N-Q_1]^*)$$

$$= \sigma_p\left(\begin{pmatrix} A & -Q_3 \\ 0 & N-Q_4 \end{pmatrix}^*\right) \cup \sigma_p([N-Q_1]^*) = \emptyset. \blacksquare$$

COROLLARY 2.6. *(i) If* $T \in L(H)$, $\sigma_p(T)$ *includes a (finite or) denumerable set* $\{\lambda_j\}_{j=1}^{\infty}$ *and* $N = (\text{diag}\{\lambda_1,\lambda_2,\ldots\})^{(\infty)}$, *then given* $\varepsilon > 0$ *there exists* K_ε *compact, with* $\|K_\varepsilon\| < \varepsilon$, *such that*

$$\sigma_p(T\oplus N - K_\varepsilon) = \sigma_p(T) \setminus \{\lambda_j\}_{j=1}^{\infty} \text{ and } \sigma_p([T\oplus N - K_\varepsilon]^*) = \sigma_p(T^*).$$

(ii) If $T \in L(H)$, $\sigma_p(T^*)^*$ *includes a (finite or) denumerable set* $\{\lambda_j\}_{j=1}^{\infty}$ *and* $N = (\text{diag}\{\lambda_1,\lambda_2,\ldots\})^{(\infty)}$, *then given* $\varepsilon > 0$ *there exists* K_ε *compact, with* $\|K_\varepsilon\| < \varepsilon$, *such that*

$$\sigma_p(T\oplus N - K_\varepsilon) = \sigma_p(T) \text{ and } \sigma_p([T\oplus N - K_\varepsilon]^*)^* = \sigma_p(T^*)^*\setminus\{\lambda_j\}_{j=1}^{\infty}.$$

PROOF. (i) The proof follows by the same argument as in Lemma 2.5. Write

$$T\oplus N = \begin{pmatrix} N & 0 \\ 0 & T \end{pmatrix},\ K_\varepsilon = \begin{pmatrix} Q_1 & Q_2 \\ 0 & 0 \end{pmatrix},\ Q_1 = \oplus_{j=1}^{\infty} Q_{1,j},\ Q_2 = \begin{pmatrix} Q_{2,1} \\ Q_{2,2} \\ \cdot \\ \cdot \end{pmatrix},$$

where Q_1, Q_2 are compact and $\|Q_1\|,\|Q_2\| < \varepsilon/2$. If $Q_{1,j}$ and $Q_{2,j}$ are cleverly chosen, then

$$T\oplus N - K_\varepsilon = \begin{pmatrix} N-Q_1 & -Q_2 \\ 0 & T \end{pmatrix}$$

has the desired properties.

(ii) This follows by applying (i) to T^* and N^*. ■

LEMMA 2.7. *If* $T \in L(H)$ *is biquasitriangular, then there exists* $K \in K(H)$ *such that*

$$\sigma_p(T - K) = \sigma_p([T - K]^*) = \emptyset.$$

PROOF. According to [3], [35] there exists $K_0 \in K(H)$ such that $\sigma(T - K_0) = \sigma_e(T - K_0) = \sigma_e(T)$.

Let $\varepsilon > 0$ be given. By using Voiculescu's theorem [36], it is possible to find $K_1 \in K(H)$, with $\|K_1\| < \varepsilon/4$, such that $T - K_0 - K_1 \simeq (T - K_0) \oplus R^{(2)}$, where $R = \rho(\tilde{T})$ for some unital *-representation $\rho: C^*(\tilde{T}) \to L(H_\rho)$ of the C*-algebra generated by $\tilde{T}$ and $\tilde{I}$ on a separable Hilbert space H_ρ. Furthermore, ρ can be chosen so that

$$R = \begin{pmatrix} \begin{pmatrix} N & 0 \\ 0 & N \end{pmatrix} & & * \\ & S & \\ O & & N \end{pmatrix},$$

where $N = (\text{diagonal}\{\lambda_1, \lambda_2, \ldots\})^{(\infty)}$ for some (finite or) denumerable dense subset $\{\lambda_n\}_{n=1}^{\infty}$ of $\sigma_e(T)$ (see [21, Chapter 4]).

Thus, $\sigma(N) = \sigma_e(N) = \sigma(R) = \sigma_e(R) = \sigma_e(T)$ and

$$(T - K_0) \oplus R^{(2)} = \begin{pmatrix} (T-K_0) & 0 \ 0 & 0 \ 0 \\ & \begin{pmatrix} N & 0 \\ 0 & N \end{pmatrix} & * \\ O & & S \\ & & & N \end{pmatrix} \oplus \begin{pmatrix} \begin{pmatrix} N & 0 \\ 0 & N \end{pmatrix} & & * \\ & S & \\ O & & N \end{pmatrix}$$

$$= \begin{pmatrix} A & C \\ 0 & N \end{pmatrix} \oplus \begin{pmatrix} N & D \\ 0 & B \end{pmatrix} = \left(\begin{array}{cc|cc} N & D & & \\ 0 & B & \multicolumn{2}{c}{O} \\ \hline & & A & C \\ \multicolumn{2}{c|}{O} & 0 & N \end{array}\right) = \left(\begin{array}{cc|cc} N & 0 & D & 0 \\ 0 & A & 0 & C \\ \hline & & B & 0 \\ \multicolumn{2}{c|}{O} & 0 & N \end{array}\right) = \begin{pmatrix} N \oplus A & D \oplus C \\ O & B \oplus N \end{pmatrix},$$

where

$$A = \begin{pmatrix} T-K_0 & 0 \ 0 \ 0 & & \\ & N & * & \\ & & N & \\ O & & & S \end{pmatrix} \quad \text{and} \quad B = \begin{pmatrix} N & & * \\ & S & \\ O & & N \end{pmatrix}$$

are biquasitriangular operators satisfying $\sigma(A) = \sigma_e(A) = \sigma(B) = \sigma_e(B) = \sigma_e(T)$.

According to [27], there exist compact operators $K_{2,A}$ and $K_{2,B}$, with $\max[\|K_{2,A}\|, \|K_{2,B}\|] < \varepsilon/8$, such that $A - K_{2,A}$ and $B - K_{2,B}$ are bitriangular operators with diagonal sequences $\{\lambda_j\}_{j=1}^{\infty}$, where $\{\lambda_j\}_{j=1}^{\infty} = \{\nu_n\}_{n=1}^{\infty}$ (as a set) and $\{j: \lambda_j = \nu_n\}$ is an infinite set for each $n = 1, 2, \ldots$.

By Lemma 2.5, there exist compact operators $K'_{2,A}$ and $K'_{2,B}$, with $\max[\|K'_{2,A}\|, \|K'_{2,B}\|] < \varepsilon/8$, such that

$$\sigma_p(N\oplus(A-K_{2,A})-K'_{2,A}) = \sigma_p([N\oplus(A-K_{2,A})-K'_{2,A}]^*)$$
$$= \sigma_p((B-K_{2,B})\oplus N-K'_{2,B}) = \sigma_p([(B-K_{2,B})\oplus N-K'_{2,B}]^*) = \emptyset.$$

It readily follows that we can find $K_2 \in K(H)$, with $\|K_2\| \le \max[\|K_{2,A}\|,\|K_{2,B}\|] + \max[\|K'_{2,A}\|,\|K'_{2,B}\|] < \varepsilon/4$, such that

$$T - K_0 - K_1 - K_2 = \begin{pmatrix} T_1 & T_{12} \\ 0 & T_2 \end{pmatrix},$$

where $\sigma_p(T_1) = \sigma_p(T_1^*) = \sigma_p(T_2) = \sigma_p(T_2^*) = \emptyset$.

By using Lemma 2.1, we conclude that if $K = K_0 + K_1 + K_2$ $(\in K(H))$, then $\sigma_p(T-K) = \sigma_p([T-K]^*) = \emptyset$. ■

REMARK 2.8. The proof of Lemma 2.6 shows that the compact operator K can be chosen so that $\|K\| \le \|K_0\| + \varepsilon/2$, where K_0 is the operator provided by J. G. Stampfli's theorem [35]. Given $A \in L(H)$, let $m_e(\lambda - A) = \min\{r \ge 0:\ r \in \sigma_e([(A-\lambda)^*(A-\lambda)]^{1/2})\}$ be the minimal essential modulus of $A-\lambda$ $(\lambda \in \mathbb{C})$, let $\Delta_\gamma(A) = \{\lambda \in \mathbb{C}:\ m_e(\lambda - A) \le \gamma\}$ $(\gamma \ge 0)$, and let

$$m_e(A;\lambda) = \min\{\gamma \ge 0:\ \operatorname{dist}[\lambda,\Delta_\gamma(A)] \le \gamma\}.$$

According to [22] (see also [4,Chapter 12]) K_0 can be chosen so that

$$\|K\| < \|K_0\| + \varepsilon/2 < \varepsilon + \max\{m_e(T;\lambda):\ \lambda \in \sigma_o(T)\}$$
$$(< \varepsilon + \tfrac{1}{2}\max\{\operatorname{dist}[\lambda,\sigma_e(T)]:\ \lambda \in \sigma_o(T)\}.$$

3. CLEANSING. SECOND PART

Let Ω be a nonempty bounded open subset of the complex plane, and let ω be a point of Ω: if $R(\Omega^-)$ denotes the uniform closure of the rational functions with poles outside Ω^-, then it is possible to find a representing measure μ for ω supported by the peak points of $R(\Omega^-)$ in $\partial(\Omega^-)$. It is not difficult to see that μ cannot have atoms. (See [17,Chapter II]. In particular, Theorem 11.6, p.56.)

Let $M(\partial\Omega) \in L(L^2(\partial\Omega,d\mu))$ be the operator "multiplication by λ". Clearly, $H^2(\partial\Omega) = L^2$-closure of $R(\Omega^-)$ is a rationally cyclic invariant subspace of $M(\partial\Omega)$; $e_0(\lambda) \equiv 1$ is a rationally cyclic vector for $M_+(\partial\Omega) = M(\partial\Omega)|H^2(\partial\Omega)$. We have

$$M(\partial\Omega) = \begin{pmatrix} M_+(\partial\Omega) & Z(\partial\Omega) \\ 0 & M_-(\partial\Omega) \end{pmatrix} \begin{matrix} H^2(\partial\Omega) \\ L^2(\partial\Omega)\ominus H^2(\partial\Omega) \end{matrix} .$$

$M(\partial\Omega)$ is normal, $\sigma_p(M(\partial\Omega)) = \emptyset$,
$M_+(\partial\Omega)$ and $M_-(\partial\Omega)^*$ are pure subnormal operators such that $\sigma(M_+(\partial\Omega)) = \sigma(M_-(\partial\Omega)) = \Omega^-$, $\sigma_e(M_+(\partial\Omega)) = \sigma_e(M_-(\partial\Omega)) = \sigma(M(\partial\Omega)) = \sigma_e(M(\partial\Omega)) = \partial(\Omega^-)$, $\mathrm{ind}(\lambda - M_-(\partial\Omega)) = \mathrm{nul}(\lambda - M_-(\partial\Omega)) = -\mathrm{ind}(\lambda - M_+(\partial\Omega)) = \mathrm{nul}(\lambda - M_+(\partial\Omega))^* = 1$ and $\mathrm{nul}(\lambda - M_+(\partial\Omega)) = \mathrm{nul}(\lambda - M_-(\partial\Omega))^* = 0$ for all $\lambda \in \mathrm{interior}(\Omega^-) = \sigma_p(M_-(\partial\Omega)) = \sigma_p(M_+(\partial\Omega)^*) = \Omega^- \setminus \partial(\Omega^-)$,
$\|f(M_+(\partial\Omega))\| = \|f(M_-(\partial\Omega))\| = \|f\|_{R(\Omega^-)}$ for all $f \in R(\Omega^-)$,
and $Z(\partial\Omega)$ is a compact (indeed, a Hilbert-Schmidt) operator.

The reader is referred to [11], [21,Chapter 3] for details.

LEMMA 3.1. *Let* Δ *be a nonempty compact subset of* $\mathbb{C}$. *There exists a probability Borel measure* ν, *with* $\mathrm{supp}(\nu) = \Delta$, *such that*

$$\int_\Delta \frac{d\nu(\lambda)}{|\zeta - \lambda|^2} = \infty$$

for all $\zeta \in \Delta$. *(Here* $|\zeta - \lambda|^{-2}$ *is defined equal to* ∞ *for* $\lambda = \zeta$.*)*

PROOF. Without loss of generality, we can assume that Δ is a subset of the interior of the unit square

$$\blacksquare^0 = \{\lambda \in \mathbb{C}: \ 0 \le \mathrm{Re}\,\lambda < 1, \ 0 \le \mathrm{Im}\,\lambda < 1\}.$$

Define $d\nu_0$ = planar Lebesgue measure on $\blacksquare^0 = dA$.

Split $\blacksquare^0$ into four pairwise disjoint squares $\blacksquare^1(1)$, $\blacksquare^1(2)$, $\blacksquare^1(3)$, $\blacksquare^1(4)$ of side 1/2. ($\blacksquare^1(1) = \{\lambda \in \mathbb{C}: \ 0 \le \mathrm{Re}\,\lambda < 1/2, \ 0 \le \mathrm{Im}\,\lambda < 1/2\}$, $\blacksquare^1(2) = \{\lambda \in \mathbb{C}: \ 1/2 \le \mathrm{Re}\,\lambda < 1/2, \ 0 \le \mathrm{Im}\,\lambda < 1/2\}$, etc. .)

If $\blacksquare^1(i_1), \ldots, \blacksquare^1(i_{r(1)})$ $(1 \le r(1) \le 4)$ are the squares that intersect Δ, then we define

$$d\nu_1 = \frac{4}{r(1)} \sum_{j=1}^{r(1)} dA|\blacksquare^1(i_j).$$

Assume we have already defined

$$d\nu_k = \sum_{j=1}^{r(k)} c^k(i_j)\, dA|\blacksquare^k(i_j),$$

where $\{\blacksquare^k(i_j)\}_{j=1}^{r(k)}$ are the squares (side 2^{-k}) in the k-th partition that intersect Δ, $c^k(i_j) \ge 1$, and $\|\nu_k\| = \sum_{j=1}^{r(k)} c^k(i_j)\, 4^{-k} = 1$.

(Split $\blacksquare^0$ as the disjoint union of 4^k pairwise disjoint squares of side 2^{-k}, and order these little squares in lexicographic order; then $i_1, i_2, \ldots, i_{r(k)}$ are the indices of those that intersect Δ.) Split $\blacksquare^k(i_j)$ into four pairwise disjoint squares $\blacksquare^{k+1}(i_j,1)$, $\blacksquare^{k+1}(i_j,2)$, $\blacksquare^{k+1}(i_j,3)$, $\blacksquare^{k+1}(i_j,4)$ (side $2^{-(k+1)}$), and let $\blacksquare^{k+1}(i_j,s_1),\ldots,\blacksquare^{k+1}(i_j,s_{r'(i_j)})$ $(1 \le r(i_j) \le 4)$ be the squares that intersect Δ; then we define

$$d\nu_{k+1} = \sum_{j=1}^{r(k)} \frac{4c^k(i_j)}{r'(i_j)} \sum_{h=1}^{r'(i_j)} dA|\blacksquare^{k+1}(i_j,s_h).$$

(Clearly, we can always write

$$d\nu_{k+1} = \sum_{j=1}^{r(k+1)} c^{k+1}(i_j)\, dA|\blacksquare^{k+1}(i_j)$$

with the above convention about lexicographic order.)

Let ν be a w*-limit point of $\{\nu_k\}_{k=0}^{\infty}$. Given $\varepsilon > 0$, $\operatorname{supp}(\nu_k) \subset \Delta_\varepsilon$ for all k large enough so that $4^{-k} < \varepsilon$. It readily follows that

$$\operatorname{supp}(\nu) \subset \cap_{\varepsilon>0} \Delta_\varepsilon = \Delta.$$

On the other, for each ζ in Δ and each $m \ge 1$, there always exist four squares with a common vertex in the partition of $\blacksquare^0$ into 4^m pairwise disjoint squares of side 2^{-m} such that ζ is *an interior point* of the union of these four squares. Let $\blacksquare^m(\zeta)$ be the square containing ζ. By induction, we can easily check that $\nu_k(\blacksquare^m(\zeta)) \ge A(\blacksquare^m(\zeta)) = 4^{-m}$ for all $k = 0,1,2,\ldots,$ and therefore

$$\nu(\blacksquare^m(\zeta)^-) \ge 4^{-m}$$

(for all $m = 1,2,\ldots$).

It follows that $\operatorname{supp}(\nu) = \Delta$.

Furthermore, if ζ is an atom of ν, then

$$\int_\Delta \frac{d\nu(\lambda)}{|\zeta-\lambda|^2} \ge \infty.\nu(\{\zeta\}) = \infty.$$

If ζ is not an atom, and $\blacksquare^n(\zeta) \subset \blacksquare^m(\zeta)$ is defined as above for $n > m$, then the regularity properties of the Borel measures indicate that

$$\nu(\blacksquare^m(\zeta)^- \setminus \blacksquare^n(\zeta)^-) > \tfrac{1}{2}\cdot 4^{-m}$$

for all $n \ge n_o(m,\zeta)$. By induction, we can construct a decreasing sequence of squares $\{\blacksquare^{m(j)}(\zeta)\}_{j=1}^{\infty}$ $(\cap_{j=1}^{\infty} \blacksquare^{m(j)}(\zeta)^- = \{\zeta\})$ such that

$$\nu(\blacksquare^{m(j)}(\zeta)^- \setminus \blacksquare^{m(j+1)}(\zeta)^-) > \tfrac{1}{2}\cdot 4^{-m(j)}$$

for all $j = 1,2,\ldots$.

Once again, we conclude that

$$\int_\Delta \frac{d\nu(\lambda)}{|\zeta-\lambda|^2} \geq \int_{\blacksquare^{m(1)}(\zeta)^-} \frac{d\nu(\lambda)}{|\zeta-\lambda|^2}$$

$$= \sum_{j=1}^\infty \int_{\blacksquare^{m(j)}(\zeta)^- \setminus \blacksquare^{m(j+1)}(\zeta)^-} \frac{d\nu(\lambda)}{|\zeta-\lambda|^2}$$

$$\geq \sum_{j=1}^\infty \frac{1}{2}\cdot 4^{-m(j)} \inf\{|\zeta-\lambda|^{-2}: \lambda \in \blacksquare^{m(j)}(\zeta)\}$$

$$\geq \sum_{j=1}^\infty \frac{1}{2}\cdot 4^{-m(j)} \times \frac{1}{2}\cdot 4^{m(j)} = \frac{1}{4}(1+1+1+\dots) = \infty.$$

In either case,

$$\int_\Delta \frac{d\nu(\lambda)}{|\zeta-\lambda|^2} = \infty. \qquad \blacksquare$$

We immediately obtain the following.

COROLLARY 3.2. *Let* Δ *and* ν *be as in the above lemma, let* $L(\Delta)$ = "multiplication by λ" *on* $L^2(\Delta,d\nu)$, *and let* f_o *be the characteristic function of* Δ.

$L(\Delta)$ *is a normal operator such that* $\sigma(L(\Delta)) = \Delta$ *and*

$$f_o \notin \cup\{\text{ran}(L(\Delta)-\lambda): \lambda \in \Delta\}.$$

LEMMA 3.3. *Let* Ω *be a nonempty bounded open subset of* $\mathbb{C}$. *There exist a normal operator* $N(\Omega)$ *and a compact operator* K *such that*

(i) $\sigma(N(\Omega)) = \sigma(N(\Omega)+K) = \sigma_e(N(\Omega)) = \partial\Omega$,

(ii) $N(\Omega)+K$ *has an invariant subspace* $M(\Omega)$ *such that*

$$N(\Omega)+K = \begin{pmatrix} N_+(\Omega) & Z'(\Omega) \\ 0 & N_-(\Omega) \end{pmatrix} \begin{matrix} M(\Omega) \\ H\ominus M(\Omega) \end{matrix},$$

$N_+(\Omega) = N(\Omega)+K|M(\Omega)$ *and* $N_-(\Omega)$ *are essentially normal operators (equivalently,* $Z'(\Omega)$ *is compact),* $\sigma(N_+(\Omega)) = \Omega^-$, $\sigma_e(N_+(\Omega)) = \partial\Omega$, $\sigma_p(N_+(\Omega)) = \emptyset$, $\sigma_p(N_+(\Omega)^*)^* = \Omega$, *and*

$$\text{ind}(N_+(\Omega)-\lambda) = -\text{nul}([N_+(\Omega)-\lambda]^*) = 1$$

for all $\lambda \in \Omega$.

Furthermore, given $\varepsilon > 0$, K *can be chosen so that*

$$\|K\| < \varepsilon.$$

PROOF. Define $N(\Omega) = M(\partial\Omega)\oplus L(\Delta)^{(\infty)}$, where $M(\partial\Omega)$ is defined as in the introduction of this section and $L(\Delta)$ is defined as in Corollary 3.2, with $\Delta = \partial[\text{interior}(\Omega^-)\setminus\Omega]^-$. Let L_a be the restriction of $L(\Delta)$ to $L^2(\Delta,d\nu_a)$, and $L_c = L(\Delta)|L^2(\Delta,d\nu_c)$, where ν_a is the atomic part of ν and $\nu_c = \nu-\nu_a$ contains no atoms; then $L(\Delta) = L_a\oplus L_c$, and we have

$$N(\Omega) = L(\Delta)\oplus\begin{pmatrix} M_+(\partial\Omega) & Z(\partial\Omega) \\ 0 & M_-(\partial\Omega)\end{pmatrix}\oplus L_c^{(\infty)}\oplus L_a^{(\infty)}$$

$$= \begin{pmatrix} M_+(\partial\Omega)\oplus L(\Delta) & Z(\partial\Omega)\oplus 0 \\ 0 & M_-(\partial\Omega)\oplus L_c^{(\infty)}\end{pmatrix} {}^{H^2(\partial\Omega)\oplus L^2(\Delta,d\nu)} \oplus L_a^{(\infty)}.$$

Define $K_1 = \frac{\varepsilon}{2}\, e_o\otimes f_o$ ($\|K_1\| = \frac{\varepsilon}{2}\,\|e_o\|\cdot\|f_o\| = \frac{\varepsilon}{2}$); then

$$N(\Omega) + K_1 = \begin{pmatrix} M_+(\partial\Omega) & \frac{\varepsilon}{2}e_o\otimes f_o & * \\ 0 & L(\Delta) & * \\ 0 & 0 & * \end{pmatrix}\oplus L_a^{(\infty)}.$$

By Lemma 2.1, $\sigma_p\left(\begin{pmatrix} M_+(\partial\Omega) & \frac{\varepsilon}{2}e_o\otimes f_o \\ 0 & L(\Delta)\end{pmatrix}\right) \subset \sigma_p(M_+(\partial\Omega)) \cup \sigma_p(L(\Delta))$ $= \sigma_p(L_a)$. On the other hand, if

$$\begin{pmatrix} M_+(\partial\Omega) & \frac{\varepsilon}{2}e_o\otimes f_o \\ 0 & L(\Delta)\end{pmatrix}^*\begin{pmatrix} x \\ y\end{pmatrix} = \begin{pmatrix} L(\Delta)^* & \frac{\varepsilon}{2}f_o\otimes e_o \\ 0 & M_+(\partial\Omega)^*\end{pmatrix}\begin{pmatrix} y \\ x\end{pmatrix}$$

$$= \begin{pmatrix} L(\Delta)^*y + \frac{\varepsilon}{2}\langle x,e_o\rangle f_o \\ M_+(\partial\Omega)^*x\end{pmatrix} = \bar{\lambda}\begin{pmatrix} y \\ x\end{pmatrix}$$

for some $\lambda \in \mathbb{C}$ ($x \in H^2(\partial\Omega)$, $y \in L^2(\Delta,\nu)$), then $\lambda \in \sigma_p(M_+(\partial\Omega)^*)^*$ $= \Omega^- \setminus \partial(\Omega^-) = \text{interior}(\Omega^-)$ and

$$[L(\Delta) - \lambda]^*y = -\frac{\varepsilon}{2}\langle x,e_o\rangle f_o \in (\vee\{f_o\}) \cap \text{ran}[L(\Delta) - \lambda]^*$$
$$= (\vee\{f_o\}) \cap \text{ran}[L(\Delta) - \lambda],$$

because ran M* = ran M for all normal operators. (Here $\vee$ denotes "the closed linear span of".)

If $\langle x,e_o\rangle \neq 0$, and $\lambda \in \Delta$, then $[L(\Delta) - \lambda]^*y = 0$ because (by Corollary 3.2) $(\vee\{f_o\}) \cap \text{ran}[L(\Delta) - \lambda] = \{0\}$.

Since e_o is a rationally cyclic vector for $M_+(\partial\Omega)$ and x is an eigenvector of $M_+(\partial\Omega)^*$, $\langle x,e_o\rangle = 0$ implies $x = 0$ (see, e.g., [4,Chapter 11]). It follows that

$$[L(\Delta) - \lambda]^*y = 0.$$

We conclude that

$$\sigma_p\left(\begin{pmatrix} M_+(\partial\Omega) & \frac{\varepsilon}{2}e_o\otimes f_o \\ 0 & L(\Delta)\end{pmatrix}^*\right) = \Omega \cup \sigma_p(L_a).$$

Therefore both,

$$\sigma_p\left(\begin{pmatrix} M_+(\partial\Omega) & \frac{\varepsilon}{2}e_o\otimes f_o \\ 0 & L(\Delta)\end{pmatrix}^*\right) \cap \Delta \quad\text{and}\quad \sigma_p\left(\begin{pmatrix} M_+(\partial\Omega) & \frac{\varepsilon}{2}e_o\otimes f_o \\ 0 & L(\Delta)\end{pmatrix}\right)$$

are *at most denumerable subsets of* $\sigma_p(L_a)$. Clearly, we can find a direct summand L_b of L_a such that $\sigma_p(L_b)$ coincides with the point spectrum of

$$\begin{pmatrix} M_+(\partial\Omega) & \frac{\varepsilon}{2}e_o \otimes f_o \\ 0 & L(\Delta) \end{pmatrix}.$$

Consider the operator

$$R = L_b^{(\infty)} \oplus \begin{pmatrix} M_+(\partial\Omega) & \frac{\varepsilon}{2}e_o \otimes f_o \\ 0 & L(\Delta) \end{pmatrix} \oplus L_a^{(\infty)}.$$

By a double application of Corollary 2.6, we can find a compact operator K_2, with $\|K_2\| < \varepsilon/2$, such that

$\sigma(R - K_2) = \Omega^-$, $\sigma_p(R - K_2) = \emptyset$ and $\sigma_p([R - K_2]^*)^* = \Omega$ $(\mathrm{nul}(R - K_2 - \lambda)^* = 1$ for all $\lambda \in \Omega)$.

Now it is completely apparent that there exists a compact operator K, with $\|K\| \leq \|K_1\| + \|K_2\| < \varepsilon$, satisfying all our requirements. ∎

Let $T \in L(H)$ and let $\{\Omega_j\}$ be an enumeration of the components of $\rho_{s-F}^-(T)$. Since the operator K of Lemma 3.3 can be chosen of arbitrarily small norm, and $\|f(M(\partial\Omega))\| = \|f(M_+(\partial\Omega))\| = \|f(M_-(\partial\Omega))\| = \|f\|_{R(\Omega^-)}$, it is not difficult to construct an operator

$$M_+(T) = \oplus_j\, N_+(\Omega_j)^{(-\mathrm{ind}(T-\omega_j))} \quad (\omega_j \in \Omega_j)$$

such that

(1) there exist $M_\ell(T)$ normal and $K_\ell(T)$ compact such that $\sigma(M_\ell(T)) = \sigma(M_\ell(T) - K_\ell(T)) = \sigma_e(M_\ell(T)) = \partial\rho_{s-F}^-(T)$, and $\|K_\ell(T)\| < \frac{\varepsilon}{2}$, and

(2) $M_+(T)$ is the restriction of $M_\ell(T) - K_\ell(T)$ to a suitable invariant subspace,

(3) $\sigma(M_+(T)) = [\rho_{s-F}^-(T)]^-$, $\sigma_{\ell e}(M_+(T)) = \partial\rho_{s-F}^-(T)$ and $\mathrm{ind}(M_+(T) - \lambda) = -\mathrm{nul}[M_+(T) - \lambda]^* = \mathrm{ind}(T - \lambda)$ for all $\lambda \in \rho_{s-F}^-(T)$, and

(4) $\sigma_p(M_+(T)) = \emptyset$ and $\sigma_p(M_+(T)^*)^* = \rho_{s-F}^-(T)$.

PROPOSITION 3.4 (Cleansing operation). *Let* $T \in L(H)$. *Given* $\varepsilon > 0$, *there exists* $K \in K(H)$, *with*

$$\|K\| < \varepsilon + \max\{m_e(T;\lambda) : \lambda \in \sigma_o(T)\},$$

such that

$$\sigma_p(T - K) = \rho_{s-F}^+(T) \quad \textit{and} \quad \sigma_p([T - K]^*)^* = \rho_{s-F}^-(T).$$

PROOF. Define $M_+(T)$ as above indicated, and

$$M_-(T) = M_+(T^*)^*.$$

($M_-(T)$ is the restriction to some invariant subspace of an operator of the form $M_r(T)-K_r(T)$, where $M_r(T)$ is a normal operator such that $\sigma(M_r(T)) = \sigma(M_r(T)-K_r(T)) = \sigma_e(M_r(T)-K_r(T)) = \partial\rho^+_{s-F}(T)$ and $K_r(T)$ is a compact operator satisfying $\|K_r(T)\| < \varepsilon/2$.)

By combining Voiculescu's theorem [36] and [8], [34], we can find $K_1 \in K(H)$, with $\|K_1\| < \varepsilon/4$, such that

$$T-K_1 \simeq T\oplus\begin{pmatrix} M_\ell(T) & * & * \\ 0 & S & * \\ 0 & 0 & M_r(T)\end{pmatrix},$$

where $\sigma(S) = \sigma_e(S) = \sigma_e(T)$, $\sigma_{\ell e}(S) = \sigma_{\ell e}(T)$ and $\sigma_{re}(S) = \sigma_{re}(T)$.

By using the results of [2], [4, Chapter 12], [22], we can find $K_2' \in K(H)$, with

$$\|K_2'\| < \frac{\varepsilon}{2} + \max\{m_e(T;\lambda):\ \lambda\in\sigma_o(T)\},$$

such that $T-K_2'$ is *smooth* in the sense of C. Apostol, that is,

$$\text{nul}(T-K_2'-\lambda) = 0 \quad \text{for all } \lambda\in\rho_{s-F}(T)\setminus\rho^+_{s-F}(T) \text{ and}$$
$$\text{nul}(T-K_2'-\lambda)^* = 0 \quad \text{for all } \lambda\in\rho_{s-F}(T)\setminus\rho^-_{s-F}(T).$$

Therefore, there exists $K_2 \simeq K_2'\oplus K_\ell(T)\oplus 0\oplus K_r(T)$ ($\|K_2\| = \|K_2'\|$!) such that

$$T-K_1-K_2 \simeq (T-K_2')\oplus\begin{pmatrix} M_\ell(T)-K_\ell(T) & * & * \\ 0 & S & * \\ 0 & 0 & M_r(T)-K_r(T)\end{pmatrix}$$

$$= \left(\begin{array}{c|ccc|c} M_+(T) & & * & & * \\ \hline & M_+'(T) & * & * & \\ 0 & 0 & S\oplus(T-K_2') & * & * \\ & 0 & 0 & M_-'(T) & \\ \hline 0 & & 0 & & M_-(T)\end{array}\right)$$

$$= \begin{pmatrix} M_+(T) & * & * \\ 0 & R' & * \\ 0 & 0 & M_-(T)\end{pmatrix},$$

where

$$\begin{pmatrix} M_+(T) & * \\ 0 & M_+'(T)\end{pmatrix} = M_\ell(T)-K_\ell(T),\quad \begin{pmatrix} M_-'(T) & * \\ 0 & M_-(T)\end{pmatrix} = M_r(T)-K_r(T)$$

and

$$R' = \begin{pmatrix} M'_+(T) & * & * \\ 0 & S\oplus(T-K'_2) & * \\ 0 & 0 & M'_-(T) \end{pmatrix}.$$

By proceeding as, for instance, in [5], or [21,Chapter 6], we can easily check that R' is a biquasitriangular operator such that $\sigma_o(R') = \emptyset$. Thus, by Lemma 2.7 and Remark 2.8, we can find K'_3 compact, with $\|K'_3\| < \varepsilon/4$, such that the operator $R = R'-K'_3$ satisfies $\sigma_p(R) = \sigma_p(R^*) = \emptyset$.

Therefore, there exists $K_3 \simeq 0\oplus K'_3\oplus 0$ (so that $K_3 \in K(H)$ and $\|K_3\| < \varepsilon/4$) such that, if $K = K_1 + K_2 + K_3$, then

$$T - K \simeq \begin{pmatrix} M_+(T) & * & * \\ 0 & R & * \\ 0 & 0 & M_-(T) \end{pmatrix}.$$

By Lemma 2.1,

$$\sigma_p(T-K) \subset \sigma_p(M_+(T)) \cup \sigma_p(R) \cup \sigma_p(M_-(T)) = \sigma_p(M_-(T)) = \rho^+_{s-F}(T)$$

and

$$\sigma_p([T-K]^*)^* \subset \sigma_p(M_-(T)^*)^* \cup \sigma_p(R^*)^* \cup \sigma_p(M_+(T)^*)^* = \sigma_p(M_+(T)^*)^* = \rho^-_{s-F}(T).$$

Since $\rho^+_{s-F}(T) \subset \sigma_p(T-K)$ and $\rho^-_{s-F}(T) \subset \sigma_p([T-K]^*)^*$, we conclude that

$$\sigma_p(T-K) = \rho^+_{s-F}(T) \quad \text{and} \quad \sigma_p([T-K]^*)^* = \rho^-_{s-F}(T).$$

It is completely apparent that $K \in K(H)$ and

$$\|K\| \le \|K_1\| + \|K_2\| + \|K_3\| < \varepsilon + \max\{m_e(T;\lambda): \lambda \in \sigma_o(T)\}. \blacksquare$$

REMARKS 3.5. With the same arguments as in Lemma 3.3, we can prove the following result:

Let $T \in L(H)$ and let Δ be a compact subset of $\mathbb{C}$. Given $\varepsilon > 0$, there exist N normal and C compact such that $\sigma(N) = \sigma_e(N) = \Delta$, $\|C\| < \varepsilon$, $\sigma(T\oplus N - C) = \sigma(T) \cup \Delta$, $\sigma_p(T\oplus N - C) = \sigma_p(T) \setminus \Delta$, $\mathrm{nul}(T\oplus N - C - \lambda) = \mathrm{nul}(T-\lambda)$ for all $\lambda \in \sigma_p(T) \setminus \Delta$, and $\mathrm{nul}(T\oplus N - C - \lambda)^* = \mathrm{nul}(T-\lambda)^*$ for all complex λ. Moreover, if T has the form "normal + compact" (or T is essentially normal), then, of course, $T\oplus N - C$ has the same form.

PROOF. Let $\{x_\nu\}$ be the set of all unit eigenvectors of T. Since H is separable, there exists a (finite or) denumerable

$\{y_j\}_{j=1}^{\infty} \subset \{x_\nu\}$ such that $\{x_\nu\} \subset \cup_{j=1}^{\infty} \{x \in H: \|y_j - x\| < \sqrt{2}\}$. It is obvious that, for each ν, $\langle x_\nu, y_j\rangle \neq 0$ (for at least one index j). Define $N = L(\Delta)^{(\infty)}$; then it is straightforward to check (as in the proof of Lemma 3.3) that if

$$T\oplus N - C_1 = \left(\begin{array}{cccccc|c} L(\Delta) & & & & & & (\varepsilon/2)f_o\otimes y_1 \\ & L(\Delta) & & 0 & & & (\varepsilon/4)f_o\otimes y_2 \\ & & L(\Delta) & & & & (\varepsilon/8)f_o\otimes y_3 \\ & & & \cdot & & & \cdot \\ & 0 & & & \cdot & & \cdot \\ & & & & & \cdot & \cdot \\ \hline & & 0 & & & & T \end{array}\right)$$

(where

$$C_1 = -\left(\begin{array}{c|c} & (\varepsilon/2)f_o\otimes y_1 \\ & (\varepsilon/4)f_o\otimes y_2 \\ & (\varepsilon/8)f_o\otimes y_3 \\ 0 & \cdot \\ & \cdot \\ & \cdot \\ \hline 0 & 0 \end{array}\right)$$

is compact and satisfies $\|C_1\| < \sum_{j=1}^{\infty} \varepsilon/2^j = \varepsilon$), then $\sigma(T\oplus N - C_1) = \sigma(T\oplus N) = \sigma(T) \cup \Delta$, $\sigma_p(T\oplus N - C_1) \subset [\sigma_p(T) \setminus \Delta] \cup \sigma_p(L(\Delta)) = [\sigma_p(T) \setminus \Delta] \cup \sigma_p(N)$, and $\mathrm{nul}([T\oplus N - C_1 - \lambda]^*) = \mathrm{nul}([T-\lambda]^*)$ for all complex $\lambda \notin \sigma_p(N)$.

By using Corollary 2.6, now we can find a compact operator C_2, with $\|C_2\| < \varepsilon$, such that $C = C_1 + C_2$ satisfies $\sigma(T\oplus N - C) = \sigma(T) \cup \Delta$, $\sigma_p(T\oplus N - C) = \sigma_p(T) \setminus \Delta$ and $\mathrm{nul}(T\oplus N - C - \lambda)^* = \mathrm{nul}(T-\lambda)^*$ for all $\lambda \in \mathbb{C}$.

On the other hand, in order to check that $\sigma_p(T\oplus N - C) \supset \sigma_p(T) \setminus \Delta$, pick any $\lambda \notin \Delta$ and observe that

$$T\oplus N - C = \begin{pmatrix} L(\Delta)^{(\infty)} & * & * \\ 0 & \lambda 1 & * \\ 0 & 0 & T_\lambda' \end{pmatrix} \begin{matrix} H_1^{(\infty)} \\ \ker(T-\lambda) \\ H \ominus \ker(T-\lambda) \end{matrix} .$$

Since $\sigma(\lambda 1) = \{\lambda\}$ is disjoint from $\sigma(L(\Delta)^{(\infty)}) = \Delta$, it follows from Rosenblum's corollary [21,Chapter 3], [33] that

$$\begin{pmatrix} L(\Delta)^{(\infty)} & * \\ 0 & \lambda 1 \end{pmatrix} \begin{matrix} H_1^{(\infty)} \\ \ker(T-\lambda) \end{matrix} \sim L(\Delta)^{(\infty)} \oplus \lambda 1_{\ker(T-\lambda)},$$

and therefore $\lambda \in \sigma_p(T \oplus N - C)$ and $\mathrm{nul}(T \oplus N - C - \lambda) \geq \mathrm{nul}(T - \lambda)$. (The inequality $\mathrm{nul}(T \oplus N - C - \lambda) \leq \mathrm{nul}(T - \lambda)$ is immediate, so that the two kernels have exactly the same dimension.) ∎

4. NEW MAKE UP. FIRST PART

As indicated in the Introduction, J. Dixmier and C. Foiaş [14] and G. K. Kalish [28], [29] have exhibited a large class of operators T with the property $\sigma(T) = \sigma_p(T)$. (In [32], M. Omladič provided a deeper analysis of some of the operators studied by Kalish. See also [19].)

In order to complete our "set of tools", we need operators satisfying the conditions $\sigma(T) = \sigma_p(T)$, $\sigma_p(T^*) = \emptyset$ and, in addition, T must be a compact perturbation of a normal operator. The construction of these operators is the core of this section.

PROPOSITION 4.1. *Let* Γ *be a nonempty compact subset of* $\mathbb{C}$. *There exists* $T(\Gamma) \in L(H)$ *such that*

$\sigma(T(\Gamma)) = \sigma_p(T(\Gamma)) = \Gamma$, $\mathrm{nul}(T(\Gamma) - \lambda) = 1$ *for all* $\lambda \in \Gamma$, *and* $\sigma_p(T(\Gamma)^*) = \emptyset$.

Furthermore, $T(\Gamma) = N + K$, *where* N *is normal and* K *is compact.*

The models constructed by J. Dixmier and C. Foiaş in [14] are based on the Sobolev space

$$W^{2,2}(\blacksquare) = \{f \in L^2(\blacksquare, dA): \text{ the distributional partial derivatives of first and second order of } f \text{ belong to } L^2(\blacksquare, dA)\},$$

where $\blacksquare = \{\lambda \in \mathbb{C}: 0 \leq \mathrm{Re}\,\lambda \leq \pi,\ 0 \leq \mathrm{Im}\,\lambda \leq \pi\}$ and dA denotes the planar Lebesgue measure.

The reader is referred to the excellent treatise of R. A. Adams [1] for everything related with Sobolev spaces. In particular, it is well-known that $W^{2,2}(\blacksquare)$ is a Hilbert space of continuous functions (by Sobolev's embedding theorem) under the norm

$$\|f\|_{W^{2,2}(\blacksquare)} = \left(\int_{\blacksquare}\left[|f(\lambda)|^2 + \left|\frac{\partial^2}{\partial s^2} f(\lambda)\right|^2 + \left|\frac{\partial^2}{\partial t^2} f(\lambda)\right|^2\right]dA\right)^{\frac{1}{2}},$$

where $\lambda = s + it$.

Instead of $W^{2,2}(\blacksquare)$, we shall use here the subspace

$$W_0^{2,2}(\blacksquare) = \{f \in W^{2,2}(\blacksquare): f|\partial\blacksquare = 0\}.$$

$W^{2,2}(\blacksquare)$ is a Banach algebra with identity (under pointwise multiplication and an equivalent norm) and contains the restrictions to $\blacksquare$ of all the functions in $C^{\infty}(\mathbb{R}^2)$ ($\mathbb{R}^2$ identified with $\mathbb{C}$ in the usual fashion). Furthermore, $W^{2,2}(\blacksquare)$ is a regular Banach algebra [31] whose maximal ideal space can be naturally identified with $\blacksquare$ via "point evaluations". For each closed subset Σ of $\blacksquare$,

$$W^{2,2}(\blacksquare,\Sigma) = \{f \in W^{2,2}(\blacksquare): \quad f|\Sigma \equiv 0\}$$

is a closed ideal. Equivalently, it is a subspace of $W^{2,2}(\blacksquare)$ invariant under multiplication by the functions in $W^{2,2}(\blacksquare)$. (In particular, $W_0^{2,2}(\blacksquare) = W^{2,2}(\blacksquare,\partial\blacksquare)$ is one of these subspaces.) Moreover, it was shown in [14] that if M_λ = "multiplication by λ" and

$$M_\lambda = \begin{pmatrix} M_\lambda | W^{2,2}(\blacksquare,\Sigma) & * \\ 0 & M_{\lambda,\Sigma} \end{pmatrix} W^{2,2}(\blacksquare,\Sigma) \quad ,$$

then

$$\sigma(M_{\lambda,\Sigma}) = \sigma_p((M_{\lambda,\Sigma})^*)^* = \Sigma.$$

Furthermore, since the maximal ideal space of $W^{2,2}(\blacksquare)$ coincides with $\blacksquare$, we can easily check that

$$\operatorname{nul}(M_{\lambda,\Sigma} - \lambda)^* = 1$$

for all $\lambda \in \Sigma$.

Without loss of generality, we can assume that the compact set Γ of Proposition 4.1 is a subset of interior $\blacksquare$. Clearly, if $\Sigma = \partial\blacksquare \cup \Gamma$, then $\partial\blacksquare \cap \Gamma = \emptyset$ and (by Riesz's decomposition theorem) $M_{\lambda,\partial\blacksquare\cup\Gamma}$ is similar to the direct sum of $M_{\lambda,\partial\blacksquare}$ and $M_{\lambda,\Gamma}$. We have

$$M_\lambda = \begin{pmatrix} M_\lambda | W^{2,2}(\blacksquare,\partial\blacksquare\cup\Gamma) & * \\ 0 & M_{\lambda,\partial\blacksquare\cup\Gamma} \end{pmatrix} W^{2,2}(\blacksquare,\partial\blacksquare\cup\Gamma)$$

$$= \left.\begin{pmatrix} M_\lambda | W^{2,2}(\blacksquare,\partial\blacksquare\cup\Gamma) & * & * \\ 0 & M^{o}_{\lambda,\Gamma} & * \\ 0 & 0 & M_{\lambda,\partial\blacksquare} \end{pmatrix}\right\} W_0^{2,2}(\blacksquare)$$

$$= \begin{pmatrix} M_\lambda | W_0^{2,2}(\blacksquare) & * \\ 0 & M_{\lambda,\partial\blacksquare} \end{pmatrix} W_0^{2,2}(\blacksquare) \quad ,$$

where $M^{o}_{\lambda,\Gamma}$ is similar to $M_{\lambda,\Gamma}$, and therefore has exactly the same spectral properties.

We want to show that $M^{0}_{\lambda,\Gamma}$ is a compact perturbation of a normal operator. The advantage of working with $W^{2,2}_{0}(\blacksquare)$ instead of $W^{2,2}(\blacksquare)$ lies in the fact that

$$W^{2,2}_{0}(\blacksquare) = W^{2,2}(\blacksquare)\text{-closure of } C^{\infty}_{0}(\blacksquare)$$

($C^{\infty}_{0}(\blacksquare) = \{g \in C^{\infty}(\mathbb{R}^2): \text{ supp}(g) \subset \text{interior } \blacksquare\}$) is the domain of the negative self-adjoint (unbounded) Laplace operator

$$\Delta : W^{2,2}_{0}(\blacksquare) \to L^2(\blacksquare, dA) \quad \left(\Delta = \frac{\partial^2}{\partial s^2} + \frac{\partial^2}{\partial t^2}\right).$$

Furthermore, Δ is *an injective linear mapping from* $W^{2,2}_{0}(\blacksquare)$ *onto* $L^2(\blacksquare,dA)$ and the $W^{2,2}(\blacksquare)$-norm, restricted to $W^{2,2}_{0}(\blacksquare)$ is actually equivalent to the norm

$$\|f\|_0 = \left(\int_{\blacksquare} |\Delta f(\lambda)|^2 \, dA\right)^{\frac{1}{2}}.$$

(The intrinsic reason being that $W^{2,2}_{0}(\blacksquare)$ contains no non-zero harmonic functions!)

Thus, we can consider Δ both as a densely defined (unbounded closed) self-adjoint operator acting on $L^2(\blacksquare,dA)$, and as the natural isometric isomorphism from the Hilbert space $W^{2,2}_{0}(\blacksquare)$ onto the Hilbert space $L^2(\blacksquare,dA)$. Indeed, Δ is a diagonal operator with respect to the orthonormal basis

$$\{e_{mn}(s,t) = \tfrac{2}{\pi} \sin ms \sin nt\}_{m,n=1}^{\infty}$$

of $L^2(\blacksquare,dA)$ ($\Delta e_{mn} = -(m^2+n^2)e_{mn}$), and the transformation from $W^{2,2}_{0}(\blacksquare)$ onto $L^2(\blacksquare,dA)$ defined by Δ can be described by

$$\Delta[-(m^2+n^2)^{-1}e_{mn}] = e_{mn}, \quad \Delta^{-1}e_{mn} = -(m^2+n^2)^{-1}e_{mn}$$

($m,n = 1,2,\ldots$; $\{-(m^2+n^2)e_{mn}\}_{m,n=1}^{\infty}$ is an orthonormal basis for $W^{2,2}_{0}(\blacksquare)$).

Let $M^{0}_{\lambda} = M_{\lambda}|W^{2,2}_{0}(\blacksquare)$ and let M^{0}_{s} and M^{0}_{t} denote the "multiplication by the first variable" and, respectively, the "multiplication by the second variable" on $W^{2,2}_{0}(\blacksquare)$. M_{λ}, M_s and M_t will indicate the analogously defined operators on $L^2(\blacksquare,dA)$. It is obvious that $M^{0}_{\lambda} = M^{0}_{s} + iM^{0}_{t}$, $M_{\lambda} = M_s + iM_t$, and that M_{λ} is normal and M_s and M_t are positive hermitian operators.

Next, we compute the adjoints of the operators M^{0}_{s} and M^{0}_{t}. Let $f,g \in C^{\infty}_{0}(\blacksquare)$. Since Δ is self-adjoint, we have

$$(M^{0}_{s}f,g)_{W^{2,2}_{0}(\blacksquare)} = \langle\Delta(M^{0}_{s}f),\Delta g\rangle_{L^2(\blacksquare,dA)} = \langle M^{0}_{s}f,\Delta^2 g\rangle$$

$$= \langle M_s f,\Delta^2 g\rangle = \langle f,M_s\Delta^2 g\rangle = \langle\Delta^{-1}(\Delta f),M_s\Delta^2 g\rangle$$

$$= \langle \Delta f, \Delta^{-1} M_s \Delta^2 g \rangle = \langle \Delta f, \Delta(\Delta^{-2} M_s \Delta^2) g \rangle$$
$$= (f, (\Delta^{-2} M_s^o \Delta^2) g).$$

Thus, $(M_s^o)^* | C_0^\infty(\blacksquare) = \Delta^{-2} M_s \Delta^2 | C_0^\infty(\blacksquare)$.

On $C_0^\infty(\blacksquare)$, we have

$$M_s^o - (M_s^o)^* = M_s^o - \Delta^{-2} M_s^o \Delta^2 = -\Delta^{-2}[M_s^o, \Delta^2] = 4\Delta^{-1} \frac{\partial}{\partial s}.$$

(Here $[A,B] = AB - BA$.)

Indeed, if $g \in C_0^\infty(\blacksquare)$, then

$$[M_s^o, \Delta^2] g(s,t) = (M_s^o \Delta^2 - \Delta^2 M_s^o) g(s,t)$$
$$= M_s^o \Delta^2 g - \left(\frac{\partial^2}{\partial s^2} + \frac{\partial^2}{\partial t^2}\right)^2 s g(s,t)$$
$$= s\Delta^2 g(s,t) - \left[4 \frac{\partial}{\partial s} \Delta g(s,t) + s\Delta^2 g(s,t)\right]$$
$$= -4 \frac{\partial}{\partial s} \Delta g(s,t) = -4\Delta \frac{\partial}{\partial s} g(s,t).$$

In order to show that $M_s^o - (M_s^o)^* = 4\Delta^{-1} \partial/\partial s$ (extended to $W_0^{2,2}(\blacksquare)$) is, indeed, a compact operator, it is convenient to work with its unitarily equivalent image in $L(L^2(\blacksquare, dA))$, restricted to the dense linear manifold $C^\infty(\blacksquare)$:

$$M_s^o - (M_s^o)^* \simeq \Delta^{-1}[M_s^o - (M_s^o)^*]\Delta = 4\Delta^{-1}(\Delta^{-1} \tfrac{\partial}{\partial s})\Delta = 4\tfrac{\partial}{\partial s} \Delta^{-1}$$

because the partial derivative operators commute on $C^\infty(\blacksquare)$.

In particular, we have

$$4\frac{\partial}{\partial s} \Delta^{-1} e_{mn} = -\frac{4m}{m^2+n^2} f_{mn} \quad (m,n = 1,2,\ldots),$$

where

$$\{f_{mn} = \tfrac{2}{\pi} \cos ms \sin nt\}_{m,n=1}^{\infty}$$

is another orthonormal basis of $L^2(\blacksquare, dA)$.

Since

$$\frac{-4m}{m^2+n^2} \to 0 \quad (m+n \to \infty),$$

we deduce that $4\frac{\partial}{\partial s} \Delta^{-1} \in K(L^2(\blacksquare, dA))$; that is, $M_s^o - (M_s^o)^* = 2i\,\mathrm{Im}\, M_s^o \in K(W_0^{2,2}(\blacksquare))$.

A similar computation shows that $M_t^o - (M_t^o)^* = 2i\,\mathrm{Im}\, M_t^o \in K(W_0^{2,2}(\blacksquare))$, and therefore $[(M_\lambda^o)^*, M_\lambda^o]$ is compact.

By using the fact that $W^{2,2}(\blacksquare)$ is a regular Banach algebra, it is possible to show that M_λ^o is a *decomposable operator* in the sense of C. Foiaş, with $\sigma(M_\lambda^o) = \blacksquare$. (Thus, $\sigma(M_\lambda^o)$ has no isolated points.) It is easily seen that $\sigma_e(M_\lambda^o) = \blacksquare$ (see, e.g., [14],

[16] for details). Since M_λ^o is essentially normal, the Brown-Douglas-Fillmore theorem implies that

$$M_\lambda^o = \text{"Normal + Compact" [10].}$$

The authors are grateful to J. Agler for the following observation.

LEMMA 4.2. *Let* $A \in L(H)$ *be an operator with compact imaginary part, and let* M *be an invariant subspace of* A*; then both* $A|M$ *and the compression of* A *to* $M^\perp$ *have compact imaginary part.*

PROOF. Let P_M be the orthogonal projection of H onto M. Since $\mathbb{I}\mathrm{m}\, A$ is compact and $(1-P_M)AP_M = 0$, we have

$$\begin{aligned}\tilde{P}_M\tilde{A} - \tilde{A}\tilde{P}_M &= \tilde{P}_M\tilde{A}(\tilde{1}-\tilde{P}_M) - (\tilde{1}-\tilde{P}_M)\tilde{A}\tilde{P}_M\\ &= \tilde{P}_M\tilde{A}^*(\tilde{1}-\tilde{P}_M) - (\tilde{1}-\tilde{P}_M)\tilde{A}\tilde{P}_M\\ &= [(\tilde{1}-\tilde{P}_M)\tilde{A}\tilde{P}_M]^* - (\tilde{1}-\tilde{P}_M)\tilde{A}\tilde{P}_M = \tilde{0},\end{aligned}$$

so that $P_MA - AP_M \in K(H)$.

Thus, we have

$$A = \begin{pmatrix} A|M & C \\ 0 & A_{M^\perp}\end{pmatrix}\begin{matrix} M \\ H\ominus M\end{matrix}, \quad P_M = \begin{pmatrix} 1 & 0 \\ 0 & 0\end{pmatrix}$$

and

$$P_MA - AP_M = \begin{pmatrix} 0 & C \\ 0 & 0\end{pmatrix} \in K(H).$$

Since both $[A^*,A]$ and C are compact operators, straightforward computations show that both

$$[(A|M)^*, A|M] \quad \text{and} \quad [(A_{M^\perp})^*, A_{M^\perp}]$$

are compact.

Since $\mathbb{I}\mathrm{m}\, A$ is compact, $\sigma_e(A) \subset \mathbb{R}$, and therefore

$$\sigma_e(A|M) \cup \sigma_e(A_{M^\perp}) \subset \mathbb{R}.$$

It readily follows (for instance, by using the Brown-Douglas-Fillmore theorem [10]) that $\mathbb{I}\mathrm{m}\, A|M$ and $\mathbb{I}\mathrm{m}\, A_{M^\perp}$ are compact operators. ∎

COROLLARY 4.3. *For each compact subset* Γ *of* interior ∎, $M_{\lambda,\Gamma}^o$ *is an essentially normal operator satisfying*

$$\sigma(M_{\lambda,\Gamma}^o) = \sigma_p((M_{\lambda,\Gamma}^o)^*) = \Gamma.$$

PROOF. It only remains to show that $[(M_{\lambda,\Gamma}^o)^*, M_{\lambda,\Gamma}^o]$ is compact. Recall that

$$M_\lambda^o = \begin{pmatrix} M_\lambda | W^{2,2}(\partial \blacksquare \cup \Gamma) & * \\ 0 & M_{\lambda,\Gamma}^o \end{pmatrix} \begin{matrix} M(\Gamma) \\ W_o^{2,2}(\blacksquare) \ominus M(\Gamma) \end{matrix},$$

where $M(\Gamma) = \{f \in W^{2,2}(\blacksquare): f|\partial\blacksquare \cup \Gamma \equiv 0\} = \{f \in W_o^{2,2}(\blacksquare): f|\Gamma \equiv 0\}$ is invariant under M_λ^o, M_s^o and M_t^o. Let $M_{s,\Gamma}^o$ and $M_{t,\Gamma}^o$ denote the compressions of M_s^o and, respectively, M_t^o to $M(\Gamma)^\perp$.

By using Lemma 4.2, we see that $M_{s,\Gamma}^o$ and $M_{t,\Gamma}^o$ are commuting operators of the form "Hermitian + Compact". Since $M_{\lambda,\Gamma}^o = M_{s,\Gamma}^o + iM_{t,\Gamma}^o$, it is straightforward to check that $[(M_{\lambda,\Gamma}^o)^*, M_{\lambda,\Gamma}^o]$ is, indeed, a compact operator. ∎

PROOF OF PROPOSITION 4.1. Let $R \in L(H)$ be an essentially normal operator such that $\sigma(R) = \sigma_p(R) = \Gamma$ and $\text{nul}(R-\lambda) = 1$ for all $\lambda \in \Gamma$. For instance, construct M_{λ,Γ^*}^o as in Corollary 4.3 and define $R = (M_{\lambda,\Gamma^*}^o)^*$. Clearly, $H_p = \vee\{\ker(R-\lambda): \lambda \in \Gamma\}$ is invariant under R. Since H_p is a separable Hilbert space, it is easily seen that

$$H_p = \vee\{\ker(R-\lambda_j)\}_{j=1}^\infty.$$

for a suitable denumerable subset $\{\lambda_j\}_{j=1}^\infty$ of Γ.

It readily follows that $R|H_p$ is a triangular operator; more precisely,

$$R|H_p = \begin{pmatrix} \lambda_1 & & & * \\ & \lambda_2 & & \\ & & \lambda_3 & \\ & & & \ddots \\ & 0 & & & \\ \end{pmatrix}.$$

By [20], [21,Chapter 3], $\sigma_p([R|H_p]^*)^*$ is a subset of $\{\lambda_j\}_{j=1}^\infty$; moreover, $\sigma(R|H_p) = \sigma_p(R|H_p) = \Gamma$. Let $M \in L(H_1)$ be a normal operator such that

$M \simeq M^{(\infty)}$, $\sigma(M) = \sigma_e(M) = \Gamma$ and $\sigma_p(M) = \sigma_p([R|H_p]^*)^*$;

then $M \simeq M \oplus M_p$, where M_p is a normal operator of uniform infinite multiplicity and purely point spectrum equal to $\sigma_p(M)$. We have

$$R \oplus M \simeq \left(\begin{array}{cc|cc} R|H_p & 0 & * & 0 \\ 0 & M_p & 0 & 0 \\ \hline & & R_{H_p^\perp} & 0 \\ & 0 & 0 & M \end{array}\right) = \begin{pmatrix} (R|H_p)\oplus M_p & * \\ 0 & R_{H_p^\perp}\oplus M \end{pmatrix}.$$

By Corollary 2.6, we can find a compact operator K_p (of

arbitrarily small norm) such that

$\sigma([(R|\mathcal{H}_p)\oplus M_p]-K_p) = \sigma_p([(R|\mathcal{H}_p)\oplus M_p]-K_p) = \sigma(R|\mathcal{H}_p) = \sigma_p(R|\mathcal{H}_p) = \Gamma$,

$\mathrm{nul}([(R|\mathcal{H}_p)\oplus M_p] - K_p - \lambda) = \mathrm{nul}(R|\mathcal{H}_p - \lambda) = \mathrm{nul}(R-\lambda) = 1$ for all $\lambda \in \Gamma$, and

$\sigma_p(\{[(R|\mathcal{H}_p)\oplus M_p] - K_p\}^*) = \emptyset$.

On the other hand, since $\sigma(R|\mathcal{H}_p) \subset \sigma(R)$ we also have

$$\sigma(R_{\mathcal{H}_p^\perp}) \subset \sigma(R) = \sigma(M) \text{ [20]},$$

whence it readily follows that $R_{\mathcal{H}_p^\perp}\oplus M$ is a biquasitriangular operator satisfying $\sigma(R_{\mathcal{H}_p^\perp}\oplus M) = \sigma_e(R_{\mathcal{H}_p^\perp}\oplus M) = \Gamma$. By Lemma 2.7, there exists a compact operator $K_p^\perp$ (of arbitrarily small norm) such that

$\sigma([R_{\mathcal{H}_p^\perp}\oplus M]-K_p^\perp) = \Gamma$

and

$\sigma_p([R_{\mathcal{H}_p^\perp}\oplus M] - K_p) = \sigma_p(\{[R_{\mathcal{H}_p^\perp}\oplus M] - K_p\}^*) = \emptyset$.

Therefore, there exists $K \in L(\mathcal{H})$ $(\mathcal{H} = \mathcal{H}_o\oplus\mathcal{H}_1)$, $K \simeq K_p\oplus K_p^\perp$, such that

$$T(\Gamma) = R\oplus M - K = \begin{pmatrix} A & C \\ 0 & B \end{pmatrix}$$

satisfies

$\sigma(T(\Gamma)) = \sigma_p(T(\Gamma)) = \sigma(A) = \sigma_p(A) = \Gamma$,

$\mathrm{nul}(T(\Gamma) - \lambda) = \mathrm{nul}(R|\mathcal{H}_p - \lambda) = \mathrm{nul}(R-\lambda) = \mathrm{nul}(A-\lambda) = 1$ for all $\lambda \in \Gamma$,

$\sigma(B) = \Gamma$ and $\sigma_p(A^*) = \sigma_p(B) = \sigma_p(B^*) = \emptyset$.

By Lemma 2.1, $\sigma_p(T(\Gamma)^*) = \emptyset$.

Finally, observe that $T(\Gamma) = R\oplus M$ - "Compact" = "Normal + Compact". ■

REMARK 4.4. Clearly, if $\Gamma \subset$ interior ■, then $\|M^o_{\lambda,\Gamma}\| \le \|M^o_\lambda\|$ and $R = (M^o_{\lambda,\Gamma^*})^*$ satisfies $\sigma(R) \subset$ ■, so that $\sigma(T(\Gamma)) = \sigma_e(T(\Gamma)) = \Gamma \subset$ ■. $T(\Gamma) = N + C$, where N is normal, $\sigma(N) = \sigma_e(N) = \Gamma$, and C is a compact operator satisfying

$$\|C\| = \|T(\Gamma) - N\| \le \|T(\Gamma)\| + \|N\| \le \|M^o_\lambda\| + 2\pi + 1 := \alpha$$

(provided $\|K_p\|, \|K_p^\perp\| \le 1$).

COROLLARY 4.5. *Let Σ be a bounded F_σ subset of $\mathbb{C}$. Given $\varepsilon > 0$, there exist a normal operator $N \in L(\mathcal{H})$ and a compact operator C, with $\|C\| < \varepsilon$, such that*

$\sigma(N+C) = \sigma_e(N+C) = \Sigma^-$, $\sigma_p(N+C) = \Sigma$,

$\mathrm{nul}(N+C-\lambda) = 1$ *for all* $\lambda \in \Sigma$, *and* $\sigma_p([N+C]^*) = \emptyset$.

PROOF. It is obvious that we can always write $\Sigma = \cup_n \Gamma_n$, where Γ_n is a compact subset of an open square $\blacksquare_n$ such that

$$\varepsilon\pi/2n\alpha > \text{side } \blacksquare_n \to 0 \ (n \to \infty).$$

By Proposition 4.1 and Remark 4.4, there exist normal operators N_n and compact operators C_n, such that

$$\varepsilon/2 > \|C_n\| \to 0 \ (n \to \infty),$$

$$\sigma(N_n + C_n) = \sigma_e(N_n + C_n) = \sigma_p(N_n + C_n) = \sigma(N_n) = \sigma_e(N_n) = \Gamma_n,$$

$\mathrm{nul}(N_n + C_n - \lambda) = 1$ for all $\lambda \in \Gamma_n$, and $\sigma_p([N_n + C_n]^*) = \emptyset$ $(n = 1,2,\ldots)$.

According to Remark 3.5, there exist normal operators N_n' and compact operators C_n' (acting on the space of $(N_n + C_n)\oplus N_n'$) such that $\|C_n'\| < \varepsilon/2n$, $\sigma(N_n') = \Gamma_n \cap (\cup_{k=1}^{n-1} \Gamma_k)$,

$$\sigma((N_n + C_n)\oplus N_n' - C_n') = \sigma_e((N_n + C_n)\oplus N_n' - C_n') = \Gamma_n,$$

$$\sigma_p((N_n + C_n)\oplus N_n' - C_n') = \Gamma_n \setminus (\cup_{k=1}^{n-1} \Gamma_k),$$

$\mathrm{nul}((N_n + C_n)\oplus N_n' - C_n' - \lambda) = 1$ for all $\lambda \in \Gamma_n \setminus (\cup_{k=1}^{n-1} \Gamma_k)$,

and $\sigma_p([(N_n + C_n)\oplus N_n' - C_n']^*) = \emptyset$ $(n = 2,3,\ldots)$.

Define $N = \oplus_{n=1}^{\infty} (N_n \oplus N_n')$ and $C = (C_1 \oplus 0)\oplus\{\oplus_{n=2}^{\infty} (C_n \oplus 0 - C_n')\}$.

It is easily seen that N is normal, C is compact, $\|C\| < \varepsilon$, and $N+C$ has all the required properties. ∎

The result of J. Dixmier and C. Foiaş [14] admits the following mild improvement. Let

$$\sigma_p^{(n)}(T) = \{\lambda \in \mathbb{C}: \mathrm{nul}(T-\lambda) \geq n\} \ (n = 1,2,\ldots).$$

PROPOSITION 4.6. *For each* n, $1 \leq n < \infty$, $\sigma_p^{(n)}(T)$ *is a bounded* F_σ *set.*

PROOF. Set $\mathcal{P}_n = \{x = (x_1, x_2, \ldots, x_n) \in \mathcal{H}^{(n)}: \|x\| \leq 1$, $Tx_i = \lambda x_i$ for some $\lambda \in \sigma_p(T)$, $i = 1,2,\ldots,n\}$, and $\mathcal{P}_o = \{x \in \mathcal{P}_n: \{x_i\}_{i=1}^n$ is linearly dependent$\}$.

It can be easily seen that $\mathcal{P}_n$ and $\mathcal{P}_o$ are weakly closed subsets of the closed unit ball of $\mathcal{H}^{(n)}$. Since the ball is weakly compact, so are $\mathcal{P}_n$ and $\mathcal{P}_o$. Moreover, the compact topology of the ball $\mathcal{B}$ is metrizable because $\mathcal{H}^{(n)}$ is separable. Let d be one of the metrics that give the weak topology on $\mathcal{B}$, and let

$$A_m = P_n \cap \{x \in B: \ d[x,P_o] \geq 1/m\} \quad (m = 1,2,\ldots).$$

A_m is a weakly compact set ($m = 1,2,\ldots$). Since P_o is closed, we have $P_n \setminus P_o = \cup_{m=1}^{\infty} A_m$, a denumerable union of weakly compact sets.

Define $e: P_n \setminus P_o \to \mathbb{C}$ by $e(x) = \lambda$ (where $x = (x_1, x_2, \ldots, x_n)$ and $Tx_i = \lambda x_i$ for all $i = 1,2,\ldots,n$). It can be easily proved that e is a continuous function with respect to the weak topology in $P_n \setminus P_o$. Moreover, since A_m is weakly compact, $\Delta_m = e(A_m)$ is a compact subset of $\sigma(T)$, and

$$\sigma_p^{(n)}(T) = e[P_n \setminus P_o] = \cup_{m=1}^{\infty} e(A_m) = \cup_{m=1}^{\infty} \Delta_m$$

is a bounded F_σ set. ■

It is convenient to remark that $\sigma_p^{(\infty)}(T) = \{\lambda \in \mathbb{C}: \ \ker(T-\lambda)$ is infinite dimensional$\}$ is an $F_{\sigma\delta}$ set (denumerable intersection of F_δ's), but not an F_σ, in general. (P_∞ is, of course, a weakly compact set, but $P_\infty \setminus \cup_{n=1}^{\infty} P_n$ is not, in general, a denumerable union of weakly compact subsets.) Indeed, Proposition 4.6 indicates that $\{\sigma_p^{(n)}(T)\}_{n=1}^{\infty}$ is a non-increasing sequence of F_σ bounded sets, and each $F_{\sigma\delta}$ subset of $\mathbb{C}$ can be $\sigma_p^{(\infty)}(T)$ for some operator T. From Corollary 4.5 we immediately obtain the following result:

COROLLARY 4.7. *Given a non-increasing sequence* $\{\Sigma_n\}_{n=1}^{\infty}$ *of bounded* F_σ *subsets of* $\mathbb{C}$, *and* $\varepsilon > 0$, *there exist* $N, C \in L(H)$ *such that* N *is normal,* $\sigma(N) = \sigma_e(N) = \Delta := (\Sigma_1)^-$, C *is compact,* $\|C\| < \varepsilon$, $\sigma_p^{(n)}(N+C) = \Sigma_n$, $\sigma_p^{(\infty)}(N+C) = \cap_{n=1}^{\infty} \Sigma_n$ *and* $\sigma_p([N+C]^*) = \emptyset$.

5. NEW MAKE UP. SECOND PART

Following C. Apostol [3], we define

$$\rho_{s-F}^{r}(T) = \{\mu \in \rho_{s-F}(T): \ \text{nul}(T-\lambda) \text{ and } \text{nul}(T-\lambda)^* \text{ are continuous on a neighborhood of } \mu\}$$

(= the *regular points* of $\rho_{s-F}(T)$), and $\rho_{s-F}^{s}(T) = \rho_{s-F}(T) \setminus \rho_{s-F}^{r}(T)$ (= the *singular points* of $\rho_{s-F}(T)$).

LEMMA 5.1. *Let* $T \in L(H)$, *and let*

$$\Phi = \rho_{s-F}(T) \cap \text{interior}[\sigma_p(T) \cap \sigma_p(T^*)^*].$$

Given $\varepsilon > 0$, *there exists* $K_\varepsilon \in K(H)$, *with* $\|K_\varepsilon\| < \varepsilon$, *such that*

$$T - K_\varepsilon = \begin{pmatrix} N(\Phi) & * \\ 0 & T_\varepsilon \end{pmatrix},$$

where $N(\Phi)$ *is a compact perturbation of a normal operator,* $\sigma(N(\Phi))$

$=\Phi^-$, $\sigma_e(N(\Phi)) = \partial\Phi$, $\sigma_p(N(\Phi)) = \sigma_p(N(\Phi)^*)^* = \Phi$, $\text{nul}(N(\Phi)-\lambda) = \text{nul}([N(\Phi)-\lambda]^*) = 1$ *for all* $\lambda \in \Phi$, $\sigma_e(T_\varepsilon) = \sigma_e(T)$, $\rho_{s-F}(T_\varepsilon) = \rho_{s-F}(T)$, $\text{ind}(T_\varepsilon - \lambda) = \text{ind}(T-\lambda)$ *for all* $\lambda \in \rho_{s-F}(T)$, *and*

$$\text{nul}(T_\varepsilon - \lambda) = \begin{cases} \text{nul}(T-\lambda) - 1, & \text{if } \lambda \in \Phi, \\ \text{nul}(T-\lambda), & \text{if } \lambda \in \rho_{s-F}(T) \setminus \Phi \end{cases}$$

and

$$\text{nul}(T_\varepsilon - \lambda)^* = \begin{cases} \text{nul}(T-\lambda)^* - 1, & \text{if } \lambda \in \Phi, \\ \text{nul}(T-\lambda)^*, & \text{if } \lambda \in \rho_{s-F}(T) \setminus \Phi. \end{cases}$$

PROOF. The Apostol's triangular representation of T indicates that T admits a 3×3 operator matrix of the form

$$T = \begin{pmatrix} T_r & * & * \\ 0 & T_o & * \\ 0 & 0 & T_\ell \end{pmatrix} \begin{matrix} H_r \\ H_o \\ H_\ell \end{matrix},$$

where $H_r = \bigvee\{\ker(T-\lambda): \lambda \in \rho^r_{s-F}(T)\}$, $H_\ell = \bigvee\{\ker(T-\lambda)^*: \lambda \in \rho^r_{s-F}(T)\}$ and $H_o = H \ominus (H_r \oplus H_\ell)$, T_r (T^*_ℓ) is a triangular operator and all the components of its spectrum intersect the interior of $\rho_{s-F}(T) \cap \sigma_p(T)$ (the interior of $\rho_{s-F}(T) \cap \sigma_p(T^*)^*$, resp.), $\sigma_p(T^*_r) = \sigma_p(T_\ell) = \emptyset$, $\rho_{s-F}(T) \subset \rho_{s-F}(T_r) \cap \rho_{s-F}(T_\ell) \cap (\mathbb{C} \setminus \sigma(T_o))$, and $\rho^s_{s-F}(T) \subset \sigma_o(T_o)$.

Moreover, if Λ is a finite subset of $\rho^s_{s-F}(T)$, then T is similar to $T_\Lambda \oplus T'_\Lambda$, where T_Λ acts on a finite dimensional space, $\sigma(T_\Lambda) = \Lambda$, and $\Lambda \cap \rho^s_{s-F}(T'_\Lambda) = \emptyset$ (see [3], [21,Theorem 3.38]).

By Voiculescu's theorem [36] (see also [21,Chapter 4]), there exists $K_o \in K(H)$, with $\|K_o\| < \varepsilon/2$, such that

$$T - K_o \simeq T \oplus \begin{pmatrix} M & * \\ 0 & S \end{pmatrix}^{(2)},$$

where $M \simeq M^{(\infty)}$ is a normal operator such that $\sigma(M) = \sigma_e(M) = \partial\Phi$, $S \simeq S^{(\infty)}$, $\sigma(S) = \sigma_e(S) = \sigma_e(T)$, $\rho^+_{s-F}(S) = \{\lambda \in \rho^+_{s-F}(T): \text{ind}(T-\lambda) = \infty\}$ and $\rho^-_{s-F}(S) = \{\lambda \in \rho^-_{s-F}(T): \text{ind}(T-\lambda) = -\infty\}$.

After replacing (if necessary) M by some arbitrarily small compact perturbation (that can be absorbed in K_o), we can directly assume that $M \simeq M \oplus M(\partial(\Phi^-))$, where $M(\partial(\Phi^-))$ is the operator defined in the introduction of Section 3.

Thus, we have

$$T - K_o \simeq \begin{pmatrix} T_r & * & * \\ 0 & T_o & * \\ 0 & 0 & T_\ell \end{pmatrix} \oplus \begin{pmatrix} M & 0 & 0 & * \\ 0 & M_+(\partial(\Phi^-)) & Z(\partial(\Phi^-)) & * \\ 0 & 0 & M_-(\partial(\Phi^-)) & * \\ 0 & 0 & 0 & S \end{pmatrix}^{(2)}$$

$$= \left(\begin{array}{cc|c|c|c|cc} M & & & & & * & 0 \\ & M_+(\partial(\Phi^-)) & O & O & Z(\partial(\Phi^-)) & * & 0 \\ \hline & & M & & & 0 & * \\ & & \begin{pmatrix} M_-(\partial(\Phi^-)) & 0 \\ Z(\partial(\Phi^-)) & M_+(\partial(\Phi^-)) \end{pmatrix} & O & O & 0 & * \\ & & & & & 0 & * \\ \hline & & & \begin{pmatrix} T_r & * & * \\ 0 & T_o & * \\ 0 & 0 & T_\ell \end{pmatrix} & 0 & O & O \\ & O & & 0 & M_-(\partial(\Phi^-)) & * & 0 \\ \hline & & & & & S & 0 \\ & & & & & 0 & S \end{array}\right)$$

$$= \begin{pmatrix} M\oplus M_+(\partial(\Phi^-)) & O & & & * \\ & \begin{pmatrix} M\oplus M_-(\partial(\Phi^-)) & 0 \\ Z'(\partial(\Phi^-)) & M_+(\partial(\Phi^-))\oplus T_r \end{pmatrix} & & & * \\ & & T_o & & \\ O & & & T_\ell\oplus M_-(\partial(\Phi^-)) & \\ & & & & S^{(2)} \end{pmatrix},$$

where $Z'(\partial(\Phi^-)) = \begin{pmatrix} 0 & Z(\partial(\Phi^-)) \\ 0 & 0 \end{pmatrix}$.

By Proposition 3.4, there exists a compact operator K_1, with $\|K_1\| < \varepsilon/2$, such that

$$L = M\oplus M_+(\partial(\Phi^-)) - K_1$$

is an essentially normal operator such that $\sigma(L) = \Phi^-$, $\sigma_e(L) = \partial\Phi$, $\sigma_p(L) = \emptyset$, $\sigma_p(L^*)^* = \Phi$, and $\mathrm{ind}(L-\lambda) = -\mathrm{nul}(L-\lambda)^* = -1$ for all $\lambda \in \Phi$.

The operator $M_+(\partial(\Phi^-))\oplus T_r$ is quasitriangular (that is, $\rho^-_{s-F}(M_+(\partial(\Phi^-))\oplus T_r = \emptyset$; see [21,Chapter 6]). Since the spectrum of this operator has no isolated points, by using the results of [25], it is not difficult to find a compact operator K_2', with $\|K_2'\| < \varepsilon/6$, such that $A_r = M_+(\partial(\Phi^-))\oplus T_r - K_2'$ is a triangular operator with diagonal entries $\{\lambda_j\}_{j=1}^{\infty}$ (with respect to some orthonormal basis $\{e_j\}_{j=1}^{\infty}$) such that $\lambda_j \in \sigma_{\ell re}(T_r) \cup \partial\Phi \subset \sigma_{\ell re}(T)$ (for all $j = 1,2,\ldots$), so that $\sigma_p(A_r^*)^* \subset \sigma_{\ell re}(A_r)$; moreover,

$$\mathrm{ind}(A_r - \lambda) = \begin{cases} \mathrm{ind}(T_r - \lambda) - 1, & \text{if } \lambda \in \Phi, \\ \mathrm{ind}(T_r - \lambda), & \text{if } \lambda \in \rho_{s-F}(T_r) \setminus \Phi^- \end{cases}$$

(same references as above, or [20]).

Now we repeat the same trick as in [7]: let P_n denote the orthogonal projection onto $\vee\{e_j\}_{j=1}^{n}$. Since $Z'(\partial(\Phi^-))$ is compact,

$$\|Z'(\partial(\Phi^-)) - P_n Z'(\partial(\Phi^-))\| \to 0 \quad (n \to \infty).$$

Thus, we can find a compact operator K_2'', with $\|K_2''\| < \varepsilon/6$, such that

$$\begin{pmatrix} M\oplus M_-(\partial(\Phi^-)) & 0 \\ Z'(\partial(\Phi^-)) & A_r \end{pmatrix} - K_2'' = \begin{pmatrix} M\oplus M_-(\partial(\Phi^-)) & 0 & 0 \\ P_n Z'(\partial(\Phi^-)) & F_{r,n} & * \\ 0 & 0 & A_{r,n} \end{pmatrix}$$

(for some n large enough), where

$$F_{r,n} = \begin{pmatrix} \lambda_1 & & & & * \\ & \lambda_2 & & & \\ & & \cdot & & \\ & & & \cdot & \\ & 0 & & & \cdot \\ & & & & \lambda_n \end{pmatrix} \begin{matrix} e_1 \\ e_2 \\ \cdot \\ \cdot \\ \cdot \\ e_n \end{matrix}$$

($= A_r|\mathrm{ran}\, P_n$) and

$$A_{r,n} = \begin{pmatrix} \lambda_{n+1} & & & * \\ & \lambda_{n+2} & & \\ & & \cdot & \\ & 0 & & \cdot \\ & & & & \cdot \end{pmatrix} \begin{matrix} e_{n+1} \\ e_{n+2} \\ \cdot \\ \cdot \\ \cdot \end{matrix} = (A_r)_{\ker P_n}$$

has exactly the same characteristics as A_r.

Observe that the spectrum of

$$R_n = \begin{pmatrix} M\oplus M_-(\partial(\Phi^-)) & 0 \\ P_n Z'(\partial(\Phi^-)) & F_{r,n} \end{pmatrix}$$

is equal to $\sigma(M) \cup \sigma(M_-(\partial(\Phi^-))) \cup \sigma(F_{r,n}) = \Phi^-$ and $\rho_{s-F}(R_n) \cap \sigma(R_n) = \rho_{s-F}^+(R_n) = \Phi$; furthermore, R_n is essentially normal and

$$\mathrm{ind}(R_n - \lambda) = 1 \quad \text{for all } \lambda \in \Phi.$$

By Proposition 3.4, there exists a compact operator K_2''', with $\|K_2'''\| < \varepsilon/6$, such that $R = R_n - K_2'''$ is essentially normal,

$\sigma(R) = \Phi^-$, $\sigma_e(R) = \partial\Phi$, $\sigma_p(R) = \Phi$, $\sigma_p(R^*) = \emptyset$, and $\text{nul}(R-\lambda) = \text{ind}(R-\lambda) = 1$ for all $\lambda \in \Phi$.

On the other hand,

$$\text{nul}\left(\begin{pmatrix} R & * \\ 0 & A_{r,n} \end{pmatrix} - \lambda\right) = \text{ind}\left(\begin{pmatrix} R & * \\ 0 & A_{r,n} \end{pmatrix} - \lambda\right)$$

$$= \text{ind}(R-\lambda) + \text{ind}(A_{r,n} - \lambda) = \text{nul}(R-\lambda) + \text{nul}(A_{r,n} - \lambda)$$

for all $\lambda \in \rho_{s-F}\left(\begin{pmatrix} R & * \\ 0 & A_{r,n} \end{pmatrix}\right) = \rho_{s-F}(T_r) \setminus \partial\Phi$.

Therefore, there exists a compact operator K_2, with $\|K_2\| \le \|K_2'\| + \|K_2''\| + \|K_2'''\| < \varepsilon/2$, such that

$$\begin{pmatrix} M \oplus M_-(\partial(\Phi^-)) & 0 \\ Z'(\partial(\Phi^-)) & T_r \end{pmatrix} - K_2 \simeq \begin{pmatrix} R & * \\ 0 & B_r \end{pmatrix},$$

where $B_r = A_{r,n}$.

Clearly, the operator $N(\Phi) = L \oplus R$ has all the desired properties.

Since $\sigma[T_\ell \oplus M_-(\partial(\Phi^-))] = \sigma(T_\ell) \cup \Phi^-$ has no isolated points, and $\text{ind}(T_\ell \oplus M_-(\partial(\Phi^-)) - \lambda) \le 0$ for all $\lambda \in \rho_{s-F}[T_\ell \oplus M_-(\partial(\Phi^-))]$, by using the results of [25] we can find a compact operator K_3, with $\|K_3\| < \varepsilon/2$, such that

$$B_\ell = T_\ell \oplus M_-(\partial(\Phi^-)) - K_3$$

is the adjoint of a triangular operator whose diagonal entries belong to $\sigma_{\ell re}[T_\ell \oplus M_-(\partial(\Phi^-))] = \sigma_{\ell re}(T_\ell) \cup \partial\Phi$. It follows that $\sigma_p(B_\ell) \subset \sigma_{\ell re}[T_\ell \oplus M_-(\partial(\Phi^-))] \subset \sigma_{\ell re}(T)$, and

$$\text{ind}(B_\ell - \lambda) = \begin{cases} \text{ind}(T_\ell - \lambda) + 1, & \text{if } \lambda \in \Phi, \\ \text{ind}(T_\ell - \lambda), & \text{if } \lambda \in \rho_{s-F}(T\ ) \setminus \Phi^- \end{cases}$$

(same reference as above, or [20]).

We conclude that there exists a compact operator $K_o' \simeq K_1 \oplus K_2 \oplus K_3 \oplus 0^{(2)}$ (so that $\|K_o'\| < \varepsilon/2$) such that

$$T - (K_o + K_o') \simeq \begin{pmatrix} N(\Phi) & & & & * \\ & B_r & & & \\ & & T_o & & \\ & O & & B_\ell & \\ & & & & S^{(2)} \end{pmatrix}.$$

Define $K_\varepsilon = K_o + K_o'$, and

$$T_\varepsilon = \begin{pmatrix} B_r & & & * \\ & T_o & & \\ & & B_\ell & \\ 0 & & & S^{(2)} \end{pmatrix}.$$

Clearly, K_ε is compact, and $\|K_\varepsilon\| \le \|K_o\| + \|K_o'\| < \varepsilon$. With the help of [3], [21,Chapter 3], it is straightforward to check that T_ε has the desired characteristics.

The proof of Lemma 5.1 is now complete. ∎

COROLLARY 5.2. *Let* T *and* Φ *be as in Lemma 5.1, and assume that* $\rho_{s-F}^{s}(T) \cap \Phi = \emptyset$. *Given a finite of denumerable subset* Λ *of* Φ *that only accumulates on* $\partial\Omega$, *and* $\varepsilon > 0$, *there exists a compact operator* K_ε', *with* $\|K_\varepsilon'\| < \varepsilon$, *such that* $K_\varepsilon' \simeq K_\varepsilon'' \oplus 0$,

$$T - (K_\varepsilon + K_\varepsilon') = \begin{pmatrix} N(\Phi) - K_\varepsilon'' & * \\ 0 & T_\varepsilon \end{pmatrix},$$

where $\sigma(N(\Phi) - K_\varepsilon'') = \partial\Phi \cup \Lambda$, $\sigma_e(N(\Phi)) = \partial\Phi$, $\sigma_p(N(\Phi) - K_\varepsilon'')$ $= \sigma_p([N(\Phi) - K_\varepsilon'']^*)^* = \Lambda$, $\mathrm{nul}(N(\Phi) - K_\varepsilon'' - \lambda) = \mathrm{nul}(N(\Phi) - K_\varepsilon'' - \lambda) = 1$ *for all* $\lambda \in \Phi$, *and* $\rho_{s-F}^{s}(T - (K_\varepsilon + K_\varepsilon')) \cap \Phi = \Lambda$.

PROOF. Since $N(\Phi)$ is a compact perturbation of a normal operator, we can find a compact operator K_1, with $\|K_1\| < \varepsilon/3$, such that

$$N(\Phi) - K_1 \simeq N(\Phi) \oplus D,$$

where D is a normal diagonal operator of uniform infinite multiplicity such that $\sigma(D) = \sigma_e(D) = \partial\Phi$.

Recall that $N(\Phi) = L \oplus R$, where $\mathrm{ind}(R - \lambda) = \mathrm{nul}(R - \lambda)$ $= \mathrm{nul}(L - \lambda)^* = -\mathrm{ind}(L - \lambda) = 1$ for all $\lambda \in \Phi$. Let

$$p(\lambda) = \Pi\,\{\lambda - \mu:\ \mu \in \Lambda \setminus (\partial\Phi)_{\varepsilon/4}\};$$

then $R|\ker p(R)$ is similar to diagonal$\{\mu_1, \mu_2, \ldots, \mu_m\} \in L(\mathbb{C}^m)$, where $\mu_1, \mu_2, \ldots, \mu_m$ denote the zeros of the polynomial p; moreover, $R_p = R_{\ker p(R)^\perp}$ is similar to R. (If $R \in L(H_R)$ and $\ker(p(R))^\perp$ $= M$, then the mapping $x \to p(R)x$ from H_R onto M implements the similarity.)

We have

$$N(\Phi) - K_1 \simeq \begin{pmatrix} R|\ker p(R) & * & 0 \\ 0 & R_p & 0 \\ 0 & 0 & L \end{pmatrix} \oplus D.$$

According to [3], [21,Chapter3], there exists a compact

K_2', with $\|K_2'\| < \varepsilon/3$, such that

$$\sigma(R_p \oplus L - K_2') = \sigma_e(R_p \oplus L - K_2') = \partial\Phi.$$

On the other hand, it is straightforward to construct a compact normal operator K_2'' commuting with D such that $\|K_2''\| < \varepsilon/3$ and

$$D - K_2'' = D' \oplus D(\Lambda,\varepsilon),$$

where $\sigma(D') = \partial\Phi$ and $D(\Lambda,\varepsilon)$ is a diagonal normal operator such that $\sigma_o(D(\Lambda,\varepsilon)) = \sigma_p(D(\Lambda,\varepsilon)) = \Lambda \cap (\partial\Phi)_{\varepsilon/4} \subset \sigma(D(\Lambda,\varepsilon)) \subset [\Lambda \cap (\partial\Phi)_{\varepsilon/4}] \cup \partial\Phi$. (Roughly speaking: K_2'' "pushes" some of the diagonal entries of D to close points in $\Lambda \cap (\partial\Phi)_{\varepsilon/4}$.)

Therefore, there exists $K_2 \simeq 0 \oplus K_2' \oplus K_2''$ (so that $\|K_2\| < \varepsilon/3$) such that

$$N(\Phi) - (K_1 + K_2) \simeq \begin{pmatrix} R|\ker p(R) & * \\ 0 & (R_p \oplus L - K_2') \oplus D' \end{pmatrix} \oplus D(\Lambda,\varepsilon).$$

Finally, according to Lemma 2.7 and Remark 2.8, there exists a compact operator K_3', with $\|K_3'\| < \varepsilon/3$, such that the operator

$$A = (R_p \oplus L - K_2') \oplus D' - K_3'$$

satisfies $\sigma(A) = \sigma_e(A) = \partial\Phi$ and $\sigma_p(A) = \sigma_p(A^*) = \emptyset$.

Define $K_3 = 0 \oplus K_3' \oplus 0$ (with respect to the above decomposition) and $K_\varepsilon'' = K_1 + K_2 + K_3$; then K_ε'' is compact, $\|K_\varepsilon''\| \le \|K_1\| + \|K_2\| + \|K_3\| < \varepsilon$, and

$$N(\Phi) - K_\varepsilon'' \simeq \begin{pmatrix} R|\ker p(R) & * \\ 0 & A \end{pmatrix} \oplus D(\Lambda,\varepsilon).$$

By Rosenblum's corollary [33], [21,Chapter 3], this operator is similar to

$$A \oplus (\mathrm{diag}\{\lambda_1,\lambda_2,\ldots,\lambda_m\}) \oplus D(\Lambda,\varepsilon),$$

whence it readily follows that $\sigma(N(\Phi) - K_\varepsilon'') = \partial\Phi \cup \Lambda$, $\sigma_e(N(\Phi) - K_\varepsilon'') = \partial\Phi$, $\sigma_p(N(\Phi) - K_\varepsilon'') = \sigma_p([N(\Phi) - K_\varepsilon'']^*) = \Lambda$ and $\mathrm{nul}(N(\Phi) - K_\varepsilon'' - \lambda) = \mathrm{nul}(N(\Phi) - K_\varepsilon'' - \lambda) = 1$ for all $\lambda \in \Lambda$.

Since $\rho_{s-F}^s(T) \cap \Phi = \emptyset$, it readily follows from [3], [21, Chapter 3] (or by a direct verification) that

$$\rho_{s-F}^s(T - (K_\varepsilon + K_\varepsilon')) \cap \Phi = \Lambda. \qquad \blacksquare$$

LEMMA 5.3. *Let* $T \in L(H)$ *and let* Φ *be a union of bounded components of* $\rho_{s-F}(T)$ *such that*

$$\Phi \cap \mathrm{interior}[\sigma_p(T) \cap \sigma_p(T^*)^*] = \emptyset.$$

Given $\varepsilon > 0$, *there exists* $K_\varepsilon \in K(H)$, *with*

$\|K_\varepsilon\| < \varepsilon + \max[\max\{m_e(T-\lambda): \lambda \in \Phi\}, \max\{m_e[(T-\lambda)^*]: \lambda \in \Phi\}]$,

such that

$$T - K_\varepsilon = \begin{pmatrix} N(\Phi) & * \\ 0 & T_\varepsilon \end{pmatrix},$$

where $N(\Phi)$ *is a compact perturbation of a normal operator*, $\sigma(N(\Phi)) = \Phi^-$, $\sigma_e(N(\Phi)) = \partial\Phi$, $\sigma_p(N(\Phi)) = \sigma_p(N(\Phi)^*)^* = \Phi$, $\text{nul}(N(\Phi)-\lambda) = \text{nul}(N(\Phi)-\lambda)^* = 1$ *for all* $\lambda \in \Phi$, $\sigma_e(T_\varepsilon) = \sigma_e(T)$, $\rho_{s-F}(T_\varepsilon) = \rho_{s-F}(T)$, *and* $\text{ind}(T_\varepsilon - \lambda) = \text{ind}(T-\lambda)$, $\text{nul}(T_\varepsilon - \lambda) = \text{nul}(T-\lambda)$ *and* $\text{nul}(T_\varepsilon - \lambda)^* = \text{nul}(T-\lambda)$ *for all* $\lambda \in \rho_{s-F}(T)$.

PROOF. According to [26], there exists $K_1 \in K(H)$, with

$\|K_1\| < \frac{\varepsilon}{2} + \max[\max\{m_e(T-\lambda): \lambda \in \Phi\}, \max\{m_e[(T-\lambda)^*]: \lambda \in \Phi\}]$,

such that

$$T - K_1 \simeq T \oplus A,$$

where $\sigma(A) = \sigma_e(T) \cup \Phi$, $\sigma_e(A) = \sigma_e(T)$, $\sigma_{\ell re}(A) = \sigma_{\ell re}(T)$, $\rho^+_{s-F}(A) = \{\lambda \in \rho^+_{s-F}(T): \text{ind}(T-\lambda) = \infty\}$, $\rho^-_{s-F}(A) = \{\lambda \in \rho^-_{s-F}(T): \text{ind}(T-\lambda) = -\infty\}$, and

$$\min[\text{nul}(A-\lambda), \text{nul}(A-\lambda)^*] = 1 \quad \text{for all } \lambda \in \Phi.$$

Now the result follows from Lemma 5.1. ■

By combining the arguments of [26] and those in the proof of Corollary 5.2 (dealing with modifications of D), we obtain the following result. (The details of the proof are left to the reader.)

LEMMA 5.4. *Let* $T \in L(H)$ *and let* Φ *be a union of (not necessarily bounded) components of* $\rho_{s-F}(T)$ *such that*

$$\Phi \cap \text{interior}[\sigma_p(T) \cap \sigma_p(T^*)^*] = \emptyset.$$

Given a bounded finite or denumerable subset Λ *of* Φ *that only accumulates of* $\partial\Phi$, *and* $\varepsilon > 0$, *there exists* $K_\varepsilon \in K(H)$, *with*

$\|K_\varepsilon\| < \varepsilon + \max[\max\{m_e(T-\lambda): \lambda \in \Lambda\}, \max\{m_e[(T-\lambda)^*]: \lambda \in \Lambda\}]$

such that

$$T - K_\varepsilon = \begin{pmatrix} N(\Lambda) & * \\ 0 & T_\varepsilon \end{pmatrix}$$

where $N(\Lambda)$ *is a compact perturbation of a normal operator*, $\sigma(N(\Lambda)) = \sigma_{\ell re}(T) \cup \Lambda$, $\sigma_e(N(\Lambda)) = \sigma_{\ell re}(T)$, $\sigma_p(N(\Lambda)) = \sigma_p(N(\Lambda)^*)^* = \Lambda$, $\text{nul}(N(\Lambda)-\lambda) = \text{nul}(N(\Lambda)-\lambda)^* = 1$ *for all* $\lambda \in \Lambda$, $\sigma_e(T_\varepsilon) = \sigma_e(T)$, $\rho_{s-F}(T_\varepsilon) = \rho_{s-F}(T)$, *and* $\text{ind}(T_\varepsilon - \lambda) = \text{ind}(T-\lambda)$, $\text{nul}(T_\varepsilon - \lambda) = \text{nul}(T-\lambda)$ *and* $\text{nul}(T_\varepsilon - \lambda)^* = \text{nul}(T-\lambda)^*$ *for all* $\lambda \in \rho_{s-F}(T)$.

6. COMPACT PERTURBATIONS AND POINT SPECTRUM

Now we are in a position to prove the main result of this article

THEOREM 6.1. *Let* $T \in L(H)$ *and let* Σ *and* Ξ *be two bounded* F_σ *subsets of* $\mathbb{C}$ *such that*

(i) $\Sigma \supset \rho^+_{s-F}(T)$ *and* $\Xi \supset \rho^-_{s-F}(T)$;

(ii) if Ω *is a component of* $\rho_{s-F}(T) \setminus \rho^+_{s-F}(T)$ $(\rho_{s-F}(T) \setminus \rho^-_{s-F}(T))$, *then either* $\Omega \subset \Sigma$ $(\Omega \subset \Xi$, *resp.*), *or* $\Omega \cap \Sigma$ $(\Omega \cap \Xi$, *resp.*) *is a finite (possibly empty) or denumerable set that only accumulates on* $\partial\rho_{s-F}(T)$; *and*

(iii) if Ω *is a component of* $\{\lambda \in \rho_{s-F}(T):\ \mathrm{ind}(T-\lambda) = 0\}$, *then* $\Omega \cap \Sigma = \Omega \cap \Xi$.

There exists $K \in K(H)$ *such that*

$$\sigma_p(T-K) = \Sigma \quad \text{and} \quad \sigma_p([T-K]^*)^* = \Xi.$$

Furthermore, if $\Phi = \rho_{s-F}(T) \cap \mathrm{interior}[\sigma_p(T) \cap \sigma_p(T^*)^*]$, $\Phi(\Sigma) = [\rho_{s-F}(T) \setminus \rho^+_{s-F}(T)] \cap \mathrm{interior}\ \Sigma$ *and* $\Phi(\Xi) = [\rho_{s-F}(T) \setminus \rho^-_{s-F}(T)] \cap \mathrm{interior}\ \Xi$, *then given* $\varepsilon > 0$, K *can be chosen so that*

$$\begin{aligned}\|K\| < \varepsilon + \max[&\max\{m_e(T;\lambda):\ \lambda \in \sigma_o(T)\},\\ &\max\{m_e(T-\lambda):\ \lambda \in \Phi(\Sigma) \setminus \Phi\},\\ &\max\{m_e([T-\lambda]^*):\ \lambda \in \Phi(\Xi) \setminus \Xi\},\\ &\max\{m_e(T-\lambda):\ \lambda \in \Sigma\setminus(\mathrm{interior}\ \Sigma \cup \Phi),\\ &\max\{m_e([T-\lambda]^*):\ \lambda \in \Xi\setminus(\mathrm{interior}\ \Xi \cup \Phi).\end{aligned}$$

On the other hand, if either Σ (Ξ) *is not a bounded* F_σ *set, or does not satisfy (i) or (ii), then* $\sigma_p(T-C) \neq \Sigma$ $(\sigma_p([T-C]^*)^* \neq \Xi$, *resp.*) *for any compact operator* C; *moreover, if* Σ *and* Ξ *do not satisfy (iii), then either* $\sigma_p(T-C) \neq \Sigma$, *or* $\sigma_p([T-C]^*)^* \neq \Xi$ *for all* C *in* $K(H)$.

REMARKS 6.2. (i) The above estimate of the norm of K is not the best possible, in general. However, our estimate *is the best possible that can be obtained without specific knowledge of the structure of* T *(other than the different parts of* $\sigma(T)$ *and the functions* $m_e(T-\lambda)$ *and* $m_e([T-\lambda]^*)$ $(\lambda \in \mathbb{C})$, *and without a careful comparison between* $\rho^s_{s-F}(T)$ *and the isolated points of* $(\Sigma \cup \Xi) \cap \rho_{s-F}(T)$ *(see* [26] *for related examples;* $m_e(T-\lambda) = 0$ *on* $\sigma_{\ell re}(T)$*).*

(ii) In particular, our estimate shows that if $\sigma_o(T)$ and the set of isolated points of $(\Sigma \cup \Xi) \cap \rho_{s-F}(T)$ are empty, then K can be chosen of arbitrarily small norm. We shall continue this analysis in the next section.

PROOF OF THEOREM 6.1. By using Voiculescu's theorem [36], we can find $K_o \in K(H)$, with $\|K_o\| < \varepsilon/9$, such that

$$T - K_o \simeq T \oplus \begin{pmatrix} N & * & * \\ 0 & S & * \\ 0 & 0 & N \end{pmatrix}^{(4)},$$

where $N \simeq N^{(\infty)}$ is a normal operator such that $\sigma(N) = \sigma_e(N) = \sigma_{\ell re}(T)$, $S \simeq S^{(\infty)}$, $\sigma(S) = \sigma_e(S) = \sigma_e(T)$, $\rho^+_{s-F}(S) = \{\lambda \in \rho_{s-F}(T): \text{ind}(T - \lambda) = \infty\}$ and $\rho^-_{s-F}(S) = \{\lambda \in \rho_{s-F}(T): \text{ind}(T - \lambda) = -\infty\}$.

Let

$$T - K_\varepsilon = \begin{pmatrix} N(\Phi) & * \\ 0 & T_\varepsilon \end{pmatrix}$$

be as described in Lemma 5.1, with $\|K_\varepsilon\| < \varepsilon/9$.

By using Proposition 3.4, Lemma 5.1 and Corollary 5.2, we can find a compact operator K_1, with $\|K_1\| < \varepsilon/9$, such that

$$\sigma_p(N(\Phi) \oplus \begin{pmatrix} N & * & * \\ 0 & S & * \\ 0 & 0 & N \end{pmatrix} - K_1) = (\Sigma \cap \Phi) \cup \rho^+_{s-F}(S)$$
$$\cup [\Sigma \cap \rho_{s-F}(T) \cap (\partial \rho_{s-F}(T))_{\varepsilon/10}]$$

and

$$\sigma_p([N(\Phi) \oplus \begin{pmatrix} N & * & * \\ 0 & S & * \\ 0 & 0 & N \end{pmatrix} - K_1]^*) = (\Xi \cap \Phi) \cup \rho^-_{s-F}(S)$$
$$\cup [\Xi \cap \rho_{s-F}(T) \cap (\partial \rho_{s-F}(T))_{\varepsilon/10}].$$

Let Σ_ε (Ξ_ε) denote the (necessarily finite) set of isolated points of $[\Sigma \cap \rho_{s-F}(T)] \setminus (\partial \rho_{s-F}(T))_{\varepsilon/10}$ $([\Xi \cap \rho_{s-F}(T)] \setminus (\partial \rho_{s-F}(T))_{\varepsilon/10}$, resp.), and let

$$A = N(\Phi) \oplus \begin{pmatrix} N & * & * \\ 0 & S & * \\ 0 & 0 & N \end{pmatrix} - K_1.$$

By combining the actions of K_o, K_ε and K_1, we have produced a compact operator K_2, with $\|K_2\| < \varepsilon/3$, such that

$$T - K_2 \simeq \begin{pmatrix} A & * \\ 0 & T_\varepsilon \end{pmatrix} \oplus \begin{pmatrix} N & * & * \\ 0 & S & * \\ 0 & 0 & N \end{pmatrix}^{(3)}.$$

By Proposition 3.4, there exists a compact operator K_3,

with

$$\|K_3\| < \frac{\varepsilon}{3} + \max\{m_e(T_\varepsilon;\lambda): \lambda \in \sigma_o(T_\varepsilon)\}$$
$$= \frac{\varepsilon}{3} + \max\{m_e(T;\lambda): \lambda \in \sigma_o(T)\},$$

such that

$$\sigma_p(T_\varepsilon - K_3) = \rho^+_{s-F}(T) \quad (\subset \Sigma)$$

and

$$\sigma_p([T_\varepsilon - K_3]^*) = \rho^-_{s-F}(T) \quad (\subset \Xi).$$

By Lemma 5.3, there exists a compact operator K_4, with

$$\|K_4\| < \frac{\varepsilon}{3} + \max[\max\{m_e(T-\lambda): \lambda \in \Phi(\Sigma)\setminus\Phi\},$$
$$\max\{m_e([T-\lambda]^*): \lambda \in \Phi(\Xi)\setminus\Phi\}]$$

such that

$$\begin{pmatrix} N & * & * \\ 0 & S & * \\ 0 & 0 & N \end{pmatrix} - K_4 \simeq \begin{pmatrix} N([\Phi(\Sigma)\cup\Phi(\Xi)]\setminus\Phi) & * \\ 0 & B_\varepsilon \end{pmatrix},$$

where $A_\varepsilon := N([\Phi(\Sigma)\cup\Phi(\Xi)]\setminus\Phi)$ is a compact perturbation of a normal operator, $\sigma(A_\varepsilon) = ([\Phi(\Sigma)\cup\Phi(\Xi)]\setminus\Phi)^-$, $\sigma_e(A_\varepsilon) = \partial([\Phi(\Sigma)\cup\Phi(\Xi)]\setminus\Phi)$, $\sigma_p(A_\varepsilon) = \sigma_p(A_\varepsilon^*)^* = [\Phi(\Sigma)\cup\Phi(\Xi)]\setminus\Phi$ $(\subset \Sigma\cap\Xi)$, $\mathrm{nul}(A_\varepsilon - \lambda) = \mathrm{nul}(A_\varepsilon - \lambda)^* = 1$ for all $\lambda \in [\Phi(\Sigma)\cup\Phi(\Xi)]\setminus\Phi$, and B_ε has the same characteristics as the operator S. By Proposition 3.4, there exists a compact operator K_4', with $\|K_4'\| < \varepsilon/3$, such that

$$\sigma_p(B_\varepsilon - K_4') = \rho^+_{s-F}(B_\varepsilon) = \rho^+_{s-F}(S) \quad (\subset \Sigma)$$

and

$$\sigma_p([B_\varepsilon - K_4']^*)^* = \rho^-_{s-F}(B_\varepsilon) = \rho^-_{s-F}(S) \quad (\subset \Xi).$$

By [3], [21,Chapter 3] and Lemma 5.4, there exists a compact operator K_5, with

$$\|K_5\| < \frac{\varepsilon}{3} + \max[\max\{m_e(T-\lambda): \lambda \in \Sigma\setminus(\text{interior }\Sigma\cup\Phi)\}$$
$$\max\{m_e([T-\lambda]^*): \lambda \in \Xi\setminus(\text{interior }\Xi\cup\Phi)\}]$$

such that

$$\begin{pmatrix} N & * & * \\ 0 & S & * \\ 0 & 0 & N \end{pmatrix} - K_5 = \begin{pmatrix} F & * \\ 0 & W \end{pmatrix},$$

where F acts on a finite dimensional space, $\sigma(F) = \Sigma_\varepsilon \cup \Xi_\varepsilon$ $(\subset \Sigma\cap\Xi)$, $\mathrm{nul}(F-\lambda) = 1$ for all $\lambda \in \sigma(F)$, $\begin{pmatrix} F & * \\ 0 & W \end{pmatrix} \sim F \oplus W$, and W has the same characteristics as the operator S.

By Proposition 3.4, there exists a compact operator K_5', with $\|K_5'\| < \varepsilon/3$, such that

$$\sigma_p(W-K_5') = \rho_{s-F}^+(W) = \rho_{s-F}^+(S) \quad (\subset \Sigma)$$

and

$$\sigma_p([W-K_5']^*)^* = \rho_{s-F}^-(W) = \rho_{s-F}^-(S) \quad (\subset \Xi).$$

Finally, by combining Corollary 4.5 and Proposition 3.4, we can find a compact operator K_6, with $\|K_6\| < 2\varepsilon/3$, such that

$$\begin{pmatrix} N & * & * \\ 0 & S & * \\ 0 & 0 & N \end{pmatrix} - K_6 = \begin{pmatrix} N'(\Sigma) & * & * \\ 0 & S_\varepsilon & * \\ 0 & 0 & N'(\Xi) \end{pmatrix},$$

where $N'(\Sigma)$ and $N'(\Xi)$ are compact perturbations of normal operators, $\sigma(N'(\Sigma)) = \sigma_e(N'(\Sigma)) = [\Sigma \cap \sigma_{\ell re}(T)]^-$, $\sigma_p(N'(\Sigma)) = \Sigma\cap\sigma_{\ell re}(T)$, $\sigma_p(N'(\Sigma)^*) = \emptyset$, $\sigma(N'(\Xi)) = \sigma_e(N'(\Xi)) = [\Xi \cap \sigma_{\ell re}(T)]^-$, $\sigma_p(N'(\Xi)) = \emptyset$, $\sigma_p(N'(\Xi)^*)^* = \Xi \cap \sigma_{\ell re}(T)$, $\sigma(S_\varepsilon) = \sigma(S)$, $\sigma_p(S_\varepsilon) = \rho_{s-F}^+(S)$ and $\sigma_p(S_\varepsilon^*)^* = \rho_{s-F}^-(S)$.

Now we put all the pieces together. We have constructed a compact operator $K = K_o + K_o'$, where

$$K_o' \simeq K_2 + 0_{(A)} \oplus K_3 \oplus (K_4 + 0_{(A_\varepsilon)} \oplus K_4') \oplus (K_5 + 0_{(F)} \oplus K_5') \oplus K_6$$

($0_{(A)}$ denotes the 0 operator in the space of A, etc.), in such a way that $\|K\| \le \|K_o\| + \|K_o'\|$ is bounded by the estimate given in the statement of the theorem, and

$$T-K \simeq \begin{pmatrix} A & * \\ 0 & T_\varepsilon - K_3' \end{pmatrix} \oplus \begin{pmatrix} A_\varepsilon & * \\ 0 & B_\varepsilon - K_4' \end{pmatrix} \oplus \begin{pmatrix} F & * \\ 0 & W-K_5' \end{pmatrix} \oplus \begin{pmatrix} N'(\Sigma) & * & * \\ 0 & S_\varepsilon & * \\ 0 & 0 & N'(\Xi) \end{pmatrix}.$$

Since $\rho_{s-F}^+(T) \subset \sigma_p(T-K)$ and $\rho_{s-F}^-(T) \subset \sigma_p([T-K]^*)^*$, with the help of Lemma 2.1, it is not difficult to check that

$$\begin{aligned}
\sigma_p(T-K) &= \sigma_p(A) \cup \sigma_p(T_\varepsilon - K_3') \cup \sigma_p(A_\varepsilon) \cup \sigma_p(B_\varepsilon - K_4') \cup \sigma_p(F) \\
&\quad \cup \sigma_p(W-K_5') \cup \sigma_p(N'(\Sigma)) \cup \sigma_p(S_\varepsilon) \cup \sigma_p(N'(\Xi)) \\
&= ([\Sigma \cap \Phi] \cup \rho_{s-F}^+(S) \cup [\Sigma \cap \rho_{s-F}(T) \cap (\partial\rho_{s-F}(T))_{\varepsilon/10}]) \\
&\quad \cup \rho_{s-F}^+(T) \cap ([\Phi(\Sigma) \cup \Phi(\Xi)] \setminus \Phi) \cup \rho_{s-F}^+(S) \cup \sigma(F) \\
&\quad \cup \rho_{s-F}^+(S) \cup [\Sigma \cap \sigma_{\ell re}(T)] \cup \rho_{s-F}^+(S) \cup \emptyset \\
&= (\Sigma \cap \Phi) \cup [\Sigma \cap \rho_{s-F}(T) \cap (\partial\rho_{s-F}(T))_{\varepsilon/10}] \\
&\quad \cup \rho_{s-F}^+(T) \cup [\Phi(\Sigma) \setminus \Phi] \cup \Sigma_\varepsilon \cup [\Sigma \cap \sigma_{\ell re}(T)] = \Sigma
\end{aligned}$$

and

$$\begin{aligned}\sigma_p([T-K]^*)^* &= \sigma_p(A^*)^* \cup \sigma_p([T_\varepsilon - K_3']^*)^* \cup \sigma_p(A_\varepsilon^*)^* \\ &\quad \cup \sigma_p([B_\varepsilon - K_4']^*)^* \cup \sigma_p(F) \cup \sigma_p([W-K_5']^*)^* \\ &\quad \cup \sigma_p(N'(\Sigma)^*)^* \cup \sigma_p(S_\varepsilon^*)^* \cup \sigma_p(N'(\Xi)^*)^* \\ &= ([\Xi \cap \Phi] \cup \rho_{s-F}^-(S) \\ &\quad \cup [\Xi \cap \rho_{s-F}(T) \cap (\partial\rho_{s-F}(T))_{\varepsilon/10}]) \cup \rho_{s-F}^-(T) \\ &\quad \cup ([\Phi(\Sigma) \cup \Phi(\Xi)] \setminus \Phi) \cup \rho_{s-F}^-(S) \cup \sigma(F) \\ &\quad \cup \rho_{s-F}^-(S) \cup \emptyset \cup \rho_{s-F}^-(S) \cup [\Xi \cap \sigma_{\ell re}(T)] \\ &= (\Xi \cap \Phi) \cup [\Xi \cap \rho_{s-F}(T) \cap (\partial\rho_{s-F}(T))_{\varepsilon/10}] \\ &\quad \cup \rho_{s-F}^-(T) \cup [\Phi(\Xi) \setminus \Phi] \cup \Xi_\varepsilon \cup [\Xi \cap \sigma_{\ell re}(T)] = \Xi.\end{aligned}$$

This shows that conditions (i), (ii) and (iii) are sufficient for the existence of a compact operator K such that $\sigma_p(T-K) = \Sigma$ and $\sigma_p([T-K]^*)^* = \Xi$.

The results mentioned in the Introduction indicate that $\sigma_p(T-C) \neq \Sigma$ $(\sigma_p([T-C]^*)^* \neq \Xi)$ for all $C \in K(H)$ unless Σ (Ξ, resp.) is a bounded F_σ satisfying (i) and (ii). Furthermore, $\{\lambda \in \rho_{s-F}(M): \ \mathrm{ind}(M-\lambda) = 0\} \cap \sigma_p(M) = \{\lambda \in \rho_{s-F}(M): \ \mathrm{ind}(M-\lambda) = 0\} \cap \sigma_p(M^*)^*$ for all M in $L(H)$, so that (iii) is also necessary.

The proof of Theorem 6.1 is now complete. ∎

In Theorem 6.1 we did not request anything special about $\mathrm{nul}(T-K-\lambda)$ for λ in Σ (or about $\mathrm{nul}(T-K-\lambda)^*$ for $\lambda \in \Xi$). With the help of Corollary 4.7, it is possible to prove a much more general result. The details of the proof are left to the reader.

THEOREM 6.3. *Let* $T \in L(H)$ *and let* $\{\Sigma_n\}_{n=1}^\infty$ *and* $\{\Xi_n\}_{n=1}^\infty$ *be two non-increasing sequences of bounded* F_σ *subsets of* $\mathbb{C}$ *such that*

(i)$_n$ $\Sigma_n \supset \{\lambda \in \rho_{s-F}^+(T): \ \mathrm{ind}(T-\lambda) \geq n\}$ *and*
$\Xi_n \supset \{\lambda \in \rho_{s-F}^-(T): \ \mathrm{ind}(T-\lambda) \leq -n\}$;

(ii)$_n$ *if* Ω *is a component of* $\{\lambda \in \rho_{s-F}(T): \ \mathrm{ind}(T-\lambda) < n\}$ *(*$\{\lambda \in \rho_{s-F}(T): \ \mathrm{ind}(T-\lambda) > -n\}$*), then either* $\Omega \subset \Sigma_n$ *(*$\Omega \subset \Xi_n$*, resp.) or* $\Omega \cap \Sigma_n$ *(*$\Omega \cap \Xi_n$*, resp.) is a finite (possibly empty) or denumerable set that only accumulates on* $\partial\rho_{s-F}(T)$*; and*

(iii)$_n$ *Let* Ω *be a component of* $\rho_{s-F}(T)$ *such that*

$\operatorname{ind}(T-\lambda) = m$ *(for some* $m \in \mathbb{Z}$*;* $\lambda \in \Omega$*). If* $m \geq 0$*, then* $\Omega \cap \Sigma_n = \Omega \cap \Xi_{n-m}$ *for all* $n > m$*; if* $m \leq 0$*, then* $\Omega \cap \Sigma_{n+m} = \Omega \cap \Xi_n$ *for all* $n > |m|$.

There exists $K \in K(H)$ *such that*

$$\sigma_p^{(n)}(T-K) = \Sigma_n \text{ and } \sigma_p^{(n)}([T-K]^*)^* = \Xi_n,$$

for all $n = 1,2,\ldots$.

On the other hand, if either Σ_n *(*Ξ_n*) is not a bounded* F_σ*, or does not satisfy (i)*$_n$ *or (ii)*$_n$*, then* $\sigma_p^{(n)}(T-C) \neq \Sigma_n$ *(*$\sigma_p^{(n)}([T-C]^*)^* \neq \Xi_n$*) for any compact operator* C*; moreover, if* $\{\Sigma_n\}_{n=1}^{\infty}$ *and* $\{\Xi_n\}_{n=1}^{\infty}$ *do not satisfy (iii)*$_n$ *for all* $n = 1,2,\ldots,$ *then either* $\sigma_p^{(n)}(T-C) \neq \Sigma_n$*, or* $\sigma_p^{(n)}([T-C]^*)^* \neq \Xi_n$ *for all* C *in* $K(H)$ *(for at least one index* n*).*

Indeed, it is also possible to obtain an upper estimate for the norm of the operator K, which involves the sup (over $n \geq 1$) of expressions of the type indicated in Theorem 6.1, with Φ replaced by $\rho_{s-F}(T) \cap \operatorname{interior}[\sigma_p^{(n)}(T) \cap \sigma_p^{(n)}(T^*)^*]$, etc., etc.. Furthermore, K can be chosen in such a way that if $\lambda \in \sigma_o(T-K)$, then the restriction of $T-K$ to the Riesz spectral subspace associated with the clopen subset $\{\lambda\}$ has a prescribed Jordan form (with spectrum equal to $\{\lambda\}$, of course!), etc., etc.. The details are left to the interested reader.

7. SMALL COMPACT PERTURBATIONS AND POINTS OF CONTINUITY OF THE FUNCTION $\bar{\sigma}_p$

Suppose $T \in L(H)$, and let Σ and Ξ be two bounded F_σ sets satisfying conditions (i), (ii) and (iii) of Theorem 6.1 (and therefore there exists a compact operator K such that $\sigma_p(T-K) = \Sigma$ and $\sigma_p([T-K]^*)^* = \Xi$).

Assume, moreover, that

(iv) $\sigma_o(T) \subset \Sigma \cap \Xi$,

(v) $\Phi(\Sigma) \cup \Phi(\Xi) \subset \Phi$, and

(vi) the isolated points of $(\Sigma \setminus \Phi) \cap \rho_{s-F}^-(T)$ and the isolated points of $(\Xi \setminus \Phi) \cap \rho_{s-F}^+(T)$ belong to $\rho_{s-F}^s(T)$.

Then minor modifications of the proof of Tbeorem 6.1 show that in this case, we still can achieve the desired result with a compact perturbation of arbitrarily small norm. The details of the necessary modifications are left to the reader. (For instance, it is not necessary to disturb the finite dimensional invariant subspaces of T associated with points of

$$[\rho^{s}_{s-F}(T) \cap (\Sigma \cup \Xi)] \setminus (\partial\rho_{s-F}(T))_{\varepsilon},$$

etc..)

On the other hand, if Σ and Ξ fail to verify some of the conditions (i) - (vi), then it is impossible to find compact operators of arbitrarily small norm such that $\sigma_p(T - K) = \Sigma$ and $\sigma_p([T - K]^*)^* = \Xi$. This is obvious if Σ and Ξ do not satisfy (i), (ii) or (iii) (use Theorem 6.1 and Remarks 6.2).

If A, B $\in L(H)$, $\lambda \in \sigma_p(A)$ and $B - \lambda$ is bounded below, then

$$\|A - B\| \geq \inf\{\|(B - \lambda)\mathbf{x}\|: \ x \in H, \ \|x\| = 1\} > 0.$$

By applying this observation to the operators and to their adjoints, it is not difficult to conclude that (iv), (v) and (vi) are also necessary conditions for K's of arbitrarily small norm.

In [12], J. B. Conway and B. B. Morrel characterized those operators that are points of continuity of the spectral function $\bar{\sigma}_p$ ($\bar{\sigma}_p(T) = \sigma_p(T)^-$). A simplified proof of this result was given in [4,Chapter 14] (see also [24]).

By combining this characterization with our previous observations, we obtain the following.

THEOREM 7.1. *$\bar{\sigma}_p$ is continuous at $A \in L(H)$ if and only if*

$$\sigma_{\ell e}(A) \subset \sigma_p(A) = [\sigma_o(A) \cup \rho^{+}_{s-F}(A)]^-.$$

Furthermore, if $\bar{\sigma}_p$ is not continuous at $R \in L(H)$, then there exists a sequence $\{K_n\}_{n=0}^{\infty}$ of compact operators such that

$$\|K_n\| \to 0 \ (n \to \infty) \text{ and } \ \sigma_p(R - K_n) = \sigma_p(R - K_0)$$

for all $n = 1,2,...,$ *but*

$$\bar{\sigma}_p(R - K_0) \neq \bar{\sigma}_p(R).$$

PROOF. It only remains to prove the second statement.

Suppose that $\bar{\sigma}_p$ is not continuous at R; then either there exists a point $\alpha \in \sigma_{\ell e}(R) \setminus \bar{\sigma}_p(R)$, or there exists a point $\beta \in \bar{\sigma}_p(R) \setminus [\sigma_o(R) \cup \sigma^{+}_{s-F}(R)]^-$. (Observe that $\sigma_p(R)$ always include $\sigma_o(R) \cup \rho^{+}_{s-F}(R)$.)

In the first case, it follows from Theorem 6.1, Remarks

6.2, and our observations in the introduction to this section that (given $\varepsilon > 0$) there exists a compact operator K_ε, with $\|K_\varepsilon\| < \varepsilon$, such that $\sigma_p(R - K_\varepsilon) = \sigma_p(R) \cup \{\alpha\}$.

It follows that

$$d_H[\bar{\sigma}_p(R - K_\varepsilon), \bar{\sigma}_p(R)] = \delta > 0$$

for some δ independent from ε. (If $\sigma_p(R) = \emptyset$, $\delta = 1$; if $\sigma_p(R) \neq \emptyset$, $\delta = \min\{1, \operatorname{dist}[\alpha, \sigma_p(T)]\}$.)

In the second one, there exists a compact operator K_ε, with $\|K_\varepsilon\| < \varepsilon$, such that $\sigma_p(R - K_\varepsilon) = \sigma_o(R) \cup \rho^+_{s-F}(R)$, and therefore

$$d_H[\bar{\sigma}_p(R - K_\varepsilon), \bar{\sigma}_p(R)] = \delta > 0,$$

where either $\delta = 1$ (if $\sigma_o(R) \cup \rho^+_{s-F}(R) = \emptyset$), or $\delta = \min\{1, \operatorname{dist}[\beta, \sigma_o(R) \cup \rho^+_{s-F}(R)]\}$ (if $\sigma_o(R) \cup \rho^+_{s-F}(R) \neq \emptyset$). In either case, δ is independent from ε.

In the same vein, it is possible to combine the results of this article with the arguments of [12], [4,Chapter 14] in order to prove the following result. The proof is left to the interested reader.

THEOREM 7.2. *(i) The spectral function*

$$T \to [\sigma_p(T) \cup \sigma_p(T^*)^*]^-$$

is continuous at $A \in L(H)$ *if and only if*

$$\sigma_{\ell re}(A) \subset [\sigma_p(A) \cup \sigma_p(A^*)^*]^- = [\sigma_o(A) \cup \rho^+_{s-F}(A) \cup \rho^-_{s-F}(A)]^-.$$

(ii) The spectral function

$$T \to [\sigma_p(T) \cap \sigma_p(T^*)^*]^-$$

is continuous at $A \in L(H)$ *if and only if*

$$\sigma_{\ell re}(A) \subset [\sigma_p(A) \cap \sigma_p(A^*)^*]^- = \sigma_o(A)^-.$$

(iii) The spectral function

$$T \to \bar{\sigma}_p(T) \cap \bar{\sigma}_p(T^*)^*$$

is continuous at $A \in L(H)$ *if and only if*

$$\sigma_{\ell re}(A) \subset \bar{\sigma}_p(A) \cap \bar{\sigma}_p(A^*)^* = [\sigma_o(A) \cup (\partial\rho^+_{s-F}(A) \cap \partial\rho^-_{s-F}(A))]^-.$$

Furthermore, if τ *denotes one of these three functions and* τ *is not continuous at* $R \in L(H)$*, then there exists a sequence* $\{K_n\}_{n=0}^\infty$ *of compact operators such that* $\|K_n\| \to 0$ *(*$n \to \infty$*), and* $\tau(R - K_n) = \tau(R - K_0)$ *for all* $n = 1,2,\ldots,$ *but* $\tau(R - K_0) \neq \tau(R)$.

As a final remark, it is worth mentioning that the tools introduced in Sections 2 - 5 can also be used to analyze the continuity properties and the behavior under small compact perturbations of spectral functions of the type $T \to [\sigma_p^{(n)}(T)]^-$, $T \to [\sigma_p^{(n)}(T) \cup \sigma_p^{(m)}(T^*)^*]^-$, $T \to [\sigma_p^{(n)}(T) \cap \sigma_p^{(m)}(T^*)^*]^-$, $T \to [\sigma_p^{(n)}]^- \cap [\sigma_p^{(m)}(T^*)^*]^-$, $T \to [\sigma_p^{(n)}(T) \setminus \sigma_p^{(m)}(T^*)^*]^-$, etc., etc..

REFERENCES

1. R. A. Adams, Sobolev spaces, Academic Press, New York-San Francisco-London, 1975.

2. C. Apostol, Matrix models for operators, Duke Math. J. 42 (1975), 779-785.

3. C. Apostol, The correction by compact perturbations of the singular behavior of operators, Rev. Roum. Math. Pures et Appl. 21 (1976), 155-175.

4. C. Apostol, L. A. Fialkow, D. A. Herrero and D. Voiculescu, Approximation of Hilbert space operators. Volume II, Research Notes in Math., vol. 102, Pitman Advanced Publ. Program, Boston-London-Melbourne, 1984.

5. C. Apostol and B. B. Morrel, On uniform approximation of operators by simple models, Indiana Univ. Math. J. 26 (1977), 427-442.

6. C. Apostol, C. M. Pearcy and N. Salinas, Spectra of compact perturbations of operators, Indiana Univ. Math. J. 26 (1977), 345-350.

7. J. Barría and D. A. Herrero, Closure of similarity orbits of Hilbert space operators. IV: Normal operators, J. London Math. Soc. (2) 17 (1978), 525-536.

8. I. D. Berg, An extension of the Weyl-von Neumann theorem to normal operators, Trans. Amer. Math. Soc. 171 (1971), 365-371.

9. I. D. Berg and K. R. Davidson, A quantitative version of the Brown-Douglas-Fillmore theorem, (Preprint).

10. L. G. Brown, R. G. Douglas and P. A. Fillmore, Unitary equivalence modulo the compact operators and extensions of C*-algebras, Proceedings of a conference on operator theory, Halifax, Nova Scotia 1973, Lect. Notes in Math., vol. 345, Springer-Verlag, Berlin-Heidelberg-New York, 1973, pp. 58-128.

11. J. B. Conway, Subnormal operators, Research Notes in Math., vol. 51, Pitman Advanced Publ. Program, Boston-London-Melbourne, 1981.

12. J. B. Conway and B. B. Morrel, Operators that are points of spectral continuity. III, Integral Equations and Operator Theory 6 (1983), 319-344.

13. J. Dixmier, Étude sur les varietés et les opérateurs de Julia avec quelques applications, Bull. Soc. Math. France 77 (1949), 11-101.

14. J. Dixmier and C. Foiaş, Sur le spectre ponctuel d'un opérateur, Colloq. Math. Soc. János Bolyai. 5. Hilbert space operators, Tihany (Hungary), 1970, pp. 127-136.

15. P. A. Fillmore and J. P. Williams, On operator ranges, Adv. in Math. 7 (1971), 254-281.

16. C. Foiaş, Une application des distributions vectorielles à la théorie spectrale, Bull. Sci. Math. 84 (1960), 147-158.

17. T. W. Gamelin, Uniform algebras, Prentice-Hall, Englewood-Cliffs, New Jersey, 1969.

18. I. C. Gohberg and M. G. Krein, Introduction to the theory of linear nonselfadjoint operators, Nauka, Moskow, 1965 (Russian); English translation, Transl. of Mathematical Monographs, vol. 18, Amer. Math. Soc., Providence, Rhode Island, 1969.

19. J. W. Helton, Operators with a representation as multiplication by x on a Sobolev space, Colloq. Math. Soc. János Bolyai. 5. Hilbert space operators, Tihany (Hungary), 1970, pp. 279-287.

20. D. A. Herrero, On the spectra of the restrictions of an operator, Trans. Amer. Math. Soc. 233 (1977), 45-58.

21. D. A. Herrero, Approximation of Hilbert space operators. Volume I, Research Notes in Math., vol. 72, Pitman Advanced Publ. Program, Boston-London-Melbourne, 1982.

22. D. A. Herrero, Economical compact perturbations. I: Erasing normal eigenvalues, J. Operator Theory 10 (1983), 289-306.

23. D. A. Herrero, On multicyclic operators. II: Two extensions of the notion of quasitriangularity, Proc. London Math. Soc. (3) 48 (1984), 247-282.

24. D. A. Herrero, Continuity of spectral functions and the Lakes of Wada, Pac. J. Math. 113 (1984), 365-371.

25. D. A. Herrero, The diagonal entries in the formula 'quasitriangular - compact = triangular', and restrictions of quasitriangularity, Trans. Amer. Math. Soc. 298

(1986), 1-42.

26. D. A. Herrero, Economical compact perturbations. II: Filling in the holes, J. Operator Theory (To appear).

27. D. A. Herrero, Most quasitriangular operators are triangular, most biquasitriangular operators are bitriangular, J. Operator Theory (To appear).

28. G. K. Kalish, On operators on separable Banach spaces with arbitrary prescribed point spectrum, Proc. Amer. Math. Soc. 34 (1972), 207-208.

29. G. K. Kalish, On operators with large point spectrum, Scripta Math. 29 (1973), 371-378.

30. T. Kato, Perturbation theory for linear operators, Springer-Verlag, New York, 1966.

31. L. H. Loomis, An introduction to abstract harmonic analysis, D. Van Nostrand, Princeton, New Jersey-Toronto-London-Melbourne, 1953.

32. M. Omladič, Some spectral properties of an operator, Operator Theory: Advances and Applications, vol. 17, Birkhäuser-Verlag, Basel, 1986, pp. 239-247.

33. M. Rosenblum, On the operator equation $BX - XA = Q$, Duke Math. J. 23 (1956), 263-269.

34. W. Sikonia, The von Neumann converse of Weyl's theorem, Indiana Univ. Math. J. 21 (1971), 121-123.

35. J. G. Stampfli, Compact perturbations, normal eigenvalues and a problem of Salinas, J. London Math. Soc. (2) 9 (1974), 165-175.

36. D. Voiculescu, A non-commutative Weyl-von Neumann theorem, Rev. Roum. Math. Pures et Appl. 21 (1976), 97-113.

Department of Mathematics,
Arizona State University,
Tempe, Arizona 85287,
U.S.A.

Operator Theory:
Advances and Applications, Vol. 32

BIMODULES OF NEST SUBALGEBRAS OF VON NEUMANN ALGEBRAS

David R. Larson*   Baruch Solel*

Dedicated to the memory of Constantin Apostol

The σ-weakly closed bimodules of nest subalgebras of σ-finite factor von Neumann algebras are characterized and structurally analysed. This generalizes work accomplished earlier by Erdos and Power for the case in which the factor is $\mathcal{B}(\mathcal{H})$. In the general case many more such bimodules exist than are given by a straightforward extension of the $\mathcal{B}(\mathcal{H})$ theory. New techniques are developed for this, including use of a partial coordinate system for bimodules, and a structural analysis of a certain boundary subspace affiliated with a lattice homomorphism of a nest.

INTRODUCTION

In this article we investigate structural properties of the σ-weakly closed linear subspaces of σ-finite factor von Neumann algebras which are closed under left and right multiplication by elements of a given nest subalgebra of the factor, so are bimodules for this nest subalgebra. In the special case where the factor is $\mathcal{B}(\mathcal{H})$, the algebra of all bounded linear transformations from a Hilbert space into itself, a complete characterization of these bimodules was given by Erdos and Power [2].

Many of the ideas and results in the Erdos-Power theory extend to the generalized setting considered in this paper. However, they do not always extend intact because in the factor context many more such bimodules exist in general than are given by the theory developed for $\mathcal{B}(\mathcal{H})$. We develop new techniques to deal with this situation, including the introduction of a partial coordinate system for subspaces of the factor affiliated with the nest, and the analysis of a certain "boundary" subspace of the factor affiliated with an order preserving mapping of a nest of projections into itself.

* This research was partially supported by grants from the National Science Foundation.

In section 1 we develop the essential aspects of our theory for the case of a continuous nest. In section 2, a complete characterization of the set of bimodules described above is given for the continuous case. In section 3, the assumption of continuity is dropped and the characterization obtained in §2 is extended to arbitrary nests in factors of types II and III.

1. Coordinates, And The Boundary Subspace.

We shall assume throughout this section that $\mathbf{M}$ is a σ-finite factor von Neumann algebra, and that $\mathcal{N}$ is a continuous nest of projections in $\mathbf{M}$ equipped with the order topology (which is equivalent to the strong operator topology, restricted to $\mathcal{N}$.) The topology on the Cartesian product $\mathcal{N} \times \mathcal{N}$ will be the product topology. It is known that $\mathcal{N}$, and therefore also $\mathcal{N} \times \mathcal{N}$, is a compact space.

If $N_1 < N_2$ and $N_i \in \mathcal{N} \setminus \{0, I\}, i = 1, 2$, then the open order interval with endpoints N_1 and N_2 is

$$(N_1, N_2) = \{N \in \mathcal{N} : N_1 < N < N_2\},$$

and we write $E(N_1, N_2) = N_2 - N_1$. Similarly, the open intervals containing 0 or I are

$$[0, N_2) = \{N \in \mathcal{N} : N < N_2\}$$

$$\text{and} \qquad (N_1, I] = \{N \in \mathcal{N} : N > N_1\},$$

and we write $E[0, N_2) = N_2$ and $E(N_1, I] = I - N_1$.

If $\mathcal{S} \subseteq \mathbf{M}$ is a set of operators we define its support, denoted **supp** $\mathcal{S}$, to be the set of all points $(N_1, N_2) \in \mathcal{N} \times \mathcal{N}$ satisfying the condition that $E(J_1)SE(J_2) \neq 0$ whenever J_i, $i = 1, 2$, is an open interval in $\mathcal{N}$ containing N_i.

LEMMA 1.1.

(1) *supp* $\mathcal{S}$ *is closed in* $\mathcal{N} \times \mathcal{N}$.

(2) *supp* $\mathbf{M} = \mathcal{N} \times \mathcal{N}$.

(3) *If* $T \in \mathbf{M}$ *and* $T \neq 0$ *then* *supp*$\{T\} \neq \phi$.

PROOF. Part (1) follows from the definition. Part (2) follows from the fact that, in a factor $\mathbf{M}$, $E\mathbf{M}F = 0$ for projections E and F in $\mathbf{M}$ only if either $E = 0$ or $F = 0$.

For part (3), suppose $supp\{T\} = \phi$. Then for every point $(N_1, N_2) \in \mathcal{N} \times \mathcal{N}$ there are open intervals J_1, J_2 containing N_1, N_2, respectively, such that $E(J_1)TE(J_2) = 0$. The open rectangles $J_1 \times J_2$ cover $\mathcal{N} \times \mathcal{N}$. By compactness, $\mathcal{N} \times \mathcal{N}$ is covered by finitely many rectangles, say

$$\mathcal{N} \times \mathcal{N} \subseteq \cup_{i=1}^{m} J_i \times K_i \quad \text{with}$$
$$E(J_i)TE(K_i) = 0 \quad , \quad 1 \leq i \leq m.$$

Let $\{F_1, F_2, ..., F_\ell\}$ be the set of the endpoints of the intervals $\{J_i\}$ and $\{K_i\}$, ordered so that $F_j < F_{j+1}$, $1 \leq j \leq \ell - 1$. Then it is easy to check that for every j and k with $1 \leq j, k \leq \ell - 1$, there is some $i \leq m$ such that $F_{j+1} - F_j \leq E(J_i)$ and $F_{k+1} - F_k \leq E(K_i)$. Also

$$\mathcal{N} \times \mathcal{N} \subseteq \cup_{\substack{1 \leq j < \ell \\ 1 \leq k < \ell}} (F_j, F_{j+1}) \times (F_k, F_{k+1}).$$

Hence $T = \Sigma_{j=1}^{\ell-1} \Sigma_{k=1}^{\ell-1} (F_{j+1} - F_j) T (F_{k+1} - F_k) = 0$. ■

We shall write $supp\ T$ for $supp\{T\}$ whenever T is an element of M. Also, if $N_1, N_2 \in \mathcal{N}$ with $N_1 < N_2$ we write $[N_1, N_2] = \{N \in \mathcal{N} : N_1 \leq N \leq N_2\}$.

LEMMA 1.2.

(1) *If $P_1 < P_2$, $Q_1 < Q_2$, $P_i, Q_i \in \mathcal{N}$ and $\mathcal{S} \subseteq$ M then*

$$supp((P_2 - P_1)\mathcal{S}(Q_2 - Q_1)) \subseteq supp\ \mathcal{S} \cap ([P_1, P_2] \times [Q_1, Q_2]).$$

(2) *If $T \in$ M and $supp\ T \cap ([P_1, P_2] \times [Q_1, Q_2]) = \phi$ then $(P_2 - P_1)T(Q_2 - Q_1) = 0$.*

PROOF. (1) Clearly $supp((P_2 - P_1)\mathcal{S}(Q_2 - Q_1)) \subseteq supp\ \mathcal{S}$. Also, if (N_1, N_2) lies in $supp((P_2 - P_1)\mathcal{S}(Q_2 - Q_1))$, then for every pair of open intervals J, K containing N_1, N_2, respectively,

$$E(J)(P_2 - P_1)\mathcal{S}(Q_2 - Q_1)E(K) \neq 0$$

and, in particular, for every pair of such intervals, $E(J)(P_2 - P_1) \neq 0$ and $(Q_2 - Q_1)E(K) \neq 0$. Since this holds for every pair of open intervals, J, K containing N_1 and N_2, respectively, we have $N_1 \in [P_1, P_2]$ and $N_2 \in [Q_1, Q_2]$.

(2) Let $T_1 = (P_2 - P_1)T(Q_2 - Q_1)$. Then *supp* $T_1 = \phi$ (using part (1)) and, thus, $T_1 = 0$ by Lemma 1.1, part 3. ■

LEMMA 1.3. *If T is a nonzero element of M and if P, Q are projections in M, then the σ-weak closure of $P\mathrm{M}T\mathrm{M}Q$ is $P\mathrm{M}Q$.*

PROOF. Since $\mathrm{M}T\mathrm{M}$ is a two-sided ideal in M and M is a factor, its σ-weak closure is equal to M. hence

$$P\mathrm{M}T\mathrm{M}Q \subseteq P\mathrm{M}Q = P(\overline{\mathrm{M}T\mathrm{M}})Q \subseteq \overline{P\mathrm{M}T\mathrm{M}Q},$$

where the closure is in the σ-weak topology. Since $P\mathrm{M}Q$ is σ-weakly closed we are done. ■

We now write $\mathcal{A}$ for the nest subalgebra associated with $\mathcal{A} \subseteq \mathrm{M}$. That is,

$$\mathcal{A} = \{T \in \mathrm{M} : (I - N)TN = 0 \text{ for all } N \in \mathcal{N}\}.$$

A linear subspace $\mathcal{U} \subseteq \mathrm{M}$ will be said to be an **$\mathcal{A}$-bimodule** if

$$\mathcal{U}\mathcal{A} \subseteq \mathcal{U} \quad \text{and} \quad \mathcal{A}\mathcal{U} \subseteq \mathcal{U}.$$

LEMMA 1.4. *Let $\mathcal{U} \subseteq \mathrm{M}$ be a σ-weakly closed $\mathcal{A}$-bimodule and let $(N_1, N_2) \in supp\mathcal{U}$. Then $N_1\mathrm{M}(I - N_2) \subseteq \mathcal{U}$.*

PROOF. Assume that $N_i \in \mathcal{N} \setminus \{0, I\}$. For every $N_i' < N_i < N_i''$, $i = 1, 2$ we have

$$(N_1'' - N_1')\mathcal{U}(N_2'' - N_2') \neq 0, \quad \text{and also}$$

$$N_1'\mathrm{M}(N_1'' - N_1') \subseteq \mathcal{A}, \; (N_2'' - N_2')\mathrm{M}(I - N_2'') \subseteq \mathcal{A}.$$

Hence

$$N_1'\mathrm{M}(N_1'' - N_1')\mathcal{U}(N_2'' - N_2')\mathrm{M}(I - N_2'') \subseteq \mathcal{U}.$$

From Lemma 1.3 we now obtain

$$N_1'\mathrm{M}(I - N_2'') \subseteq \mathcal{U}.$$

This holds for any $N_1' < N_1$, $N_2'' > N_2$. Hence, since $\mathcal{U}$ is σ-weakly closed and $\mathcal{N}$ is continuous, $N_1 \mathrm{M}(I - N_2) \subseteq \mathcal{U}$. The proof for the case where either N_1 or N_2 are in $\{0, I\}$ is similar. ∎

Now let $\mathcal{U}$ be a σ-weakly closed $\mathcal{A}$-bimodule in M and define a map $\theta : \mathcal{N} \to \mathcal{N}$ by

$$\theta(N) = \vee\{P : (P, N) \in \mathit{supp}\ \mathcal{U}\}$$

if the set $\{P : (P, N) \in \mathit{supp}\ \mathcal{U}\}$ is nonempty, and $\theta(N) = 0$ otherwise.

It is clear that for each $N \in \mathcal{N}$, $(I - \theta(N))\mathcal{U}N = 0$. In particular, $\theta(I) = 0$ if and only if $\mathcal{U} = \{0\}$.

LEMMA 1.5.

(1) *If* $\{P : (P, N) \in \mathit{supp}\ \mathcal{U}\}$ *is nonempty, then* $P \leq \theta(N)$ *if and only if* $(P, N) \in \mathit{supp}\ \mathcal{U}$, *and in particular,* $(\theta(N), N) \in \mathit{supp}\ \mathcal{U}$.

(2) $N_1 \leq N_2 \Rightarrow \theta(N_1) \leq \theta(N_2)$ *for* $N_i \in \mathcal{N}, i = 1, 2$.

(3) $\theta(N)\mathrm{M}(I - N) \subseteq \mathcal{U}$ *for all* $N \in \mathcal{N}$

(4) *If* $\{N_a\}$ *is a subset of* $\mathcal{N}$, *then* $\wedge_a \theta(N_a) = \theta(\wedge_a N_a)$. *(This says that* θ *is* **right-continuous.**)

(5) *If* $\{N_a\}$ *is a subset of* $\mathcal{N}$ *with* $\vee_a N_a = I$, *then* $\vee_a \theta(N_a) = \theta(I)$. *(This says that* θ *is left-continuous at* I.)

PROOF. (1) Suppose $(P, N) \in \mathit{supp}\ \mathcal{U}$. Then, by the definition of $\theta(N)$, we have $P \leq \theta(N)$. Note also that $(\theta(N), N) \in \mathit{supp}\ \mathcal{U}$ since $\mathit{supp}\ \mathcal{U}$ is a closed subset of $\mathcal{N} \times \mathcal{N}$. Suppose $P < \theta(N)$ and that J_1 is an open interval containing P and J_2 is an open interval containing N. If $\theta(N) \in J_1$ then $E(J_1)\mathcal{U}E(J_2) \neq 0$ and we are done. If $\theta(N) \notin J_1$ then there is an open interval J in $\mathcal{N}$ containing $\theta(N)$ whose left endpoint is larger than the right endpoint of J_1 and, therefore, $E(J_1)\mathrm{M}E(J) \subseteq \mathcal{A}$. Since $E(J)\mathcal{U}E(J_2) \neq 0$ and M is a factor,

$$E(J_1)\mathcal{U}E(J_2) \supseteq E(J_1)\mathrm{M}E(J)\mathcal{U}E(J_2) \neq 0.$$

Hence $(P, N) \in \mathit{supp}\ \mathcal{U}$.

(2) We may assume $\theta(N_1) \neq 0$. Then the hypothesis of part (1) is satisfied, so $(\theta(N_1), N_1) \in \mathit{supp}\,\mathcal{U}$. We will show that $(\theta(N_1), N_2) \in \mathit{supp}\,\mathcal{U}$, and hence $\theta(N_1) \leq \theta(N_2)$.

Let J be any open interval containing $\theta(N_1)$ and let J_2 be any open interval containing N_2. We may assume J_2 is sufficiently small so that $N_1 \notin J_2$. Now let J_1 be an open interval containing N_1 which is disjoint from N_2. Then $E(J_1)\mathrm{M}E(J_2) \subseteq \mathcal{A}$. Also, $E(J)\mathcal{U}E(J_1) \neq 0$ since $(\theta(N_1), N_1) \notin \mathit{supp}\,\mathcal{U}$. Thus $E(J)\mathcal{U}E(J_1)\mathrm{M}E(J_2) \neq 0$ since M is a factor. But this set is contained in $E(J)\mathcal{U}E(J_2)$. Thus $(\theta(N_1), N_2) \in \mathit{supp}\,\mathcal{U}$, as desired.

(3) If $\theta(N) = 0$ the result is trivial, and for $\theta(N) \neq 0$ this follows from Lemma 1.4 since $(\theta(N), N) \in \mathit{supp}\,\mathcal{U}$.

(4) The inequality $\theta(\wedge_a N_a) \leq \wedge_a \theta(N_a)$ follows from part (2). For the other direction, if some $\theta(N_a) = 0$ the result is trivial, and if not, note that $\wedge_a \theta(N_a) \leq \theta(N_{a_0})$ for every a_0, and therefore $(\wedge_a \theta(N_a), N_{a_0}) \in \mathit{supp}\,\mathcal{U}$ for every a_0. Since $\mathit{supp}\,\mathcal{U}$ is closed and $\wedge_a N_a$ is a limit point of $\{N_a\}$, $(\wedge_a \theta(N_a), \wedge_a N_a) \in \mathit{supp}\,\mathcal{U}$. Hence by part (1), $\wedge_a \theta(N_a) \leq \theta(\wedge_a N_a)$.

(5) For each $a, \theta(N_a) \leq \theta(I)$ from part (2). So $\vee_a \theta(N_a) \leq \theta(I)$. Let $P = \vee_a \theta(N_a)$. For each a we have $(I - \theta(N_a))\mathcal{U}N_a = 0$, and hence $(I - P)\mathcal{U}N_a = 0$. So since $\vee_a N_a = I$, we have $(I - P)\mathcal{U} = 0$. So if $Q \in \mathcal{N}$ and $Q > P$, then $(Q, I) \notin \mathit{supp}\,\mathcal{U}$. This shows that $\theta(I) \leq P$. Thus $P = \theta(I)$. ■

Given a nonzero map $\theta : \mathcal{N} \to \mathcal{N}$ satisfying:

(i) $N_1 \leq N_2 \Rightarrow \theta(N_1) \leq \theta(N_2)$, and

(ii) $\wedge_a \theta(N_a) = \theta(\wedge_a N_a)$ for every subset $\{N_a\} \subseteq \mathcal{N}$, and

(iii) $\vee_a \theta(N_a) = \theta(I)$ for every subset $\{N_a) \subseteq \mathcal{N}$ with $\vee_a N_a = I$ we define

$$\mathcal{J}(\theta) = \{T \in \mathrm{M} : (I - \theta(N))TN = 0 \text{ for every } N \in \mathcal{N}\}.$$

REMARK. In the special case $\mathrm{M} = \mathcal{B}(\mathcal{H})$ (or more generally a type I factor) the theory of Erdos and Power [2] states that for an arbitrary σ-weakly closed $\mathcal{A}$-bimodule $\mathcal{U}$ we have $\mathcal{U} = \mathcal{J}(\theta)$, where $\theta : \mathcal{N} \to \mathcal{N}$ is the **left** continuous map assigned to $\mathcal{U}$ defined by $\theta(N) = \mathrm{proj}[\mathcal{U}NH], N \in \mathcal{N}$, and $\mathcal{J}(\theta) = \{T : (I - \theta(N))TN = 0, N \in \mathcal{N}\}$. This is different from the **right** continuous induced map θ we define via a consideration of *supp* $\mathcal{U}$. In the theory we develop, the right-continuous map is the more natural one. For $\mathrm{M} = \mathcal{B}(\mathcal{H})$) (and so for M of type I), the theories are easily verified to be equivalent in that always $\mathcal{U} = \mathcal{J}(\theta)$ for the right continuous map as well. For M not of type I **not** every σ-weakly closed bimodule has the form $\mathcal{J}(\theta)$ for any order homomorphism $\theta : \mathcal{N} \to \mathcal{N}$. Thus our theory is necessarily more technically involved than [2]. In an arbitrary factor situation, $\mathcal{U}$ may be properly contained in $\mathcal{J}(\theta)$, so a fine structural analysis is needed

LEMMA 1.6.

(1) $\mathcal{J}(\theta)$ *is a* σ*-weakly closed* $\mathcal{A}$*-bimodule.*

(2) *If* $\tilde{\theta}$ *is the map assigned to* $\mathcal{J}(\theta)$ *as in Lemma 1.5, then* $\tilde{\theta} = \theta$.

(3) *supp* $\mathcal{J}(\theta) = \{(P, N) \in \mathcal{N} \times \mathcal{N} : N \geq N_0$ *and* $P \leq \theta(N)\}$,
where $N_0 = \wedge\{N \in \mathcal{N} : \theta(N) > 0\}$.

(4) $\mathcal{J}(\theta) = \{T \in \mathrm{M} :$ *supp* $T \subseteq$ *supp* $\mathcal{J}(\theta)\}$.

PROOF. If $R, S \in \mathcal{A}$ then $SN = NSN$ and $(I - \theta(N))R = (I - \theta(N))R(I - \theta(N))$. Hence if $T \in \mathcal{J}(\theta)$ then

$$(I - \theta(N))RTSN = (I - \theta(N))R(I - \theta(N))TNSN = 0.$$

Hence $\mathcal{J}(\theta)$ is an $\mathcal{A}$-bimodule. It is clearly σ-weakly closed.

Now write $B = \{(P, N) \in \mathcal{N} \times \mathcal{N} : P \leq \theta(N))\}$. For $M \in \mathcal{N}$ and $Q_1 > \theta(M)$, $Q_2 < M$ we have

$$B \cap ([Q_1, I] \times [0, Q_2]) = \phi$$

(because if not then there is some (P, N) such that $P \leq \theta(N)$, $P \geq Q_1 > \theta(M)$, $N \leq Q_2 < M$ and therefore $P > \theta(M) \geq \theta(N)$). Using Lemma 1.2, for every T with *supp* $T \leq B$, $(I - Q_1)TQ_2 = 0$. By continuity of $\mathcal{N}$, $(I - \theta(M))TM = 0$ for every $M \in \mathcal{N}$. Hence if *supp* $T \subseteq B$ then $T \in \mathcal{J}(\theta)$.

For every $T \in \mathrm{M}$ and $N \in \mathcal{N}$, by Lemma 1.2 (1) we have *supp*$(\theta(N)TN) \subseteq [0, \theta(N)] \times [0, N]$ and *supp*$(T(I - N)) \subseteq \mathcal{N} \times [N, I]$. Hence, if $T \in \mathcal{J}(\theta)$ then $T = T(I - N) + \theta(N)TN$ and *supp* $T \subseteq \cap_N([0, \theta(N)] \times [0, N]) \cup (\mathcal{N} \times [N, I])$. Thus, if

$(P,N_1) \in$ *supp* T then for every $N > N_1$, $P \le \theta(N)$. By right continuity $P \le \theta(N_1)$, hence $(P,N_1) \in B$. We conclude that

$$\mathcal{J}(\theta) = \{T \in \mathrm{M} : \textit{supp } T \subseteq B\}.$$

Now let $N_0 = \wedge\{N \in \mathcal{N} : \theta(N) > 0\}$, and let $B_1 = \{(P,N) \in \mathcal{N} \times \mathcal{N} : N \ge N_0 \text{ and } P \le \theta(N)\}$. To complete the proof of (2), (3) and (4) it will suffice to show that *supp* $\mathcal{J}(\theta) = B_1$. We first show that *supp* $\mathcal{J}(\theta) \supseteq B_1$.

Let $N \in \mathcal{N}$ with $\theta(N) > 0$, and let $P \in \mathcal{N}$ with $0 \le P \le \theta(N)$, so $(P,N) \in B_1$. First assume $P \ne 0$ and that $N \ne I$. Choose $P' < P < P''$, and $N' < N < N''$. Then $(P''-P')\mathcal{J}(\theta)(N''-N') \supseteq (P''-P')\theta(N)\mathrm{M}(I-N)(N''-N') \supseteq (P''-P)\mathrm{M}(N''-N) \ne 0$. Hence $(P,N) \in$ *supp* $\mathcal{J}(\theta)$. For the case where $N = I$ and $P \ne 0$, use left continuity of θ at I and the fact that *supp* $\mathcal{J}(\theta)$ is closed to conclude that $(P,I) \in$ *supp* $\mathcal{J}(\theta)$. For the case $P = 0$, now use closure of *supp* $\mathcal{J}(\theta)$ to obtain $(0,N) \in$ *supp* $\mathcal{J}(\theta)$. Finally, for the case $N = N_0$ (note that $\theta(N_0)$ **may** be 0) simply use right continuity of θ at N_0 and closure of *supp* $\mathcal{J}(\theta)$ to obtain $(P,N_0) \in$ *supp* $\mathcal{J}(\theta)$, $0 \le P \le \theta(N_0)$.

We have shown above that $B_1 \subseteq$ *supp* $\mathcal{J}(\theta) \subseteq B$. If $(P,N) \in B$ and $(P,N) \notin B_1$, then $P \le \theta(N)$ and $N < N_0$. So $\theta(N) = 0$. This implies that $\mathcal{J}(\theta)N = 0$. Moreover, since N is strictly less that N_0 there exists an open interval J containing N with right endpoint N' contained in N_0. We also have $\theta(N') = 0$, so $\mathcal{J}(\theta)N' = 0$, and so $\mathcal{J}(\theta)E(J) = 0$. It follows that (P,N) fails to be in *supp* $\mathcal{J}(\theta)$. We conclude that $B_1 =$ *supp* $\mathcal{J}(\theta)$. The proof of the lemma is complete. ■

PROPOSITION 1.7. *Let $\mathcal{U}$ be a σ-weakly closed $\mathcal{A}$-bimodule and let θ be the associated map. Then $\mathcal{U} \subseteq \mathcal{J}(\theta)$ and supp $\mathcal{U}$ = supp $\mathcal{J}(\theta)$.*

PROOF. Lemma 1.6 (3) implies that *supp* $\mathcal{U}$ = *supp* $\mathcal{J}(\theta)$. Since $\mathcal{U} \subseteq \{T \in \mathrm{M} : \textit{supp } T \subseteq \textit{supp } \mathcal{U}\} = \{T \in \mathrm{M} : \textit{supp } T \subseteq \textit{supp } \mathcal{J}(\theta)\}$, by Lemma 1.6 (4) we have $\mathcal{U} \subseteq \mathcal{J}(\theta)$. ■

Given a map $\theta : \mathcal{N} \to \mathcal{N}$ as above we can also associate another subspace, $\mathcal{K}(\theta)$, defined to be the σ-weakly closed subspace generated by $\cup\{\theta(N)\mathrm{M}(1-N) : N \in \mathcal{N}\}$.

REMARK. Let θ_0 be the identity map on $\mathcal{N}$, and let $\mathcal{K}_0 = \mathcal{K}(\theta_0)$. It follows from the characterization of the Jacobson radical of a nest subalgebra of a von Neumann algebra in [3], that since M is a factor, rad($\mathcal{A}$) is the **norm** - closed linear subspace spanned by $\vee_{N \epsilon \mathcal{N}}(N\mathrm{M}(I-N)$. In the special case $\mathrm{M} = \mathcal{B}(\mathcal{H})$ it is known that the σ-weak closure of the radical of an arbitrary nest algebra is the set of all operators in the nest algebra for which $PAP = 0$ for every minimal interval projection for the nest. So if the nest is continuous (as is assumed in this section) it follows that $\mathcal{K}_0 = \mathcal{J}(\theta_0) = \mathcal{A}$. On the other hand, if M is a type II_1 factor, there is a σ-weakly continuous expectation from M onto $\mathcal{D}$ whose restriction to the nest subalgebra A is a homomorphism [1]. This annihilates $N\mathrm{M}(I-N)$ for every $N \epsilon \mathcal{N}$, so annihilates $\mathcal{K}_0$. Thus $\mathcal{K}_0$ has trivial intersection with $\mathcal{D}$, and so the inclusion $\mathcal{K}_0 \subseteq \mathcal{J}(\theta_0)$ is proper. For continuous nests in type II_∞ factors, many possibilities can occur for $\mathcal{K}_0 \cap \mathcal{D}$ between the extremes $\{0\}$ and $\mathcal{D}$. For a "bimodule" map $\theta : \mathcal{N} \to \mathcal{N}, \mathcal{K}(\theta)$ is an $\mathcal{A}$-bimodule which plays an essential role in our theory. If $\mathcal{U}$ is a bimodule then $\mathcal{U}$ will lie between $\mathcal{K}(\theta)$ and $\mathcal{J}(\theta)$, and many possibilities can occur.

LEMMA 1.8.

(1) $\mathcal{K}(\theta)$ *is a* σ*-weakly closed* $\mathcal{A}$*-bimodule.*

(2) *supp* $\mathcal{K}(\theta) =$ *supp*$\mathcal{J}(\theta)$.

PROOF. (1) For R, S in $\mathcal{A}$ we have $S\theta(N) = \theta(N)S\theta(N)$ and $(I-N)R = (I-N)R(I-N)$ for every $N \epsilon \mathcal{N}$. Hence

$$S\theta(N)\mathrm{M}(I-N)R = \theta(N)S\theta(N)\mathrm{M}(I-N)R(I-N) \subseteq \theta(N)\mathrm{M}(I-N).$$

Therefore $S\mathcal{K}(\theta)R \subseteq \mathcal{K}(\theta)$.

(2) It is easy to check that $\mathcal{K}(\theta) \subseteq \mathcal{J}(\theta)$ and, therefore, *supp* $\mathcal{K}(\theta) \subseteq$ *supp* $\mathcal{J}(\theta)$. The argument given in the proof of Lemma 1.6 to show that $\{(P,N) : N \geq N_0,\ P \leq \theta(N)\} \subseteq$ *supp* $\mathcal{J}(\theta)$ in fact shows that this set is contained in *supp* $\mathcal{K}(\theta)$, and we are done. ■

THEOREM 1.9. *Let* $\mathcal{U}$ *be a* σ*-weakly closed* $\mathcal{A}$*-bimodule and let* θ *be the associated map. Then*

$$\mathcal{K}(\theta) \subseteq \mathcal{U} \subseteq \mathcal{J}(\theta)$$

and supp $\mathcal{K}(\theta) =$ *supp* $\mathcal{U} =$ *supp* $\mathcal{J}(\theta)$.

PROOF. It remains only to show that $\mathcal{U}$ contains $\mathcal{K}(\theta)$, but this follows from Lemma 1.5 (3). ∎

For a map $\theta : \mathcal{N} \to \mathcal{N}$ as above we write

$$\partial\theta = \{T \in \mathrm{M} : \text{ for every } N \in \mathcal{N}, (I - \theta(N))TN = \theta(N)T(I - N) = 0\}.$$

Clearly $\partial\theta$ is a $\mathcal{D}$-bimodule, where $\mathcal{D} = \mathrm{M} \cap \mathcal{N}'$. We will call $\partial\theta$ the **boundary subspace** associated with θ.

LEMMA 1.10. *Let* $\theta : \mathcal{N} \to \mathcal{N}$ *be a map as above. Write* $B(\theta) = \{(P,Q) \in \mathcal{N} \times \mathcal{N} : \vee_{N<Q}\theta(N) \leq P \leq \theta(Q)\}$. *(In particular, if* θ *is also left continuous, then* $B(\theta) = \{(\theta(N), N) : N \in \mathcal{N}\}$*). Then* $\partial\theta = \{T \in \mathrm{M} :$ *supp* $T \in B(\theta)\}$.

PROOF. If *supp* $T \subseteq B(\theta)$ then, by Lemma 1.6, $T \in \mathcal{J}(\theta)$; hence $(I - \theta(N))TN = 0, N \in \mathcal{N}$.

For $P \in \mathcal{N}$ and $N < \theta(P)$, $F > P$ (F and N in $\mathcal{N}$) we have $B(\theta) \cap ([0, N] \times [F, I]) = \phi$, since if $(P_1, Q_1) \in B(\theta)$ and $P_1 \leq N$, $Q_1 \geq F$ then $\vee_{Q \leq Q_1}\theta(Q) \leq P_1 \leq N < \theta(P)$ and $Q_1 \geq F > P$, so that $\theta(P) < \theta(P)$, which is a contradiction. Now Lemma 1.2 implies that if *supp* $T \subseteq B(\theta)$, then

$$NT(I - F) = 0$$

whenever $N < \theta(P)$ and $F > P$. Since $\mathcal{N}$ is continuous,

$$\theta(P)T(I - P) = 0.$$

This shows that if *supp* $T \subseteq B(\theta)$ then $T \in \partial\theta$.

For the other direction, note that, if $T \in \partial\theta$ then $T = \theta(N)TN + (I - \theta(N))T(I - N)$. Using Lemma 1.2 (1), we have

$$\textit{supp}\ T \subseteq \cap_N ([0, \theta(N)] \times [0, N]) \cup ([\theta(N), I] \times [N, I]).$$

Hence if $(P,Q) \in supp\ T$ then for every $N < Q$ we have $\theta(N) \leq P$, and for every $N > Q$ we have $\theta(N) \geq P$. Hence

$$\vee_{N<Q}\theta(N) \leq P \leq \wedge_{N>Q}\theta(N) = \theta(Q).$$

That is, $(P,Q) \in B(\theta)$. ∎

LEMMA 1.11. *Let $\mathcal{D} = \mathrm{M} \cap \mathcal{N}' = \mathcal{A} \cap \mathcal{A}^*$. Then*

(1) $\partial\theta\mathcal{D}(\partial\theta)^* \subseteq \mathcal{D}$ *and* $(\partial\theta)^*\mathcal{D}\partial\theta \subseteq \mathcal{D}$.

(2) *If $T \in \partial\theta$ and $T = V|T|$ is its polar decomposition (or $T = |T^*|V$) then $V \in \partial\theta$ and $|T|, |T^*| \in \mathcal{D}$.*

PROOF. (1) Fix $T \in \partial\theta$, $S \in \partial\theta$ and $A \in \mathcal{D}$. Then $T = NT\theta(N) + (I-N)T(I-\theta(N))$; hence $NT = \theta(N)$. Similarly, $S^*N = \theta(N)S^*$. Thus $NTAS^* = T\theta(N)AS^* = TA\theta(N)S^* = TAS^*N$; i.e. $TAS^* \in \mathrm{M} \cap \mathcal{N}' = \mathcal{D}$.

(2) The fact that $|T|, |T^*| \in \mathcal{D}$ follows from (1). For $T \in \partial\theta$ we have $(I-N)V\theta(N)|T| = (I-N)V|T|\theta(N) = (I-N)T\theta(N) = 0$. Since V^*V is the range projection of $|T|$ we have $(I-N)V\theta(N) = (I-N)V\theta(N)V^*V = 0$. Similarly $NV(I-\theta(N)) = 0$. Hence $V \in \partial\theta$. ∎

Now apply Zorn's Lemma to obtain a maximal family of partial isometries $\{v_\gamma\}$, in $\partial\theta$, such that $\{v_\gamma v_\gamma^*\}$ and $\{v_\gamma^* v_\gamma\}$ are orthogonal families of projections. Let $\nu_1 = \Sigma v_\gamma$.

LEMMA 1.12. *Suppose V is a partial isometry in $\partial\theta$ and p is a projection in $\mathcal{D}$ such that $VV^*p \neq 0$ (resp. $V^*Vp \neq 0$). Then there is a non zero partial isometry $V_1 \in \partial\theta$ such that $V_1V_1^* \leq p$ (resp. $V_1^*V_1 \leq p$) and $V_1^*V_1 \leq V^*V$ (resp. $V_1V_1^* \leq VV^*$).*

PROOF. Assume $VV^*p \neq 0$ (the other case is similar). By the comparison theorem there is a projection $z \in Z(\mathcal{D})$ and partial isometries $v_1, v_2 \in \mathcal{D}$ such that $v_1^*v_1 = zp, v_1v_1^* \leq zVV^*$, $v_2^*v_2 = (I-z)VV^*$, $v_2v_2^* \leq (1-z)p$. If $v_1 \neq 0$, let V_1 be v_1^*zV. Then $V_1V_1^* = v_1^*v_1 \leq p$ and $V_1^*V_1 = V^*zv_1v_1^*zV \leq V^*V$. If $v_2 \neq 0$ let $V_1 = v_2(I-z)V$. Then $V_1V_1^* = v_2v_2^* \leq p$, $V_1^*V_1 = V^*(I-z)v_2^*v_2(I-z)V \leq V^*V$. ∎.

Now write

$$f = \sup\{VV^* : V \text{ is a partial isometry in } \partial\theta\}.$$

If $U \in \mathcal{D}$ is a unitary operator then,

$$UfU^* = \sup\{UVU^*UV^*U^* : V \in \partial\theta\} \leq f,$$

as UVU^* is a partial isometry in $\partial\theta$. Hence $f \in Z(\mathcal{D})$.

LEMMA 1.13. *Let $\{\omega_\alpha\}$ be a maximal family of partial isometries in $\partial\theta$ satisfying:*

(1) $\{\omega_\alpha\omega_\alpha^*\}$ is an orthogonal family of projections (in $\mathcal{D}$).

(2) $\forall\alpha \quad \omega_\alpha\omega_\alpha^* \leq f - \nu_1\nu_1^*$

(3) $\forall\alpha \quad \omega_\alpha^*\omega_\alpha \leq \nu_1^*\nu_1$.

Then $\Sigma\omega_\alpha\omega_\alpha^* = f - \nu_1\nu_1^*$.

PROOF. Assume $f_1 = f - \nu_1\nu_1^* - \Sigma\omega_\alpha\omega_\alpha^* \neq 0$. Then there is a non zero partial isometry $\nu \in \partial\theta$ such that $\nu\nu^* f_1 \neq 0$ (by the definition of f). Using Lemma 1.12, there is a partial isometry $V_1 \in \partial\theta$ s.t. $V_1V_1^* \leq f_1$ and $V_1^*V_1 \leq \nu^*\nu$. If $V_1^*V_1 \leq 1 - \nu_1^*\nu_1$ we obtain a contradiction to the maximality of ν_1. Thus $V_1^*V_1\nu_1^*\nu_1 \neq 0$ and, using Lemma 1.12 again, we find a non zero partial isometry $\omega \in \partial\theta$ such that $\omega^*\omega \leq \nu_1^*\nu_1$ and $\omega\omega^* \leq V_1V_1^* \leq f_1$. This contradicts the maximality of $\{\omega_\alpha\}$ and, hence, $f = \Sigma\omega_\alpha\omega_\alpha^*$ ■

Let $\{\omega_\alpha\}$ be the set defined above and enumerate it as $\{\nu_i\}_{i=2}^{\infty}$ (M is σ-finite). Then $\Sigma_{i=1}^{\infty}\nu_i\nu_i^* = f$ and $\nu_i^*\nu_i \leq \nu_1^*\nu_1$ for all i.

LEMMA 1.14.

(1) $\partial\theta = \Sigma_{k=1}^{\infty}\nu_k\mathcal{D} = \mathcal{D}\nu_1\mathcal{D}$.

(2) If $\mathcal{D}$ is abelian, then $\partial\theta = \nu_1\mathcal{D}$.

PROOF. (1) Clearly $\Sigma_k\nu_k\mathcal{D} \subseteq \partial\theta$. For $x \in \partial\theta$, $x = \nu|x|$ and ν is a partial isometry in $\partial\theta$. Hence $x = fx$. Also $x = |x^*|\nu$. Hence $x = |x^*|\nu = f|x^*|\nu = \Sigma\nu_k\nu_k^*|x^*|\nu = \Sigma\nu_k(\nu_k^*|x^*|\nu) \in \mathcal{D}$. Now note that, since $\nu_k^*\nu_k \leq \nu_1^*\nu_1$ for $k > 1$, we have $\nu_k = \nu_k\nu_1^*\nu_1 \in \mathcal{D}\nu_1$. Thus $\nu_k\mathcal{D} \subseteq \mathcal{D}\nu_1\mathcal{D}$ and consequently, $\partial\theta = \mathcal{D}\nu_1\mathcal{D}$.

(2) If $\mathcal{D}$ is abelian, for $k > 1$ we have

$$\nu_k^*\nu_k \leq \nu_1^*\nu_1 \qquad \text{and} \qquad \nu_k\nu_k^* \leq I - \nu_1\nu_1^*.$$

Let $z_k = \nu_1\nu_k^*$. Then $z_k z_k^* = \nu_1\nu_k^*\nu_k\nu_1^* \leq \nu_1\nu_1^*$ and $z_k^* z_k = \nu_k\nu_1^*\nu_1\nu_k^* \leq \nu_k\nu_k^* \leq I - \nu_1\nu_1^*$. Since $z_k \in \mathcal{D}$ and $\mathcal{D}$ is abelian, this implies that $\nu_1\nu_k^* = 0$ and therefore $\nu_k = \nu_k\nu_1^*(\nu_1\nu_k^*)\nu_k = 0$ and $\partial\theta = \nu_1\mathcal{D}$. ■

LEMMA 1.15. *If $\mathcal{E} \subseteq \partial\theta$ is a sub-$\mathcal{D}$-bimodule, then there is a projection $P \in Z(\mathcal{D})$ such that $\mathcal{E} = P\partial\theta$.*

PROOF. Consider $\mathcal{E}\mathcal{E}^* \subseteq \mathcal{D}$ where $\mathcal{E}\mathcal{E}^*$ is the σ-weakly closed subspace spanned by $\{ab^* : a, b \in \mathcal{E}\}$. Then $\mathcal{E}\mathcal{E}^*$ is an ideal in $\mathcal{D}$ and thus $\mathcal{E}\mathcal{E}^* = P\mathcal{D}$ for some projection $P \in Z(\mathcal{D})$. For $T \in \mathcal{E}$, $(I - P)T \in \mathcal{E}$ and $(I - P)TT^*(I - P) \in \mathcal{E}\mathcal{E}^* = P\mathcal{D}$. Hence $(I - P)T = 0$. That is, $\mathcal{E} \subseteq P\partial\theta$. Now note that $\mathcal{E}\mathcal{E}^*\partial\theta = \mathcal{E}(\mathcal{E}^*\partial\theta) \subseteq \mathcal{E}$ (as $\mathcal{E}^*\partial\theta \subseteq \mathcal{D}$). Hence $P\mathcal{D}\partial\theta \subseteq \mathcal{E}$, and consequently, $P\partial\theta \subseteq \mathcal{E}$. Therefore $\mathcal{E} = P\partial\theta$. ■

2. The Bimodules $\mathcal{K}(\theta)$ and $\mathcal{J}(\theta)$.

In the present section we study the "difference" between the bimodules $\mathcal{J}(\theta)$ and $\mathcal{K}(\theta)$ defined in section 2 and, in particular, the question of when we have $\mathcal{K}(\theta) = \mathcal{J}(\theta)$. Recall that we always have $\mathcal{K}(\theta) \subseteq \mathcal{J}(\theta)$. We maintain the assumption that nests are continuous in this section.

For $\theta_0 = id$ (the identity map) we have $\mathcal{J}(\theta_0) = \mathcal{A}$. We shall write $\mathcal{K}_0$ for $\mathcal{K}(\theta_0)$; i.e. $\mathcal{K}_0$ is the σ-weakly closed subspace spanned by $\vee_{N \in \mathcal{N}} N\mathbf{M}(I - N)$. In this case $\partial\theta_0$ is just $\mathbf{M} \cap \mathcal{N}' = \mathcal{A} \cap \mathcal{A}^*$, which was denoted by $\mathcal{D}$.

The nest $\mathcal{N}$, in $\mathbf{M}$, gives rise to a strongly continuous one parameter unitary group $\{U_t : t \in \mathbf{R}\} \subseteq \mathbf{M}$, and the action on $\mathbf{M}$, defined by $\alpha_t(T) = U_t T U_t^*, T \in \mathbf{M}$, has the property that $\mathbf{M}^\alpha[0, \infty] = \mathcal{A}$. (See [7] for details.)

If the factor $\mathbf{M}$ has a separating vector in $\mathcal{H}$ (where we assume $\mathbf{M}$ is represented on $\mathcal{H}$) then every σ-weakly closed linear subspace $\mathcal{S} \subseteq \mathbf{M}$ is **reflexive** in $\mathcal{H}$ in the sense that whenever $T \in \mathcal{B}(\mathcal{H})$ satisfies the condition that for each $x \in \mathcal{H}, Tx$ is in the closure of $\mathbf{M}x$, where $[\cdot]$ denotes closure, then necessarily $T \in \mathcal{S}$. This is in [8], and will be useful.

LEMMA 2.1. *Suppose* M *has a separating vector in* $\mathcal{H}$ *and* $\mathcal{K}_0 \cap \mathcal{D} = \{0\}$. *Then for every* $0 \leq D \in \mathcal{D}$ *there is a vector* $y \in \mathcal{H}$ *satisfying:*

(i) $\langle \alpha_t(T)y, y\rangle = \langle Ty, y\rangle$, $t \in \mathbf{R}$, $T \in \mathrm{M}$; *and*

(ii) $\langle Dy, y\rangle \neq 0$.

PROOF. Since $D \notin \mathcal{K}_0$ and $\mathcal{K}_0$ is reflexive in $\mathcal{H}$, there is some $0 \neq x \in \mathcal{H}$ such that $D^{3/2}x \notin [\mathcal{K}_0 x]$. Let Q be the projection onto $[\mathcal{K}_0 x]$ and let $y = (I - Q)Dx$. As $(I - Q)D^{3/2}x \neq 0$, $D^{1/2}y \neq 0$ and (ii) follows (note that D commutes with Q since $\mathcal{K}_0$ is a $\mathcal{D}$-bimodule). We have $[\mathcal{K}_0 Q D x] \subseteq [\mathcal{K}_0 Q(\mathcal{H})] = [\mathcal{K}_0[\mathcal{K}_0 x]] \subseteq [\mathcal{K}_0 x] = Q(\mathcal{H})$. Also, since $\mathcal{K}_0$ is a $\mathcal{D}$-bimodule, $\mathcal{K}_0 Dx \subseteq [\mathcal{K}_0 x]$. Hence $[\mathcal{K}_0 y] \subseteq [\mathcal{K}_0 x]$ and, therefore, $\langle Ky, y\rangle = 0$ for all $K \in \mathcal{K}_0$. For $K \in \mathcal{K}_0 + \mathcal{K}_0^*$ we have $\alpha_t(K) \in \mathcal{K}_0 + \mathcal{K}_0^*$ and, thus,

$$\langle Ky, y\rangle = 0 = \langle \alpha_t(K)y, y\rangle.$$

For $A \in \mathcal{D}$ we have $\alpha_t(A) = A$ and $\langle \alpha_t(A)y, y)\rangle = \langle Ay, y\rangle$. Since the σ-weak closure of $\mathcal{D} + \mathcal{K}_0 + \mathcal{K}_0^*$ is M ([4, Proposition 2.1]), (i) follows. ■

With α as defined above there is a unique projection Q_0 (in $Z(\mathcal{D})$, the center of $\mathcal{D}$) satisfying:

(i) There is a normal faithful expectation ϕ from $Q_0 M Q_0$ onto $Q_0\mathcal{D}$ such that $\phi \circ \alpha_t = \phi$, $t \in \mathbf{R}$, on $Q_0 \mathrm{M} Q_0$; and

(ii) The zero expectation is the only expectation from $(I - Q_0)\mathrm{M}(I - Q_0)$ into $(I - Q_0)\mathcal{D}$ that is α-invariant (see [7, Remark 3.4]).

Note that $\mathcal{K}_0 \cap \mathcal{D}$ is a two-sided ideal in the von Neumann alegbra $\mathcal{D}$. Since it is σ-weakly closed, there is some projection $P_0 \in Z(\mathcal{D})$ satisfying

$$\mathcal{K}_0 \cap \mathcal{D} = P_0\mathcal{D}.$$

THEOREM 2.2. *With the above notation,*

$$P_0 = I - Q_0.$$

PROOF. Write $F = (I - P_0)(I - Q_0)$, $\mathrm{M}_F = F\mathrm{M}F$, $\mathcal{D}_F = \mathcal{D}F$, $\mathcal{N}_F = \{NF : N \in \mathcal{N}\}$ and let $\mathcal{K}_F$ be the closure in the σ-weak topology of

$F\mathcal{K}_0F = \vee_{N\epsilon\mathcal{N}} FN\mathrm{M}(I-N)F = \vee_{N'\epsilon\mathcal{N}_F} N'\mathrm{M}_F(I-N')$. We have $\mathcal{K}_F \cap \mathcal{D}_F = \{0\}$ as $F \leq I - P_0$. Using Lemma 2.1 and [5, Proposition 1] we see that there is a normal faithful expectation ψ_0 from $F\mathrm{M}F$ onto $\mathcal{D}F$ with $\psi_0 \circ \alpha_t = \psi_0$. Define $\psi : (I-Q_0)\mathrm{M}(I-Q_0) \to (I-Q_0)\mathcal{D}$ by $\psi(T) = \psi_0(FTF)$. The properties of Q_0 now imply that $\psi = 0$; hence $F = 0$ and $P_0 \geq I - Q_0$.

Now write G for P_0Q_0 and replace M by $G\mathrm{M}G$, $\mathcal{N}$ by $\mathcal{N}_G = \{NG : N \,\epsilon\, \mathcal{N}\}$, $\mathcal{K}_0$ by $G\mathcal{K}_0G$ and $\mathcal{D}$ by $\mathcal{D}G$ to get:

(1) There is a normal faithful expectation $\phi : \mathrm{M} \to \mathcal{D}$ satisfying $\phi \circ \alpha_t = \phi$, $t \,\epsilon\, \mathbb{R}$; and

(2) $\mathcal{D} \subseteq \mathcal{K}_0$.

Then, for $T \,\epsilon\, N\mathrm{M}(I-N)$, $N \,\epsilon\, \mathcal{N}$, we have $\phi(T) = N\phi(T)(I-N) = \phi(T)N(I-N) = 0$; hence $\phi|\mathcal{K}_0 = 0$. Since $\mathcal{D} \subseteq \mathcal{K}_0$, $\mathcal{D} = \phi(\mathcal{D}) = \{0\}$. This implies that $G = 0$ and $P_0 \leq I - Q_0$. ■

LEMMA 2.3. *Suppose* $T \,\epsilon\, \mathrm{M}$ and P_1, P_2, Q_1, Q_2 *are in* $\mathcal{N}$ *with* $P_1 < P_2$, $Q_1 < Q_2$. *Then*

$$supp(T - (P_2 - P_1)T(Q_2 - Q_1)) \subseteq supp\ T \setminus ((P_1, P_2) \times (Q_1, Q_2)).$$

PROOF. We have $T - (P_2 - P_1)T(Q_2 - Q_1) = P_1T + (P_2 - P_1)TQ_1 + (P_2 - P_1)T(I - Q_2) + (I - P_2)T$. Now use Lemma 1.2 (1) to complete the proof. (Note that $supp(T_1 + T_2) \subseteq supp\ T_1 \cup supp\ T_2$.)

PROPOSITION 2.4. *If* $T \,\epsilon\, \mathcal{J}(\theta)$ *and* $supp\ T \cap B(\theta) = \phi$ *(where* $B(\theta) = \{(P,Q) \,\epsilon\, \mathcal{N} \times \mathcal{N} : \vee_{N<Q}\theta(N) \leq P \leq \theta(Q)\}$*) then* $T \,\epsilon\, \mathcal{K}(\theta)$.

PROOF. Since $T \,\epsilon\, \mathcal{J}(\theta)$ and $supp\ T \cap B(\theta) = \phi$ we have,

$$supp\ T \subseteq \{(P,Q) : P \leq \theta(Q)\} \setminus B(\theta) \subseteq \{(P,Q) : P < \theta(N) \text{ for some } N < Q\}$$
$$= \cup_{N\epsilon\mathcal{N}} [0, \theta(N)) \times (N, I].$$

Since $supp\ T$ is compact, there is a finite set $\{N_i\}_{i=1}^m \subseteq \mathcal{N}$ such that

$$supp\ T \subseteq \cup_{i=1}^m [0, \theta(N_i)) \times (N_i, I] \text{ and } N_1 < N_2 < N_3 < \dots < N_m.$$

Write $S = \theta(N_1)T(I-N_1) + (\theta(N_2)-\theta(N_1))T(I-N_2) + \ldots$ $+(\theta(N_m)-\theta(N_{m-1}))T(I-N_m)$. By repeatedly using Lemma 2.3 we find that $supp(T-S) = \phi$; hence $T = S \in K(\theta)$. ■

PROPOSITION 2.5.

(1) *For every* θ *as above,* $\mathcal{J}(\theta)P_0 = \mathcal{K}(\theta)P_0$.

(2) $\mathcal{A}P_0 = \mathcal{K}_0 P_0$ *and* $P_0\mathcal{A} = P_0\mathcal{K}_0$.

(3) $\mathcal{K}_0 + \mathcal{D}Q_0 = \mathcal{A}$ *and* $\mathcal{K}_0 \cap \mathcal{D}Q_0 = \{0\}$.

PROOF. Since P_0 was defined by the equation $P_0\mathcal{D} = \mathcal{K}_0 \cap \mathcal{D}$ we have $P_0 \in \mathcal{K}_0$. If $S \in \mathrm{N}\mathrm{M}(I-N)$ for some $N \in \mathcal{N}$ then, for every $T \in \mathcal{J}(\theta), TS = TNS(I-N) = \theta(N)TNS(I-N) \in \mathcal{K}(\theta)$.

Since P_0 is in the σ-weak closure of $\vee_{N \in \mathcal{N}} \mathrm{NM}(I-N)$ this shows that $TP_0 \in \mathcal{K}(\theta)$ for every $T \in \mathcal{J}(\theta)$. This proves (1), and when we take $\theta = id$ we obtain $\mathcal{A}P_0 = \mathcal{K}_0P_0$. The proof of the equation $P_0\mathcal{A} = P_0\mathcal{K}_0$ is similar, as $ST = NS(I-N)T = NS(I-N)T(I-N) \in \mathcal{K}_0$ for $T \in \mathcal{A}$, $S \in \mathrm{NM}(I-N)$. This completes the proof of (2).

For part (3) let T be in $\mathcal{A}$. Then

$$\begin{aligned} T &= Q_0TQ_0 + P_0TQ_0 + TP_0 \\ &= \phi(Q_0TQ_0) + (Q_0TQ_0 - \phi(Q_0TQ_0)) + P_0TQ_0 + TP_0, \end{aligned}$$

where ϕ is the normal faithful conditional expectation from $Q_0\mathrm{M}Q_0$ onto $Q_0\mathcal{D}$. By (2) we have $P_0TQ_0 + TP_0 \in \mathcal{K}_0$. Also $\phi(Q_0TQ_0) \in Q_0\mathcal{D}$. It remains to show that $S = Q_0TQ_0 - \phi(Q_0TQ_0)$ lies in $\mathcal{K}_0$. For this note that S lies in the σ-weak closure of $Q_0\mathcal{K}_0Q_0 + Q_0\mathcal{D}$ ([4, Proposition 2.1 (i)]); hence $S = \sigma\text{-}w - \lim_\alpha(S_\alpha + D_\alpha)$, $S_\alpha \in Q_0\mathcal{K}_0Q_0$, $D_\alpha \in \mathcal{D}$. Applying ϕ, we have $0 = \phi(S) = \sigma\text{-}w - \lim_\alpha(\phi(S_\alpha) + D_\alpha)$. But $\phi|\mathcal{K}_0 = 0$ (see the proof of Theorem 2.2). Hence $\sigma\text{-}w - \lim_\alpha D_\alpha = 0$ and $S = \sigma\text{-}w - \lim_\alpha S_\alpha \in \mathcal{K}_0$. The fact that $\mathcal{K}_0 \cap \mathcal{D}Q_0 = \{0\}$ follows from the fact that $\phi|\mathcal{K}_0 = 0$. ■

LEMMA 2.6. *Let* θ *be as above. Then*

$$\mathcal{J}(\theta) = \mathcal{K}(\theta) + \partial\theta.$$

PROOF. The proof is a modification of the proof of Proposition 2.1 in [4], where this result was proven for $\theta = id$.

We can find a sequence $\{\mathcal{N}_k\}$ of finite subnests of $\mathcal{N}$ with $\mathcal{N}_k = \{N_{ik}\}_{i=0}^k$, $N_{0k} = 0, N_{kk} = I, N_{ik} < N_{(i+1)k}$ such that $\mathcal{N}_k \subseteq \mathcal{N}_{k+1}, k \geq 1$, and $\cup_k \mathcal{N}_k$ is dense in $\mathcal{N}$. We write $E_{ik} = N_{ik} - N_{(i-1)k}$, $1 \leq i \leq k$, and $F_{ik} = \theta(N_{ik}) - \theta(N_{(i-1)k})$. Then $\Sigma_{i=1}^k E_{ik} = \Sigma_{i=1}^k F_{ik} = I$ and, for $T \in \mathcal{J}(\theta)$, $T = \left(\Sigma_{j=1}^k F_{jk}\right) T \left(\Sigma_{i=1}^k E_{ik}\right) = \Sigma_{i=1}^k F_{ik} T E_{ik} + \Sigma_{i=1}^k \Sigma_{j<i} F_{jk} T E_{ik}$. (For $j > i$, $F_{jk} T E_{ik} = 0$ as $T \in \mathcal{J}(\theta)$.) Now write $A_k = \Sigma_{i=1}^k F_{ik} T E_{ik}$ and $R_k T - A_k = \Sigma_{i=1}^k \Sigma_{j<i} F_{jk} T E_{ik} \in \mathcal{K}(\theta)$. As $\{A_n\}$ is a bounded sequence, there is a subsequence A_{n_j} that converges σ-weakly to an operator $A_\infty \in \mathrm{M}$. For every i, k, $A_k N_{ik} = \theta(N_{ik}) A_k$ and, since $\mathcal{N}_k \subseteq \mathcal{N}_m$ for $m \geq k$ we have for $m \geq k$,

$$A_m N_{ik} = \theta(N_{ik}) A_m.$$

Consequently $A_\infty N_{ik} = \theta(N_{ik}) A_\infty$ and, thus, $A_\infty \in \partial\theta$. Since $R_{n_j} = T - A_{n_j} \to T - A_\infty$ and $R_{n_j} \in \mathcal{K}(\theta)$ for all j, we are done. ■

THEOREM 2.7. *Given $\theta : \mathcal{N} \to \mathcal{N}$ as above (i.e. monotone and right continuous), we have*

$$\mathcal{J}(\theta) = \mathcal{K}(\theta) + (\partial\theta)Q_0.$$

Moreover, the sum is direct; that is, $\mathcal{K}(\theta) \cap (\partial\theta)Q_0 = \{0\}$.

PROOF. Using Lemma 2.6 and Proposition 2.5 (1) we see that $\mathcal{J}(\theta) = \mathcal{K}(\theta) + (\partial\theta)Q_0$. It remains to show that $\mathcal{K}(\theta) \cap (\partial\theta)Q_0 = \{0\}$. For this, fix $T \in \mathcal{K}(\theta) \cap (\partial\theta)Q_0$. Let ν be a partial isometry in $\partial\theta$. Then, for $N \in \mathcal{N}$, $\nu^*\theta(N)\mathrm{M}(I - N) \subseteq N\mathrm{M}(I - N)$ and, thus, $\nu^*(\mathcal{K}(\theta) \subseteq \mathcal{K}_0$. Also, by Lemma 1.11 (1), $\nu^*\partial\theta \subseteq \mathcal{D}$. Hence for such ν, $\nu^*T \in \mathcal{K}_0 \cap \mathcal{D}Q_0 = \{0\}$ (Proposition 2.5 (3)). Write $f = \sup\{\nu\nu^* : \nu \text{ is a partial isometry in } \partial\theta\}$. Then $(I - f)\partial\theta = 0$ (Lemma 1.14 (1)) and, by the above, $fT = 0$. Hence $T = fT + (I - f)T = 0$. ■

COROLLARY 2.8. For every $\mathcal{A}$-bimodule $\mathcal{U} \subseteq \mathrm{M}$ there exists a map $\theta : \mathcal{N} \to \mathcal{N}$ (monotone and right continuous) and a projection $e \in Z(\mathcal{D})$, $e \leq Q_0$, such that $\mathcal{U}$ is the direct sum $\mathcal{K}(\theta) + (\partial\theta)e$. ■

A subspace $\mathcal{U} \subseteq \mathrm{M}$ is said to be **M-reflexive** [6] if it contains every $T \in \mathrm{M}$ that satisfies the condition that $QTP = 0$ whenever Q and P are projections in M with

$\mathcal{QUP} = 0$. Nest subalgebras of M are M-reflexive, in particular.

PROPOSITION 2.9. *If* $\mathcal{K}_0$ *is* M*-reflexive then* $\mathcal{K}_0 = \mathcal{A}$ *and, consequently,* $\mathcal{K}(\theta) = \mathcal{J}(\theta)$ *for every monotone and right-continuous map* $\theta : \mathcal{N} \to \mathcal{N}$.

PROOF. If $\mathcal{K}_0 \neq \mathcal{A}$ and $\mathcal{K}_0$ is M-reflexive then there are projections P and Q in M such that $Q\mathcal{A}P \neq \{0\}$ and $Q\mathcal{K}_0P = \{0\}$. Since both $\mathcal{K}_0$ and $\mathcal{A}$ are $\mathcal{D}$-bimodules, it is not hard to see that the projections P and Q can be chosen in $\mathrm{M} \cap \mathcal{D}' = Z(\mathcal{D})$.

Since $N\mathrm{M}(I - N) \subseteq \mathcal{K}_0$ for all $N \in \mathcal{N}$ we have

$$NQ\mathrm{M}P(I - N) = \{0\} \quad , \quad N \in \mathcal{N}.$$

Since $(I - N)Q\mathcal{A}PN \subseteq (I - N)\mathcal{A}N = \{0\}$, $N \in \mathcal{N}$, we have $Q\mathcal{A}P \subseteq \mathcal{D}$. So $Q\mathcal{A}P = Q\mathcal{D}P = QP\mathcal{D}$. But $Q\mathcal{A}P \neq \{0\}$; hence $F = QP \neq 0$. Suppose that for every $N \in \mathcal{N}$ either $NF = 0$ or $(I - N)F = 0$. Then $N_0 = \sup\{N \in \mathcal{N} : NF = 0\}$ satisfies $N_0F = 0$ and $(I - N_0)F = 0$, contradicting the fact that $F \neq 0$. Hence, for some $N \in \mathcal{N}, NF \neq 0$ and $(I - N)F \neq 0$. Since M is a factor,

$$NQ\mathrm{M}P(I - N) \supseteq NF\mathrm{M}F(I - N) \neq \{0\},$$

contradicting the fact that $Q\mathcal{K}_0P = \{0\}$. ∎

3. Nests That Are Not Necessarily Continuous.

So far in this article we have assumed that the nest $\mathcal{N}$ is continuous. In this section we adapt our results to arbitrary nests in type II and III factors.

The theory in [2] gives a complete characterization of the σ-weakly closed bimodules of an arbitrary nest algebra in $\mathcal{B}(\mathcal{H})$, so covers the type I case, including nests with finite dimensional gaps. To extend our work to this case would be duplication (although some technical devices are different in the approaches). Thus we restrict attention.

Let M be a factor of type II or III, let $\mathcal{N}$ be a nest in M, and let $\mathcal{A}$ be the associated nest subalgebra. Applying Zorn's lemma, there is a maximal nest $\tilde{\mathcal{N}} \subseteq \mathrm{M}$ containing $\mathcal{N}$. Since M is not of type I, $\tilde{\mathcal{N}}$ is continuous. Write $\tilde{\mathcal{A}}$ for the associated nest subalgebra of M. So

$$\tilde{\mathcal{A}} = \{T \in \mathrm{M} : (I-P)TP = 0,\ P \in \tilde{\mathcal{N}}\}.$$

Then $\tilde{\mathcal{A}} \subseteq \mathcal{A}$. Let $\mathcal{U}$ be an $\mathcal{A}$-bimodule in M. Then $\mathcal{U}$ is also an $\tilde{\mathcal{A}}$-bimodule. Let $\theta : \tilde{\mathcal{N}} \to \tilde{\mathcal{N}}$ be the monotone, right-continuous (and left-continuous at I) map associated with $\mathcal{U}$ as in Lemma 1.5.

LEMMA 3.1. $\tilde{\theta}(\tilde{\mathcal{N}}) \subseteq \mathcal{N}$.

PROOF. Fix $N \in \tilde{\mathcal{N}}$. We may assume $\tilde{\theta}(N) \neq 0$. Let $Q = \vee\{E \in \mathcal{N} : E \leq \tilde{\theta}(N)\}$ and $P = \wedge\{E \in \mathcal{N} : E \geq \tilde{\theta}(N)\}$. Then $Q, P \in \mathcal{N}$. If $\tilde{\theta}(N) \notin \mathcal{N}$, then $P - Q$ is a gap projection for $\mathcal{N}$. (i.e. there is no $E \in \mathcal{N}$ with $Q < E < P$.) Since $\mathcal{A} = \{T \in \mathrm{M} : (I-E)TE = 0,\ E \in \mathcal{N}\}$, we then have $(P-Q)\mathcal{A}(P-Q) = (P-Q)\mathrm{M}(P-Q)$.

First consider the case $N \neq I$. Fix $N' < N < N''$, $N', N'' \in \tilde{\mathcal{N}}$. Then

$$(P-Q)\mathcal{U}(N''-N') \neq \{0\}$$

as $(\tilde{\theta}(N), N) \in \mathit{supp}\ \mathcal{U}$ (Lemma 1.5 (1)). Hence for every projection $0 \neq F < P-Q$, $F\mathcal{A}(P-Q)\mathcal{U}(N''-N') = F\mathrm{M}(P-Q)\mathcal{U}(N''-N') \neq \{0\}$. As $\mathcal{U}$ is an $\mathcal{A}$-bimodule, $F\mathcal{U}(N''-N') \neq \{0\}$. Since this holds for every $0 \neq F < P-Q$ and N', N'' as above, it follows that $(P, N) \in \mathit{supp}\ \mathcal{U}$. But this contradicts the fact that $P > \tilde{\theta}(N)$. Hence $\tilde{\theta}(N) \in \mathcal{N}$.

For the case $N = I$, the above argument adapts intact with $N'' = N = I$, since an arbitrary basic order neighborhood of I has the form $(N', I]$. ∎

LEMMA 3.2. *Suppose P and Q are in $\mathcal{N}$, $P < Q$ and there is no $E \in \mathcal{N}$ satisfying $P < E < Q$. Then for every $N \in \tilde{\mathcal{N}}$ that satisfies $P < N < Q$ we have $\tilde{\theta}(N) = \tilde{\theta}(P)$.*

PROOF. Suppose $\tilde{\theta}(N) > \tilde{\theta}(P)$. So, in particular, $\tilde{\theta}(N) > 0$, and we know that $(\tilde{\theta}(N), N) \in \mathit{supp}\ \mathcal{U}$. Hence for every $E' < \tilde{\theta}(N) < E''$ we have

$$(E''-E')\mathcal{U}(Q-P) \neq \{0\}.$$

Since (P, Q) is a gap in $\mathcal{N}$, $(Q-P)\mathrm{M}(Q-P) \subseteq \mathcal{A}$ and, using the fact that $\mathcal{U}$ is an $\mathcal{A}$-bimodule we have,

$$(E'' - E')\mathcal{U}F \supseteq (E'' - E')\mathcal{U}(Q - P)\mathcal{A}(Q - P)F = (E'' - E')\mathcal{U}(Q - P)\mathrm{M}F \neq \{0\}$$

for every projection $0 \neq F \leq Q - P$. Hence $(\tilde{\theta}(N), P)$ lies in *supp* $\mathcal{U}$ and we have $\tilde{\theta}(N) \leq \tilde{\theta}(P)$, a contradiction. ■

Define, for $N \in \tilde{\mathcal{N}}$, $N \neq 0$, $N_- = \vee\{P \in \mathcal{N} : P < N\}$.

LEMMA 3.3. *For an $\mathcal{A}$-bimodule $\mathcal{U}$ and $\tilde{\theta}$ as above we have the following:*

(1)
$$\begin{aligned}\mathcal{J}(\tilde{\theta}) &= \{T \in \mathrm{M} : (I - \tilde{\theta}(N_-))TN = 0,\ N \in \tilde{\mathcal{N}}\}\\ &= \{T \in \mathrm{M} : (I - \tilde{\theta}(N_-))TN = 0,\ N \in \mathcal{N}\}.\end{aligned}$$

In particular $\mathcal{J}(\tilde{\theta})$ is an $\mathcal{A}$-bimodule.

(2)
$$\partial\tilde{\theta} = \{T \in \mathrm{M} : (I - \tilde{\theta}(N_-))TN = \tilde{\theta}(N)T(I - N) = 0,\ N \in \mathcal{N}\}.$$

In particular $\partial\tilde{\theta}$ is a $\mathcal{D}$-bimodule (where $\mathcal{D} = \mathrm{M} \cap \mathcal{N}'$).

(3) $\mathcal{K}(\tilde{\theta})$ is the σ-weakly closed subspace spanned by $\cup\{\tilde{\theta}(N)\mathrm{M}(I - N) : N \in \mathcal{N}\}$. In particular, $\mathcal{K}(\tilde{\theta})$ is an $\mathcal{A}$-bimodule.

(4) For $N \in \mathcal{N}, (\partial\theta)(N - N_-) = (\tilde{\theta}(N) - \tilde{\theta}(N_-))\partial\theta = \{0\}$.

PROOF. (1) Recall that $\mathcal{J}(\tilde{\theta}) = \{T \in \mathrm{M} : (I - \tilde{\theta}(N))TN = 0,\ N \in \tilde{\mathcal{N}}\}$. Write

$$\begin{aligned}\mathcal{J}_1 &= \{T \in \mathrm{M} : (I - \tilde{\theta}(N_-))TN = 0,\ N \in \tilde{N}\} \quad \text{and}\\ \mathcal{J}_2 &= \{T \in \mathrm{M} : (1 - \tilde{\theta}(N_-))TN = 0,\ N \in \mathcal{N}\}.\end{aligned}$$

Clearly $\mathcal{J}_1 \subseteq \mathcal{J}(\tilde{\theta})$ (as $\tilde{\theta}(N_-) \leq \tilde{\theta}(N)$) and $\mathcal{J}_1 \subseteq \mathcal{J}_2$. Suppose T lies in $\mathcal{J}(\tilde{\theta})$. If $N = N_-$ then $(I - \tilde{\theta}(N_-))TN = 0$. If $N \in \mathcal{N}$ then, by Lemma 2.11, $\tilde{\theta}(N) = \tilde{\theta}(N_-)$ and, again, $(I - \tilde{\theta}(N_-))TN = 0$. Let N be in $\mathcal{N}$ and $N \neq N_-$. Then $N = \vee\{P \in \tilde{N} : N_- < P < N\}$ (as $\tilde{\mathcal{N}}$ is continuous), and for every $N_- < P < N$ we have $\tilde{\theta}(P) = \tilde{\theta}(N_-)$ and, thus, $(I - \tilde{\theta}(N_-))TP = 0$. Therefore $(I - \tilde{\theta}(N_-))TN = 0$. This proves $\mathcal{J}(\tilde{\theta}) = \mathcal{J}_1$. Now fix $T \in \mathcal{J}_2$ and $N \in \tilde{\mathcal{N}}$. Write $N_+ = \wedge\{P \in \mathcal{N} : P > N\}$. Then $N_+ \in \mathcal{N}$ and $(I - \tilde{\theta}((N_+)_-))TN_+ = 0$. But $(N_+)_- = N_-$. Hence $T \in \mathcal{J}(\tilde{\theta})$ and $J_2 = \mathcal{J}(\tilde{\theta})$.

For part (2), suppose $T \in \mathrm{M}$ and $\tilde{\theta}(N)T(I-N) = 0$ for every $N \in \mathcal{N}$. Then for $N \notin \mathcal{N}, \tilde{\theta}(N) = \tilde{\theta}(N_-)$ (Lemma 2.11) and $I - N \leq I - N_-$. Since $N_- \in \mathcal{N}$, $\tilde{\theta}(N_-)T(I-N_-) = 0$; hence $\tilde{\theta}(N)T(I-N) = 0$ for every $N \in \tilde{\mathcal{N}}$. Combined with part (1), this completes the proof of (2). For part (3), it is to be shown that for every $N \notin \mathcal{N}$, $\tilde{\theta}(N)\mathrm{M}(I-N) \subseteq \cup\{\tilde{\theta}(P)\mathrm{M}(I-P) : P \in \mathcal{N}\}$. But this follows from the fact that $I - N \leq I - N_-$, $\tilde{\theta}(N) = \tilde{\theta}(N_-)$ and $N_- \in \mathcal{N}$. Part (4) follows from (2). ■

We are now in a position to show that Corollary 2.8 holds for every complete nest in a factor M. For such a nest $\mathcal{N}$ and a monotone, right continuous map $\theta : \mathcal{N} \to \mathcal{N}$ we define $\mathcal{K}(\theta)$ to be the σ-weakly closed subspace spanned by

$$\cup\{\theta(N)\mathrm{M}(I-N) : N \in \mathcal{N}\} \text{ and}$$

$$\partial\theta = \{T \in \mathrm{M} : (I - \theta(N_-))TN = \theta(N)T(I-N) = 0, \ N \in \mathcal{N}\}.$$

(For a continuous nest these coincide with our previous definitions.)

COROLLARY 3.4. *For every $\mathcal{A}$-bimodule $\mathcal{U} \subseteq \mathrm{M}$ there is a monotone and right continuous map $\theta : \mathcal{N} \to \mathcal{N}$ and a projection $e \in Z(\mathcal{D})$ such that $\mathcal{U}$ is the direct sum $\mathcal{K}(\theta) + (\partial\theta)e$.*

PROOF. Let $\tilde{\mathcal{N}}$ and $\tilde{\theta}$ be as above and define θ to be the restriction of $\tilde{\theta}$ to $\mathcal{N}$. (Lemma 3.1 shows that $\tilde{\theta}(\mathcal{N}) \subseteq \mathcal{N}$). As Lemma 3.3 shows $\mathcal{K}(\theta) = \mathcal{K}(\tilde{\theta})$ and $\partial\theta = \partial\tilde{\theta}$, Corollary 2.8 now implies that $\mathcal{U}$ is the direct sum $\mathcal{K}(\theta) + (\partial\theta)e$ where e is a projection in the centre of $\tilde{\mathcal{D}} = \mathrm{M} \cap \tilde{\mathcal{N}}'$. In our case at hand we know (Lemma 3.3 (4)) that we can replace e by eF where

$$F = I - \sum_{N \in \mathcal{N}} (N - N_-).$$

It is easy to check that $F\mathcal{D} = F\tilde{\mathcal{D}}$. Hence $eF \in Z(\mathcal{D})$ and the proof is complete. ■

REFERENCES

1. W.B. Arveson, Analyticity in operator algebras, Amer. J. Math. 89 (1967), 578-642.

2. J. Erdos and S. Power, Weakly closed ideals of nest algebras, J. Operator Theory 7 (1982), 219-235.

3. F. Gilfeather and D.R. Larson, Nest subalgebras of von Neumann algebras, Adv. in Math. 46 (1982), 171-199.

4. F. Gilfeather and D.R. Larson, Structure in reflexive subspace lattices, J. London Math. Soc. 26 (1982), 117-131.

5. I. Kovacs and J. Szücs, Ergodic type theorems in von Neumann algebras, Acta Sci. Math. Szeged, 27 (1966), 233-246.

6. D.R. Larson and Baruch Solel, Nests and inner flows, J. Operator Theory 16 (1986), 157-164.

7. R. Loebl and P.S. Muhly, Analyticity and flows in von Neumann algebras, J. Funct. Anal. 29 (1978), 214-252.

8. A. Loginov and V. Sulman, Hereditary and intermediate reflexivity of w*-algebras, Izv. Akad. Nauk SSSR 39 (1975), 1260-1273; USSR-Isv. 9 (1975), 1189-1201.

D.R. Larson
Department of Mathematics
Texas A&M University
College Station, Texas 77843

B. Solel
Department of Mathematics
and Computer Science
University of Haifa
Haifa, Israel

Operator Theory:
Advances and Applications, Vol. 32

STABILITY OF INVARIANT LAGRANGIAN SUBSPACES I.

André C.M. Ran and Leiba Rodman*

Dedicated to the memory of Constantin Apostol

CONTENTS

0. INTRODUCTION.
1. STABILITY AND STRONG STABILITY.
   1.1. Definitions and notation.
   1.2. The complex case.
   1.3. The case $\xi = \eta = 1$.
   1.4. Stability of invariant subspaces of general real matrices.
   1.5. Strong stability.
2. GENERAL THEORY.
3. THE CASE (-1,-1).
   3.1. The canonical form.
   3.2. Existence and uniqueness of invariant Lagrangian subspaces.
   3.3. Stability of invariant Lagrangian subspaces.
   3.4. Isolatedness of invariant Lagrangian subspaces

0. INTRODUCTION

Recently, the problems of stability of invariant subspaces of matrices and operators, i.e. the behaviour of invariant subspaces under small perturbations of the matrix or the operator, attracted much attention (see [BGK, GR, GLR1, AFS]). The main motivation to consider these problems comes from factorizations of matrix and operator functions, where invariant subspaces appear as the main tool in describing the factorizations (for this approach to factorization see, e.g., [BGK, GLR2]).

In engineering applications the rational matrix functions play a prominent role, because they appear as transfer functions for linear systems with constant coefficients. Moreover, often these functions possess various symmetry properties, e.g., real symmetric, or real skew-symmetric, etc. Factorizations of such functions that reflect the symmetries are of special importance. In the language of invariant subspaces such factorizations

---

* Partially supported by an NSF grant, and by Summer Research Grant from the College of Liberal Arts, Arizona State University.

correspond to invariant subspaces with certain symmetries. Hence, stability of invariant subspaces with symmetries bears directly onto stability of symmetric factorizations of rational matrix functions. It turns out that stability of invariant subspaces with symmetries is fundamental in other problems as well, such as behaviour of solutions of certain quadratic matrix equations, symmetric boundary value problems, etc.

This paper is devoted to the study and various applications of stability properties of important classes of invariant subspaces with symmetries that we call *lagrangian* (largely reflecting the standard use of this term). These subspaces are invariant for real matrices which are either symmetric or skew-symmetric with respect to a real quadratic form (which itself may be symmetric or skew-symmetric), and they are maximal isotropic with respect to the same quadratic form.

To make the presentation easier, the paper is divided into 3 parts. In the first part, we set up the framework, prove general results which will be used over and over again in the first two parts, and study in detail the cases when the real matrix and the quadratic form are either both symmetric or both skew-symmetric (for the symmetric case many results were obtained in [RR1]). We introduce and study here also new concepts of stability (conditional stability and strong stability) which seem to be important; in particular, the concept of strong stability should be more useful from the point of numerical analysis than the usual concept of stability developed in [BGK, CD]. For the precise definition of the various concepts of stability used in this paper see Sections 1.1 and 1.5. In the second part we study stability of lagrangian subspaces in the case when the matrix for which the subspace is invariant, and the quadratic form have opposite symmetries (i.e., one is symmetric and the other is antisymmetric). Here for the first time we encounter a situation where strong stability is not the same as stability. In the third part numerous applications of the results obtained in the first two parts will be given

1. STABILITY AND STRONG STABILITY

In this section we set forth the framework for our investigation, and bring together for the reader's convenience the basic facts on stable invariant subspaces which will be used subsequently.

1.1. Definitions and notation

For given $\xi = \pm 1$, $\eta = \pm 1$ introduce the class $L_n(\xi,\eta)$ of all pairs (A,H) of $n \times n$ real matrices such that H is invertible and $H^T = \xi H$, $HA = \eta A^T H$ (the superscript "T" denotes the transposed matrix). To avoid trivialities (i.e. the cases when $L_n(\xi,\eta)$ is empty) it will be assumed that n is even whenever ξ is -1.

Given $(A,H) \in L_n(\xi,\eta)$, let $J(A,H)$ be the class of all A-invariant lagrangian subspaces $M \subset \mathbb{R}^n$, i.e. such that $Ax \in M$ for every $x \in M$, $x^T Hy = 0$ for every $x,y \in M$ and $\dim M = \frac{n}{2}$. The term "lagrangian" is usually used in the case $\xi = -1$ only. However, we shall extend this terminology to the case when $\xi = 1$. Obviously, a necessary condition for non-emptiness of $J(A,H)$ is that n is even, and moreover, in the case $\xi = 1$, the signature (i.e. difference between the number of positive eigenvalues and the number of negative eigenvalues, multiplicities counted) of the real symmetric matrix H is zero. (As a side remark observe that if $\xi = -1$ then the signature of the hermitian matrix iH is always 0.) However, this necessary condition is generally not sufficient, and precise criteria for non-emptiness of $J(A,H)$ will be given in due course.

The real vector space $\mathbb{R}^n$ of n-dimensional columns will be considered with the standard inner product

$$\left\langle \begin{pmatrix} x_1 \\ x_2 \\ \vdots \\ x_n \end{pmatrix}, \begin{pmatrix} y_1 \\ y_2 \\ \vdots \\ y_n \end{pmatrix} \right\rangle = \sum_{i=1}^{n} x_i y_i .$$

Given two subspaces $M,N \subset \mathbb{R}^n$, the gap $\theta(M,N)$ between M and N is defined as follows:

$$\theta(M,N) = \|P_M - P_N\|,$$

where P_M (resp. P_N) is the orthogonal projection on M (resp. N). Here and elsewhere the norm of a matrix is the matrix norm induced by the euclidean vector norm. It is well-known (see, e.g., the expositions in [GLR1, GLR2]) that $\theta(M,N)$ is a metric on the set $G(\mathbb{R}^n)$ of all subspaces in $\mathbb{R}^n$ which makes this set into a compact metric space. Moreover, the connected components in $G(\mathbb{R}^n)$ consist precisely of all subspaces in $\mathbb{R}^n$ of fixed dimension. In the sequel we shall use the following description of convergence in $G(\mathbb{R}^n)$ (in the gap metric).

PROPOSITION 1.1. *Let* $\{M_m\}_{m=1}^{\infty}$ *be a sequence in* $G(\mathbb{R}^n)$ *such that* $\lim_{m\to\infty} M_m = M$, $M \in G(\mathbb{R}^n)$. *Then* M *consists of precisely those vectors* x *for which there is a sequence* $\{x_m\}_{m=1}^{\infty}$ *such that* $x_m \in M_m$ *for* $m = 1,2,\ldots$ *and* $\lim_{m\to\infty} x_m = x$.

For the proof of Proposition 1.1 see, for instance, [GLR1], [GLR2].

Let $(A,H) \in L_n(\xi,\eta)$. A subspace $M \in J(A,H)$ is called *conditionally stable* if for every $\varepsilon > 0$ there is $\delta > 0$ such that for any pair $(B,G) \in L_n(\xi,\eta)$ with $J(B,G) \neq \emptyset$ and

$$\|G-H\| + \|B-A\| < \delta$$

there is $M' \in J(B,G)$ satisfying

$$\theta(M',M) < \varepsilon.$$

If in the above definition "$J(B,G) \neq \emptyset$ and" is omitted we obtain the definition of *unconditionally stable* M. A subspace $M \in J(A,H)$ is called *Lipschitz conditionally stable* if there are positive constants ε and K such that for every pair $(B,G) \in L_n(\xi,\eta)$ with $J(B,G) \neq \emptyset$ and

$$\|G-H\| + \|B-A\| < \varepsilon$$

there is $M' \in J(B,G)$ satisfying

$$\theta(M',M) \leq K(\|G-H\| + \|B-A\|).$$

If "$J(B,G) \neq \emptyset$ and" is omitted, the subspace M will be called *Lipschitz unconditionally stable*.

In this paper we describe these and other classes of stable subspaces, for the cases when at least one of ξ and η is -1. For the case $\xi = \eta = 1$, these classes at least the non-trivial ones, have been described in [RR1]; for completeness we shall quote these results from [RR1] in subsection 1.3. The present part further considers the case $\xi = \eta = -1$.

For a complex $n \times n$ matrix X, the partial multiplicities of X corresponding to its eigenvalue λ_0 are, by definition the sizes of Jordan blocks with eigenvalue λ_0 that appear in the Jordan normal form of X. This definition applies in particular, to real $n \times n$ matrices. Thus, for example, the matrix

$$\begin{pmatrix} 0 & 1 & 0 & 0 \\ -1 & 0 & 0 & 0 \\ 0 & 0 & 0 & 1 \\ 0 & 0 & -1 & 0 \end{pmatrix}$$

has two eigenvalues i and $-i$ with partial multiplicities 1,1 for each. The set of all eigenvalues of A is called the spectrum of A and is denoted $\sigma(A)$. Throughout the paper we use $J_k(\lambda)$ to designate the $k \times k$ lower Jordan blocks with eigenvalue λ:

$$J_k(\lambda) = \begin{pmatrix} \lambda & & & & 0 \\ 1 & \lambda & & & \\ & 1 & \ddots & & \\ & & \ddots & \ddots & \\ 0 & & & 1 & \lambda \end{pmatrix}.$$

For real numbers a,b,c,d we also denote

$$J_k\begin{pmatrix} a & b \\ c & d \end{pmatrix} = \begin{pmatrix} a & b & \cdots & \cdots & 0 & 0 \\ c & d & \cdots & \cdots & 0 & 0 \\ 1 & 0 & & & \vdots & \vdots \\ 0 & 1 & & & \vdots & \vdots \\ \vdots & & & & \vdots & \vdots \\ 0 & 0 & \cdots & 1 \;\; 0 & a & b \\ 0 & 0 & \cdots & 0 \;\; 1 & c & d \end{pmatrix};$$

the size of $J_k\begin{pmatrix} a & b \\ c & d \end{pmatrix}$ is $2k \times 2k$. The block diagonal matrix

$$\begin{pmatrix} Z_1 & 0 & \cdots & 0 \\ 0 & Z_2 & \cdots & 0 \\ & \vdots & & \vdots \\ 0 & 0 & \cdots & Z_p \end{pmatrix}$$

will be denoted $Z_1 \oplus \ldots \oplus Z_p$ or $\mathrm{diag}[Z_1,\ldots,Z_p]$.
The notation e_k stands for the vectors all whose coordinates (except for the k-th) are zeros and the k-th coordinate is 1 (the dimension of e_k will be clear from the context).

1.2. The complex case

We recall briefly some basic facts concerning pairs of *complex* $n \times n$ matrices (A,H) such that $H = H^*$ is invertible and $HA = A^*H$ (here X^* is the conjugate transpose of X). Denote by L_n^c the set of such pairs.

THEOREM 1.2. *Given* $(A,H) \in L_n^c$, *there exists an invertible complex matrix* S *such that the matrices* $A_0 = S^{-1}AS$ *and* $H_0 = S^*HS$ *have the following form:*

(1.1) $$A_0 = J_{r_1}(\lambda_1) \oplus \dots \oplus J_{r_p}(\lambda_p) \oplus J_{\ell_1}(\mu_1 + i\nu_1) \oplus J_{\ell_1}(\mu_1 - i\nu_1) \oplus \dots \oplus J_{\ell_q}(\mu_q + i\nu_q) \oplus J_{\ell_q}(\mu_q - i\nu_q);$$

(1.2) $$H_0 = \kappa_1 E_{r_1} \oplus \dots \oplus \kappa_p E_{r_p} \oplus E_{2\ell_1} \oplus \dots \oplus E_{2\ell_q},$$

where $\lambda_1,\dots,\lambda_p$, $\mu_1,\dots,\mu_q$ *are real numbers,* $\nu_n,\dots,\nu_q$ *are positive numbers,* $\kappa_1,\dots,\kappa_p$ *are signs* (±1) *and*

$$E_k = \begin{bmatrix} & 0 & & 1 \\ & & \cdot^{\cdot^{\cdot}} & \\ & 1 & & 0 \\ 1 & & & \end{bmatrix}$$

is the $k \times k$ *matrix. The form* (1.1) *and* (1.2) *is canonical, i.e. uniquely determined by* (A,H) *up to simultaneous perturbation of blocks in* (1.1) *and* (1.2).

This result is well-known (see, e.g., the expositions in [GLR2,GLR3]). The form (1.1) is obviously the Jordan form for A. The signs $\kappa_1,\dots,\kappa_p$ are called the *sign characteristic* of (A,H); so there is a sign in the sign characteristic attached to each partial multiplicity of A corresponding to a real eigenvalue. Other descriptions of the sign characteristic are available (see [GLR4, GLR5, RR2]).

For $(A,H) \in L_n^c$, introduce the class $J^c(A,H)$ of all A-invariant lagrangian (complex) subspaces $M \subset \mathbb{C}^n$; thus, $AM \subset M$; $x^*Hy = 0$ for all $x,y \in M$ and $\dim M = \frac{n}{2}$. One can show (see, e.g., Theorem 5.1 in [RR3]):

PROPOSITION 1.3. $J^c(A,H) \neq \emptyset$ *if and only if for each real eigenvalue* λ_0 *of* A *(if any) the number of odd partial multiplicities of* A *corresponding to* λ_0 *is even (in particular, this number may be zero) and exactly half of them have sign* +1 *in the sign characteristic of* (A,H) *(so the other half have sign* -1).

As in subsection 1.1, we introduce the notions of conditionally stable, unconditionally stable, Lipschitz conditionally stable and Lipschitz unconditionally stable subspaces in $J^c(A,H)$. (The metric in the set of all subspaces in $\mathbb{C}^n$ is introduced by the same formula (1.0), where the orthogonality is understood with respect to the standard inner product in $\mathbb{C}^n$.) These classes are decribed as follows (see [RR2, RR3]).

THEOREM 1.4. *Let* $(A,H) \in L_n^c$. *There is a conditionally stable subspace in* $J^c(A,H)$ *if and only if for each real eigenvalue* λ_0 *of* A *(if any) all partial multiplicities of* A *corresponding to* λ_0 *are even and all signs in the sign characteristic of* (A,H) *corresponding to* λ_0 *are all equal. In this case the following statements are equivalent for* $M \in J^c(A,H)$:

(a) M *is conditionally stable;*

(b) M *is isolated as an element in* $J^c(A,H)$ (*in the topology induced by the gap metric*);

(c) *for each* $\lambda_0 \in \sigma(A) \setminus \mathbb{R}$ *such that* A *has more than one partial multiplicity corresponding to* λ_0, *either* $M \supset \operatorname{Ker}(A-\lambda_0 I)^n$ *or* $M \cap \operatorname{Ker}(A-\lambda_0 I)^n = \{0\}$ *holds.*

Here $\operatorname{Ker}(A-\lambda_0 I)^n$ is, of course, the *root subspace* of A corresponding to λ_0, i.e. the largest subspace with the property that the restriction of A to that subspace has the only eigenvalue λ_0.

THEOREM 1.5. *Let* $(A,H) \in L_n^c$ *and* $M \in J^c(A,H)$. *The following statements are equivalent:*

(a) M *is Lipschitz conditionally stable;*

(b) M *is Lipschitz unconditionally stable;*

(c) M *is a direct sum of root subspace of* A;

(d) *the spectrum of the restriction* $A|_M$ *does not contain any pair of non-real complex conjugate numbers.*

In particular, there is a Lipschitz conditionally stable subspace in $J^c(A,H)$ *if and only if* A *has no real eigenvalues.*

In fact, Theorems 1.4 and 1.5 (apart from equivalence of (a) and (b) in Theorem 1.5) were proved in [RR3].

The equivalence of (a) and (b) in Theorem 1.5 follows from the fact that, given the condition (c), for every matrix A' sufficiently close to A there is a subspace $M' \subset \mathbb{C}^n$ which is direct sum of root subspaces of A' such that

$$\theta(M',M) \le K\|A'-A\|,$$

where the positive constant K depends on A only.

The unconditionally stable subspaces are described as follows.

THEOREM 1.6. *Let* $(A,H) \in L_n^c$. *There is an unconditionally stable subspace in the class* $J^c(A,H)$ *if and only if* $\sigma(A) \cap \mathbb{R} = \emptyset$. *In this case* $M \in J^c(A,H)$ *is unconditionally stable if and only if it is conditionally stable.*

PROOF. If $\sigma(A) \cap \mathbb{R} = \emptyset$, then the sum of the root subspaces of A corresponding to the eigenvalues in the open upper half-plane belongs to $J^c(A,H)$ and is unconditionally stable, as follows from Theorem 1.4. Conversely, assume that $M \in J^c(A,H)$ is an unconditionally stable subspace. By Theorem 1.4 for each real eigenvalue λ_0 of A (if any) all partial multiplicities of A at λ_0 are even with equal signs. We have to prove that in fact A has no real eigenvalues. To this end for each $\varepsilon > 0$ we produce a pair $(A_\varepsilon, H) \in L_n^c$ such that $\|A - A_\varepsilon\| < \varepsilon$ and some partial multiplicities of A_ε corresponding to its real eigenvalues are odd. Without loss of generality it will be assumed that A is the nilpotent Jordan block of size n, and $H = E_n$ is given by (1.3). Now we can take

$$A_\varepsilon = \begin{pmatrix} 0 & 1 & 0 & \cdots & 0 \\ 0 & 0 & 1 & \cdots & 0 \\ \vdots & \vdots & \vdots & & \vdots \\ 0 & 0 & 0 & \cdots & 1 \\ \varepsilon & 0 & 0 & \cdots & 0 \end{pmatrix}.$$

Next, assume $\sigma(A) \cap \mathbb{R} = \emptyset$, and let $M \in J^c(A,H)$ be conditionally stable. By Proposition 1.3 $J^c(A',H') \neq \emptyset$ for every pair $(A',H') \in L_n^c$ such that $\|A' - A\|$ is small enough, and hence M is unconditionally stable. □

A pair $(A,H) \in L_n^c$ is said to satisfy the *sign condition* if for each real eigenvalue λ_0 of A the signs in the sign characteristic of (A,H) corresponding to odd partial multiplicities are all the same, and also the signs corresponding to even partial multiplicities are all the same. Note that by Theorem 1.4 and Proposition 1.3 a pair $(A,H) \in L_n^c$ with $J^c(A,H) \neq \emptyset$ satisfies the sign condition if and only if there exists a conditionally stable subspace in $J^c(A,H)$.

1.3. The case $\xi = \eta = 1$

We describe the stability classes for pairs in $L_n(1,1)$. The main results here were obtained in [RR1].

We start with the criterium for $J(A,H) \neq \emptyset$.

THEOREM 1.7. (see [RR1]). *Let* $(A,H) \in L_n(1,1)$. *Then* $J(A,H) \neq \emptyset$ *if and only if the following conditions hold*

(i) *for every* $\lambda_0 \in \sigma(A) \cap \mathbb{R}$ *the number of odd partial multiplicities of* A *corresponding to* λ_0 *is even, and precisely half of them have signs* +1 *in the sign characteristic of* (A,H) (*so the other half have signs* -1);

(ii) *for every* $\lambda_0 \in \sigma(A) \setminus \mathbb{R}$ *the number of odd partial multiplicities of* A *corresponding to* λ_0 *is even.*

Moreover, J(A,H) *consists of only one element if and only if* $J(A,H) \neq \emptyset$ *and in addition the following holds: for every real* $\lambda_0 \in \sigma(A)$ *all the partial multiplicities of* A *corresponding to* λ_0 *are even and all corresponding signs in the sign characteristic are equal, and every non-real eigenvalue of* A *has geometric multiplicity one.*

Next, we describe the conditional stability.

THEOREM 1.8 (see [RR1]). *Let* $(A,H) \in L_n(1,1)$, *and let* $M \in J(A,H)$. *The following statements are equivalent:*

(a) *M is conditionally stable;*

(b) *M is the only member of* J(A,H);

(c) *for every non-real eigenvalue* λ_0 *of* A *we have only one partial multiplicity and this partial multiplicity is even and for every real eigenvalue* λ_0 *of* A *the partial multiplicities of* A *corresponding to* λ_0 *are all even and the signs in the sign characteristic of* (A,H) *corresponding to* λ_0 *are all equal.*

It turns out that the three other classes (unconditionally stable, Lipschitz conditionally stable and Lipschitz unconditionally stable) are empty for every $(A,H) \in L_n(1,1)$. For the Lipschitz conditionally stable class this was proved in [RR1], which obviously suplies the emptiness of Lipschitz unconditionally stable class. To prove this statement for the unconditionally stable class, in view of Theorem 3.1 in [RR1] and Theorem 1.8(c) we can assume without loss of generality that one of the 2 cases holds: 1) $\sigma(A) = \{\lambda_0, \bar{\lambda}_0\}$ wiht non-real λ_0, and A has only one partial multiplicity at λ_0 which is even; 2) $\sigma(A) = \{\lambda_0\}$ with real λ_0, all partial multiplicities of A are even and all signs in the sign characteristic are equal. Using the canonical form of (A,H) under the transformation $(A,H) \to (S^{-1}AS, S^THS)$ where S is real and invertible (this canonical form is given, for instance, in [U,DPWZ], see also [RR1]) we can further restrict our attention to the following cases:

1) $A = J_{\frac{n}{2}}\begin{pmatrix} a & b \\ -b & a \end{pmatrix}$, where $b > 0$, a real; $H = E_n$ (see (1.3) for the definition); and n^2 is a multiple of 4;

2) $A = J_{r_1}(\lambda_0) \oplus \ldots \oplus J_{r_p}(\lambda_0)$, where λ_0 is real and all r_j are even; $H = \pm(E_{r_1} \oplus \ldots \oplus E_{r_p})$.

In the case 1), let $A(\varepsilon) = A + \varepsilon Q$, where $\varepsilon > 0$ is close to zero and Q is the $n \times n$ matrix with 2×2 identity matrix in the upper right corner and zeros elsewhere. It is not difficult to check that $HA(\varepsilon) = A(\varepsilon)^TH$ and that the eigenvalues of $A(\varepsilon)$ in the open upper halfplane are $a + ib + n_j$, $j = 1,\ldots,n$, where

$\eta_1,\dots,\eta_n$ are the n-th roots of ε. By Theorem 1.7, $J(A(\varepsilon),H) = \emptyset$. In the case 2) an analogous analysis shows also that $J(A(\varepsilon),H) = \emptyset$, where

$$A(\varepsilon) = (J_{r_1}(\lambda_0) + \varepsilon Q) \oplus J_{r_2}(\lambda_0) \oplus \dots \oplus J_{r_p}(\lambda_0),$$

and Q is $r_1 \times r_1$ matrix with 1 in the upper right corner and zeros elsewhere. So in both cases the unconditionally stable class is empty.

1.4. Stability of invariant subspaces of general real matrices

Let A be $n \times n$ real matrix, and let Inv(A) be the set of all its invariant subspaces in $\mathbb{R}^n$. A subspace $M \in \mathrm{Inv}(A)$ will be called *unconditionally stable* (*in the class* Inv) if for every $\varepsilon > 0$ there is $\delta > 0$ such that every real matrix B with $\|A - B\| < \delta$ has an invariant subspace N with $\theta(M,N) < \varepsilon$. We quote the description of this class of subspaces obtained in [BGK]. For a real eigenvalue λ of A we denote by $R(A;\lambda)$ the root subspace of A corresponding to λ, i.e. $\mathrm{Ker}(A - \lambda I)^n$. For a pair of non-real complex conjugate eigenvalues $\lambda \pm i\mu$ ($\mu \neq 0$) of A we define the root subspace by

$$R(A;\lambda \pm i\mu) = \mathrm{Ker}\,((A - \lambda I)^2 + \mu^2 I)^n.$$

THEOREM 1.9. *The subspace* $M \in \mathrm{Inv}(A)$ *is unconditionally stable in the class* Inv *if and only if the following conditions hold:*

(i) *the subspace* $M \cap R(A;\lambda)$ *is even dimensional for every real eigenvalue* λ *of* A *that has only one partial multiplicity and for which this partial multiplicity is even;*

(ii) *either* $M \supset R(A;\lambda)$ *or* $M \cap R(A;\lambda) = \{0\}$ *for every real eigenvalue* λ *of* A *that has more than one partial multiplicity;*

(iii) *either* $M \supset R(A;\lambda \pm i\mu)$ *or* $M \cap R(A;\lambda \pm i\mu) = \{0\}$ *for every non-real eigenvalue* $\lambda + i\mu$ *of* A *that has more than one partial multiplicity.*

Further, $M \in \mathrm{Inv}(A)$ *is an isolated element in* Inv(A) (*in the topology induced by the gap metric*) *if and only if the conditions* (ii) *and* (iii) *hold true.*

As in subsection 1.1, one introduces the notion of Lipschitz unconditional stability in the class Inv. It turns out that $M \in \mathrm{Inv}(A)$ is Lipschitz unconditionally stable in the class Inv if and only if M is a (necessarily direct) sum of root subspaces. Indeed, the part "if" of this statement is easily seen. To verify the part "only if" observe first that without loss of generality we can restrict our attention to the case when either $\sigma(A) = \{\lambda_0\}$ with λ_0 real or $\sigma(A) = \{\lambda_0 \pm i\mu_0\}$, where λ_0,μ_0 are real and $\mu_0 > 0$. Further, since obviously Lipschitz unconditional stability implies unconditional stability, in view of Theorem 1.9 we can further assume that either $A = J_n(\lambda_0)$ (λ_0 real) and M is even dimensional for even n, or $A = J_{\frac{n}{2}}\begin{pmatrix} \lambda_0 & \mu_0 \\ -\mu_0 & \lambda_0 \end{pmatrix}$. In the former case use Lemma 4.3 of [RR3] (see also the proof of Theorem 4.7 in [KMR]) to show that the only Lipschitz unconditionally stable subspaces are the trivial ones ($M = \{0\}$ or $M = \mathbb{R}^n$), in the latter case a similar argument yields the desired result.

We shall define now conditional stability in the class Inv(A). Let $\lambda_1 < \dots < \lambda_p$ be all the real eigenvalues of the $n \times n$ real matrix A, and let Γ_j be the circle with center λ_j and radius η where $\eta = \min\{\frac{1}{3}(\lambda_2 - \lambda_1), \dots, \frac{1}{3}(\lambda_p - \lambda_{p-1})\}$. For a given A-invariant subspace M introduce the set $\Gamma(M)$ of all $n \times n$ real matrices X with the following properties:

(i) $\sigma(X) \cap \Gamma_j = \emptyset$, $j = 1,\dots,p$;

(ii) for each $j = 1,\dots,p$, there exists a q_j-dimensional X-invariant subspace in the sum of all root subspaces of X corresponding to the eigenvalues of X inside Γ_j, where $q_j = \dim(M \cap R_{\lambda_j}(A))$.

In connection with this definition observe that (i) holds for all X sufficiently close to A but (ii) generally does not. It is not difficult to see that the property (ii) (for fixed j) holds for all X sufficiently close to A if and only if either $\dim R_{\lambda_j}$ is odd or both $\dim R_{\lambda_j}$ and q_j are even (cf. the proof of Lemma 9.5 in [BGK]). We say that $M \in \mathrm{Inv}(A)$ is *conditionally stable* in the class Inv(A) if for every $\varepsilon > 0$ there is $\delta > 0$ such that any $n \times n$ matrix $B \in \Gamma(M)$ for which $\|B - A\| < \delta$ has an invariant subspace N satisfying $\theta(M,N) < \varepsilon$. So this property is generally weaker than unconditional stability because here B must satisfy the additional requirement that $B \in \Gamma(M)$.

THEOREM 1.10. *The following statements are equivalent for a subspace* $M \in \mathrm{Inv}(A)$:

(a) M *is conditionally stable (in the class* Inv);

(b) M *is isolated as an element in* Inv(A);

(c) *the conditions* (ii) *and* (iii) *of* Theorem 1.7 *are satisfied.*

PROOF. Arguing as in the proof of Lemmas 8.5 and 8.6 from [BGK] we can assume that either $\sigma(A) = \{\lambda_0\}$, λ_0 real or $\sigma(A) = \{\lambda_0 \pm i\mu_0\}$, λ_0 real and $\mu_0 > 0$. In view of Theorem 1.9 and the remark thereafter the only statement which remains to be proven is the following. Let $A = J_n(\lambda_0)$ where λ_0 real and n is even. Then every odd dimensional A-invariant subspace M is conditionally stable. This can be done by repeating the arguments from the proof of Theorem 8.2 in [BGK] (see also the proof of Theorem S4.9 in [GLR2]). □

1.5. Strong stability

We introduce here another concept of stability of subspaces, which will be studied in this paper.

Let $(A,H) \in L_n(\xi,\eta)$. Choose sufficiently small positively oriented contours $\Gamma_1,\dots,\Gamma_k$ in such a way that $\Gamma_i \cap \Gamma_j = \emptyset$ for $i \neq j$, inside Γ_j there is either precisely one distinct real eigenvalue of A or precisely one pair of distinct non-real complex conjugate eigenvalues of A, and each eigenvalue of A is inside some contour Γ_j. Let $\Gamma = \bigcup_{j=1}^{k} \Gamma_j$.

The subspace $M \in J(A,H)$ will be called *conditionally strongly stable* if for any sequence $(A_m,H_m) \in L_n(\xi,\eta)$ with $J(A_m,H_m) \neq \emptyset$, $A_m \to A$ and $H_m \to H$ and *any* sequence $M_m \in J(A_m,H_m)$ with

$$(1.4) \qquad \dim P_{mj}M_m = \dim P_jM, \quad j = 1,\dots,k$$

where P_{mj} (resp. P_j) denotes the spectral projection of A_m (resp. A) with respect to the eigenvalues inside Γ_j, we have

$$(1.5) \qquad \theta(M_m,M) \to 0 \quad \text{as } m \to \infty.$$

(Observe that $\sigma(A_m) \cap \Gamma = \emptyset$ for m large enough, so the projections P_{mj} are well defined at least for m sufficiently large.) Replacing here the condition (1.5) by

$$(1.6) \qquad \theta(M_m,M) \leq K(\|A_m - A\| + \|H_m - H\|),$$

where $K > 0$ depends on A, H and M only, we obtain the definition of *conditional Lipschitz strong stability*. Further, $M \in J(A,H)$ will be called *unconditionally strongly stable* if for any sequence $(A_m,H_m) \in L_n(\xi,\eta)$ with $A_m \to A$ and $H_m \to H$ we have $J(A_m,H_m) \neq \emptyset$ and (1.5) holds for *any* sequence $M_m \in J(A_m,H_m)$ satisfying (1.4). Finally, replacing in this definition (1.5) by (1.6) the notion of *unconditional Lipschitz strong stability* is obtained

Clearly, a strongly stable subspace is stable.

We shall see later that in many, but not all, cases the notions of stability and strong stability actually coincide.

From computational point of view it seems to us that strong stability is a more desirable property than merely stability. Indeed, as we shall show in the second part of the paper, it is possible that a subspace M is conditionally stable and still for any sequence $(A_m,H_m) \in L_n(-1,1)$ $A_m \to A$, $H_m \to H$ there is a sequence $M_m \in J(A_m,H_m)$ such that (1.4) holds, but $\theta(M_m,M) > \varepsilon > 0$ where ε does not depend on M_m.

In the following theorem "strong stability" and "stability" means one of the 4 classes (conditional, unconditional, Lipschitz conditional, Lipschitz unconditional). The notions of various classes of strong stability for $(A,H) \in L_n^c$ are introduced in the same way as for $(A,H) \in L_n(\xi,\eta)$.

THEOREM 1.11. *Let* $(A,H) \in L_n^c$. *A subspace* $M \in J^c(A,H)$ *is strongly stable if and only if* M *is stable (here* $M \subset \mathbb{C}^n$*).*

THEOREM 1.12. *Let* $(A,H) \in L_n(1,1)$. *Then* $M \in J(A,H)$ *is conditionally strongly stable if and only if* M *is conditionally stable. All other classes of strong stability are empty.*

The proofs of Theorems 1.11 and 1.12 will be given after some preparation.

First observe the following lemma, the part (a) of which is an immediate consequence of the proof of Theorem 8.2 in [BGK] (see also Lemma 5.4 in [GR]), and the part (b) can be proved analogously.

LEMMA 1.13.(a). *Let* $M \subset \mathbb{C}^n$ *be the* p*-dimensional invariant subspace of an* $n \times n$ *complex matrix* A *with one-point spectrum* $\sigma(A) = \{\lambda_0\}$ *and with* $\dim \operatorname{Ker}(\lambda_0 I - A) = 1$. *Then for every sequence* B_m *of matrices such that* $\|B_m - A\| \to 0$ *as* $m \to \infty$ *and for every* p*-dimensional* B_m*-invariant subspace* M_m *we have* $\theta(M_m,M) \to 0$ *as* $m \to \infty$.
(b) *Let* $M \subset \mathbb{R}^n$ *be* p*-dimensional invariant subspace of an* $n \times n$ *real matrix* A *with either* $\sigma(A) = \{\lambda_0,\bar{\lambda}_0\}$, λ_0 *non-real, or* $\sigma(A) = \{\lambda_0\}$, λ_0 *real, and with* $\dim \operatorname{Ker}(\lambda_0 I - A) = 1$. *Then for every sequence* B_m *of real matrices such that* $\|B_m - A\| \to 0$ *as* $m \to \infty$ *and such that there is a* p*-dimensional* B_m*-invariant subspace for all* m, *we have* $\theta(M_m,M) \to 0$ *as* $m \to \infty$, *where* M_m *is any* p*-dimensional* B_m*-invariant subspace.*

PROOF OF THEOREM 1.11. For the Lipschitz class Theorem 1.11 follows from Theorem 1.5 because the property of being a direct sum of root subspaces is easily seen to be strongly stable. For the case of conditional stability Theorem 1.11 follows from the proof of Theorem 5.2 in [RR3] combined with Lemma 1.13. Analogously one proves the theorem for the unconditional stability (see Theorem 1.6). □

PROOF OF THEOREM 1.12. The last statement follows from the emptiness of corresponding classes of stability (see end of Section 1.3). Assume now

$M \in J(A,H)$ is conditionally stable. We can assume without loss of generality that either $\sigma(A) = \{\lambda_0, \overline{\lambda}_0\}$, λ_0 non-real or $\sigma(A) = \{\lambda_0\}$, λ_0 real (see Theorem 2.2 in the next section). In the former case by Theorem 1.8 the geometric multiplicity of λ_0 as an eigenvalue of A is 1 and the algebraic multiplicity m is even. As M is the only member of $J(A,H)$ it is easily seen that the characteristic polynomial of $A|_M$ is $((\lambda - \lambda_0)(\lambda - \overline{\lambda}_0))^{\frac{m}{2}}$. Now Lemma 1.13(b) ensures that M is conditionally stronly stable. Analogously one proves the conditional strong stability of M in the case $\sigma(A) = \{\lambda_0\}$, λ_0 real. □

2. GENERAL THEORY

Here we present general results concerning stability of invariant lagrangian subspaces. Some of these results and their proofs are essentially not new, but others are new. The general principles presented here will be used all over in the subsequent sections.

Everywhere in this section (except for Theorem 2.4) "stability" and "stable" means one of the 8 classes (conditional and unconditional stability, Lipschitz conditional and unconditional stability, as well as the classes of strong stability) introduced in Section 1. The results apply to all 8 classes (except for Theorem 2.4).

We start with the notion of H-stability. Let $(A,H) \in L_n(\xi,\eta)$. A subspace $M \in J(A,H)$ will be called *H-stable* if in the definition of stability (given in Section 1) we replace the letter "G" by "H". Informally speaking, M is H-stable if and only if it is stable under the allowed perturbations of (A,H) where only A may change and H is kept fixed. It is clear from the definition that stability implies H-stability. The converse is also true:

THEOREM 2.1. *A subspace* $M \in J(A,H)$ *is stable if and only if it is H-stable.*

PROOF. Use the proof of Theorem 3.2 in [RR3] together with the following fact: Given invertible real $n \times n$ matrix H with $H^T = \xi H$, there exist positive constants ε and K such that for any G with $G^T = \xi G$ and

$$(2.1) \qquad \|G - H\| < \varepsilon$$

there is an invertible real matrix S such that $S^T G S = H$ and

$$\max\{\|I - S\|, \|I - S^{-1}\|\} \leq K\|G - H\|.$$

To verify this fact in the case $\xi = 1$ observe that for $\varepsilon > 0$ small enough every real symmetric G with the property (2.1) is invertible and has the same signature as H; then use the Lagrange's algorithm for reduction of a bilinear form to the sum of squares, see, e.g., [LT]. For the case $\xi = -1$ see, e.g., [J] Section V 10. □

The following general principle is the possibility to localize the property of stability, as we explain below. It follows from the canonical form of $(A,H) \in L_n(\xi,\eta)$ (for the case when at least one of ξ and η is equal to -1 the canonical form is given in the corresponding section) that in case $\eta = -1$ the spectrum of A is symmetric relative to the imaginary axis. Of course, in any case $\sigma(A)$ is symmetric relative to the real axis, because A is a real matrix. Here symmetry with respect to the real axis means that if $\lambda \in \sigma(A)$ then $\bar{\lambda} \in \sigma(A)$ and the partial multiplicities of A corresponding to $\bar{\lambda}$ coincide with those corresponding to λ. Symmetry with respect to the imaginary axis is understood in a similar way.

Given $(A,H) \in L_n(\xi,\eta)$ define *local subspaces* R of A as follows:

(i) if $\eta = 1$, then either $R = R(A;\lambda)$ for some real eigenvalue λ of A, or $R = R(A;\lambda \pm i\mu)$ for some conjugate pair $\lambda \pm i\mu$ of non-real eigenvalues of A;

(ii) if $\eta = -1$, then one of the four possibilities occur:

(a) $R = R(A;0)$ (if 0 is an eigenvalue of A);

(b) $R = R(A;\lambda) + R(A;-\lambda)$ for a non-zero real eigenvalue λ of A;

(c) $R = R(A;\pm i\mu)$ for a pair of non-zero pure imaginary eigenvalues $\pm i\mu$ of A;

(d) $R = R(A;\lambda \pm i\mu) + R(A;-\lambda \pm i\mu)$ for a quadruple of non-real and non-pure imaginary eigenvalues $\pm\lambda \pm i\mu$ of A.

One checks without difficulties, using the canonical form for instance, that $(A|_R, PH|_R) \in L_m(\xi,\eta)$ for every $(A,H) \in L_n(\xi,\eta)$ and every local subspace of A of dimension m. Here P is the orthogonal projection $(P^2 = P = P^T)$ onto R, and the transformations $A|_R$, $PH|_R$ are understood as matrices written in some orthonormal basis in R.

THEOREM 2.2. *Let* $(A,H) \in L_n(\xi,\eta)$. *A subspace* $M \in J(A,H)$ *is stable if and only if for each local subspace* R *of* A *the subspace* $M \cap R \in J(A|_R, PH|_R)$ *is stable (here* P *is the orthogonal projection on* R*).*

The proof is analogous to the proof of Theorem 3.1 in [RR4] (see also Theorem 3.1 in [RR3]).

Next, there is a natural group action of J(A,H). Given $(A,H) \in L_n(\xi,\eta)$, let G(A,H) be the group of all real invertible $n \times n$ matrices S such that

$S^{-1}AS = A$, $S^{T}HS = H$. Clearly, $SM \in J(A,H)$ for every $M \in J(A,H)$ and $S \in G(A,H)$. Thus, the group $G(A,H)$ generates action on $J(A,H)$. The next theorem asserts that this action preserves stability.

THEOREM 2.3. *If* $M \in J(A,H)$ *is stable, then* SM *is also stable for every* $S \in G(A,H)$.

PROOF. We prove for the stability only (the strong stability is dealt with analogously). Consider first the case of unconditioned stability (so "stable" means "unconditionally stable"). Suppose SM is not stable. By definition of stability and by Theorem 2.1 there is $\varepsilon > 0$ and a sequence $A_m \to A$ (as $m \to \infty$) such that $(A_m,H) \in L_n(\xi,\eta)$, and for all $N_m \in J(A_m,H)$ the inequality

$$(2.2) \qquad \theta(N_m,SM) > \varepsilon$$

holds. Now $S^{-1}A_mS \to A$ and $(S^{-1}A_mS,H) \in L_n(\xi,\eta)$. By the stability of M, for each m there is $\tilde{M}_m \in J(S^{-1}A_mS,H)$ such that $\theta(\tilde{M}_m,M) \to 0$ as $m \to \infty$. Clearly $S\tilde{M}_m \in J(A_m,H)$, and one can show (see, e.g., [GLR1]) that

$$(2.3) \qquad \theta(S\tilde{M}_m,SM) \le K\theta(\tilde{M}_m,M),$$

where the positive constant K depends on M and S only (and is independent of m). We obtained a contradiction with (2.2).

In the case of conditioned stability the proof is analogous, with the obvious modification that $J(A_m,H) \neq \emptyset$ is required, and then also $J(S^{-1}A_mS,H) \neq \emptyset$.

Assume now that the stability under consideration is Lipschitz unconditional stability. Again arguing by contradiction, suppose S is not stable. So there is a sequence such that $(A_m,H) \in L_n(\xi,\eta)$,

$$\|A_m - A\| < \frac{1}{m}, \quad m = 1,2,\ldots$$

but

$$\theta(N_m,SM) > m\,\|A_m - A\|$$

for every $N_m \in J(A_m,H)$ and $m = 1,2,\ldots$. Now use the stability of M with respect to the sequence $(S^{-1}A_mS,H)$ and the inequality (2.3) to obtain a contradiction. The case of Lipschitz unconditional stability is done analogously. □

We conclude this section with the following useful observations.

THEOREM 2.4. *Let* $(A,H) \in L_n(\xi,\eta)$, *and assume that* $J(A,H)$ *consists of one element* M *only. Then* M *is conditionally strongly stable.*

The proof is the same as that of Theorem 3.3 in [RR3].

PROPOSITION 2.5. *Let* $(A_i,H_i) \in L_{n_i}(\xi,\eta)$, $i=1,2$, *and* $M_i \in J(A_i,H_i)$. *Suppose* $M = M_1 \oplus M_2$ *is stable as an element of* $J(A,H)$ *where* $A = A_1 \oplus A_2$, $H = H_1 \oplus H_2$. *Then* M_i *is stable as an element of* $J(A_i,M_i)$.

PROOF. We shall give the proof for unconditional stability only, the proof for the other classes of stability is similar. Suppose M_1 is not stable. Then there is an $\varepsilon > 0$ and a sequence $A_{m1} \to A_1$ and $(A_{m1},H_1) \in L_{n_1}(\xi,\eta)$ such that $\text{gap}(M_{m1},M_1) > \varepsilon$ for all $M_{m1} \in J(A_{m1},H_1)$. Choose a sequence $A_{m2} \to A_2$ such that $(A_{m2},H_2) \in L_{n_2}(\xi,\eta)$ and such that $\sigma(A_{m1}) \cap \sigma(A_{m2}) = \emptyset$. (This is always possible.) Since M is stable there is an element $M_m \in J(A_{m1} \oplus A_{m2},H)$ such that $\text{gap}(M_m,M) \to 0$. But since $\sigma(A_{m1}) \cap \sigma(A_{m2}) = \emptyset$ we have $M_m = M_{m1} + M_{m2}$ for some $M_{mi} \in J(A_{mi},H_i)$, and then

$$\text{gap}(M_{m1},M_1) \le \text{gap}(M_m,M) \to 0,$$

which contradicts the assumption. □

The converse of Proposition 2.5 is false in general, but holds in case $\sigma(A_1) \cap \sigma(A_2) = \emptyset$ as one readily checks.

3. THE CASE $\xi = -1$, $\eta = -1$

3.1. The canonical form

We start this section by recalling the canonical form for the case $\xi = -1$, $\eta = -1$ (see [DPWZ]). So let $(A,H) \in L_n(-1,-1)$. Then there is an invertible S such that $S^{-1}AS$ and S^THS are block diagonal matrices with the diagonal blocks (A_i,H_i) of one of the following four types:

$$(3.1)\qquad A_i = J_{2n_1}(0) \oplus \dots \oplus J_{2n_p}(0) \oplus J_{2n_{p+1}+1}(0) \oplus -(J_{2n_{p+1}+1}(0))^T \oplus \dots \oplus J_{2n_{p+q}+1}(0) \oplus -(J_{2n_{p+q}+1}(0))^T,$$

$$H_i = x_1F_{2n_1} \oplus \dots \oplus x_pF_{2n_p} \oplus \begin{pmatrix} 0 & I_{2n_{p+1}+1} \\ -I_{2n_{p+1}+1} & 0 \end{pmatrix} \oplus \dots \oplus \begin{pmatrix} 0 & I_{2n_{p+q}+1} \\ -I_{2n_{p+q}+1} & 0 \end{pmatrix},$$

where F_{2n} is the $2n \times 2n$ matrix given by

$$F_{2n} = \begin{bmatrix} & & & & 1 \\ & & & -1 & \\ & & \cdot^{\cdot^{\cdot}} & & \\ & 1 & & & \\ -1 & & & & \end{bmatrix}$$

and κ_j is 1 or -1.

(3.2) $$A_i = J_{n_1}(a) \oplus (-J_{n_1}(a))^T \oplus \dots \oplus J_{n_p}(a) \oplus (-J_{n_p}(a))^T$$

$$H_i = \begin{bmatrix} 0 & I_{n_1} \\ -I_{n_1} & 0 \end{bmatrix} \oplus \dots \oplus \begin{bmatrix} 0 & I_{n_p} \\ -I_{n_p} & 0 \end{bmatrix},$$

where a > 0.

(3.3) $$A_i = J_{n_1}\begin{pmatrix} 0 & b \\ -b & 0 \end{pmatrix} \oplus \dots \oplus J_{n_p}\begin{pmatrix} 0 & b \\ -b & 0 \end{pmatrix} \quad (b > 0)$$

$$H_i = \kappa_1 \begin{bmatrix} 0 & & & \begin{pmatrix} 0 & 1 \\ -1 & 0 \end{pmatrix}^{n_1} \\ & & -\begin{pmatrix} 0 & 1 \\ -1 & 0 \end{pmatrix}^{n_1} & \\ & \cdot^{\cdot^{\cdot}} & & \\ (-1)^{n_1-1}\begin{pmatrix} 0 & 1 \\ -1 & 0 \end{pmatrix}^{n_1} & & & 0 \end{bmatrix} \oplus \dots \oplus \kappa_p \begin{bmatrix} 0 & & & \begin{pmatrix} 0 & 1 \\ -1 & 0 \end{pmatrix}^{n_p} \\ & & \cdot^{\cdot^{\cdot}} & \\ & \cdot^{\cdot^{\cdot}} & & \\ (-1)^{n_p-1}\begin{pmatrix} 0 & 1 \\ -1 & 0 \end{pmatrix}^{n_p} & & & 0 \end{bmatrix}$$

where κ_j is +1 or -1.

(3.4) $$A_i = J_{n_1}\begin{pmatrix} a & b \\ -b & a \end{pmatrix} \oplus \left(-J_{n_1}\begin{pmatrix} a & b \\ -b & a \end{pmatrix}\right)^T \oplus \dots \oplus J_{n_p}\begin{pmatrix} a & b \\ -b & a \end{pmatrix} \oplus \left(-J_{n_p}\begin{pmatrix} a & b \\ -b & a \end{pmatrix}\right)^T$$

$$(a,b > 0)$$

$$H_i = \begin{bmatrix} 0 & I_{2n_1} \\ -I_{2n_1} & 0 \end{bmatrix} \oplus \dots \oplus \begin{bmatrix} 0 & I_{2n_p} \\ -I_{2n_p} & 0 \end{bmatrix}.$$

3.2. Existence and uniqueness of invariant lagrangian subspaces

In this section we shall prove the following two theorems.

THEOREM 3.1. *Suppose* $(A,H) \in L_n(-1,-1)$. *Then* $J(A,H) \neq \emptyset$ *if and only if for all pure imaginary non-zero eigenvalues* ib $(b > 0)$ *of* A *(if any) the number of odd partial multiplicities of* A *corresponding to* ib *is even, and the sum of the signs* κ_j *corresponding to these odd multiplicities is zero.*

THEOREM 3.2. *Let* $(A,H) \in L_n(-1,-1)$ *and let* M_+ *be the* A*-invariant subspace corresponding to the eigenvalues of* A *in the open right half plane. Then for each* A*-invariant subspace* $N \subset M_+$ *there exists a unique* $M \in J(A,H)$ *with* $M \cap M_+ = N$ *if and only if for each pure imaginary eigenvalue* ib *of* A $(b \geq 0)$ *all partial multiplicities of* A *corresponding to* ib *are even, and for* $b > 0$ *all the signs* κ_j $(j = 1,\dots,p)$ *corresponding to* ib *are equal, for* $b = 0$ *all the signs* $(-1)^{n_j}\kappa_j$ $(j = 1,\dots,p)$ *are equal, where* $2n_1,\dots,2n_p$ *are the partial multiplicities of* A *corresponding to zero.*

The proof is partly based on the observation that $(iA,iH) \in L_n^c$ if $(A,H) \in L_n(-1,-1)$. So there is a sign characteristic of (iA,iH) (see Section 1.2). We shall compute this sign characteristic in terms of the canonical form of (A,H) as given in the previous section.

PROPOSITION 3.3.(i) *Suppose* (A,H) *is as in* (3.1). *Then the sign characteristic of* (iA,iH) *is given by*

$$(-1)^{n_1}\kappa_1,\dots,(-1)^{n_p}\kappa_p,\underbrace{1,-1,\dots,1,-1}_{q \text{ times}}.$$

(ii) *Suppose* (A,H) *is as in* (3.3). *Then the sign characteristic of* (iA,iH) *at the eigenvalue* b *of* iA *is given by*

$$\kappa_1,\dots,\kappa_p,$$

and the sign characteristic of (iA,iH) *at the eigenvalue* $-b$ *of* iA *is given by*

$$(-1)^{n_1}\kappa_1,\dots,(-1)^{n_p}\kappa_p.$$

PROOF. (i) We only have to consider the cases $A = J_{2n}(0)$, $H = \kappa F_{2n}$ and $A = J_n(0) \oplus -J_n(0)^T$, $H = \begin{pmatrix} 0 & I_n \\ -I_n & 0 \end{pmatrix}$ where n is odd. In the first case $i^{2n-1}e_{2n}, i^{2n-2}e_{2n-1},\dots,e_1$ forms a Jordan chain for iA at zero. The the sign corresponding to such a block in the sign characteristic of (iA,iH) is determined by the sign of $\langle iHi^{2n-1}e_{2n},e_1\rangle = (-1)^n\langle He_{2n},e_1\rangle = (-1)^n\kappa$. In the second case note that there exists an iA-invariant lagrangian subspace. Since the partial multiplicities of iA at zero are n, twice, and n is odd Theorem 5.1 in [RR3] (see also Section 1.2) shows that the signs in the sign characteristic of (iA,iH) are $+1$ and -1. This proves (i).

(ii). We only have to consider the case

$$A = J_n\begin{pmatrix} 0 & b \\ -b & 0 \end{pmatrix} \ (b > 0), \quad H = \kappa\begin{pmatrix} & & & \begin{pmatrix} 0 & 1 \\ -1 & 0 \end{pmatrix}^n \\ & & -\begin{pmatrix} 0 & 1 \\ -1 & 0 \end{pmatrix}^n & \\ & \cdot^{\cdot^{\cdot}} & & \\ (-1)^{n-1}\begin{pmatrix} 0 & 1 \\ -1 & 0 \end{pmatrix}^n & & & \end{pmatrix}$$

It is easily checked that $e_1 - ie_2, i(e_3 - ie_4), \ldots, i^{n-1}(e_{2n-1} - ie_{2n})$ is a Jordan chain at b for iA, and $e_1 + ie_2, i(e_3 + ie_4), \ldots, i^{n-1}(e_{2n-1} + ie_{2n})$ is a Jordan chain at -b for iA.

We consider separately the cases $n = j \bmod 4$ $(j = 0,1,2,3)$. If n = n = 0 mod 4 we have

$$H = \kappa\begin{pmatrix} & & & 1 & 0 \\ & & & 0 & 1 \\ & & \cdot^{\cdot^{\cdot}} & & \\ & 1 \ \ 0 & & & \\ & 0 \ \ 1 & & & \\ -1 \ \ 0 & & & & \\ 0 \ \ -1 & & & & \end{pmatrix}$$

Since $i^{n-1} = -i$ we have

$$\langle iHi^{n-1}(e_{2n-1} - ie_{2n}), e_1 - ie_2\rangle = \langle He_{2n-1} - ie_{2n}, e_1 - ie_2\rangle =$$
$$= \langle He_{2n-1}, e_1\rangle + \langle He_{2n}, e_2\rangle = 2\kappa$$

So the sign at b in the sign characteristic of (iA,iH) is κ. Likewise

$$\langle iHi^{n-1}(e_{2n-1} + ie_{2n}), e_1 + ie_2\rangle = \langle He_{2n-1}, e_1\rangle + \langle He_{2n}, e_2\rangle = 2\kappa,$$

so the sign at -b in the sign characteristic of (iA,iH) is κ.

If n = 1 mod 4 we have

$$H = \kappa\begin{pmatrix} & & & 0 & 1 \\ & & & -1 & 0 \\ & & \cdot^{\cdot^{\cdot}} & & \\ & 0 \ \ -1 & & & \\ & 1 \ \ 0 & & & \\ 0 \ \ 1 & & & & \\ -1 \ \ 0 & & & & \end{pmatrix}$$

Using $i^{n-1} = 1$ we have

$$\langle iHi^{n-1}(e_{2n-1} - ie_{2n}), e_1 - ie_2\rangle = \langle He_{2n}, e_1\rangle - \langle He_{2n-1}, e_2\rangle = 2\kappa$$

and

$$\langle iHi^{n-1}(e_{2n-1} + ie_{2n}), e_1 + ie_2\rangle = -\langle He_{2n}, e_1\rangle + \langle He_{2n-1}, e_2\rangle = -2\kappa.$$

So the sign at b in the sign characteristic of (iA,iH) is κ, the sign at -b is $-\kappa$.

If n = 2 mod 4 we have

$$H = \kappa \begin{pmatrix} & & & & -1 & 0 \\ & & & & 0 & -1 \\ & & & \cdot^{\cdot^{\cdot}} & & \\ & & -1 & 0 & & \\ & & 0 & -1 & & \\ 1 & 0 & & & & \\ 0 & 1 & & & & \end{pmatrix}$$

A computation like in the case n = 0 mod 4 shows that the sign at both b and -b in the sign characteristic of (iA,iH) is κ.

In case n = 3 mod 4 we have

$$H = \kappa \begin{pmatrix} & & & & 0 & -1 \\ & & & & 1 & 0 \\ & & & \cdot^{\cdot^{\cdot}} & & \\ & & 0 & 1 & & \\ & & -1 & 0 & & \\ 0 & -1 & & & & \\ 1 & 0 & & & & \end{pmatrix}$$

Then a computation as in the case n = 1 mod 4 shows that the sign at b in the sign characteristic of (iA,iH) is κ, the sign at -b is $-\kappa$. □

Nex, we shall prove Theorems 3.1 and 3.2.

PROOF OF THEOREM 3.1. Because of the localization mentioned in Section 2 we can assume (A,H) is in one of the four forms (3.1) - (3.4). In case A has no pure imaginary eigenvalues existence of an A-invariant lagrangian subspace is evident: one can just take the spectral subspace of A corresponding to the open right half plane. In case $\sigma(A) = \{0\}$ and (A,H) are given by (3.1) we construct and $M \in J(A,H)$ as follows:

(3.5) $$M = N_1 \dotplus \dots \dotplus N_p \dotplus M_{p+1} \dotplus \dots \dotplus M_{p+q}.$$

Here the subspaces N_i and M_i are given by:

(3.6) $\quad N_i = \text{span}\{e_{i2n_i},\dots,e_{in_i+1}\} \quad i=1,\dots,p$

(3.7) $\quad M_i = \text{span}\{e_{i1},\dots,e_{i\,2n_i+1}\} \quad i=p+1,\dots,p+q,$

where e_{ij} denotes the vector with 1 on the $2n_1+\dots+2n_{i-1}+j$'th coordinate and zeros elsewhere for $i=1,\dots,p$ and with 1 on the $2n_1+\dots+2n_p+4n_{p+1}+2+\dots+4n_{i-1}+2+j$'th coordinate and zeros elsewhere for $i=p+1,\dots,p+q$.

Now suppose $\sigma(A) = \{\pm bi\}$, and suppose $J(A,H) \neq \emptyset$. Then $J^c(iA,iH) \neq \emptyset$ and by Theorem 5.1 in [RR3] (see also Section 1) the number of odd partial multiplicities of iA at $\pm b$ is even and the signs in the sign characteristic corresponding to these odd multiplicities sum to zero. From Proposition 3.3 one sees that this is equivalent to: the sum of the κ_j's corresponding to odd partial multiplicities of A at ib is zero, and the number of this odd partial multiplicities is even. Note that the partial multiplicities at -ib coincide with those at ib.

Conversely, suppose the condition of Theorem 3.1 is satisfied, and let

$$A = J_{2n_1}\begin{pmatrix}0 & b\\ -b & 0\end{pmatrix} \oplus \dots \oplus J_{2n_p}\begin{pmatrix}0 & b\\ -b & 0\end{pmatrix} \oplus J_{2n_{p+1}+1}\begin{pmatrix}0 & b\\ -b & 0\end{pmatrix} \oplus \dots \oplus J_{2n_{p+2q}+1}\begin{pmatrix}0 & b\\ -b & 0\end{pmatrix}.$$

Let H be as in (3.3). Let e_{ij} denote the vector with 1 on the $m_1+\dots+m_{i-1}+j$'th coordinate and zeros elsewhere, where $m_i = 2n_i$ for $i=1,\dots,p$ and $m_i = 2n_i+1$ for $i=p+1,\dots,p+2q$. Put

(3.8) $\quad N_i = \text{span}\{e_{i4n_i},\dots,e_{i2n_i+1}\} \quad i=1,\dots,p,$

(3.9) $\quad N_i = \text{span}\{e_{i2m_i},\dots,e_{im_i+2}\} \quad i=p+1,\dots,p+q,$

(3.10) $\quad M_{p+i} = N_{2i-1+p} \dotplus N_{2i+p} \dotplus \text{span}\{x_i,y_i\} \quad i=1,\dots,q$

where

$$x_i = e_{\alpha,m_\alpha+1} + e_{\alpha+1,m_{\alpha+1}+1} \qquad y_i = e_{\alpha,m_\alpha} + e_{\alpha+1,m_{\alpha+1}}$$

where $\alpha = p+2i-1$, $i=1,\dots,q$.

Then one checks that

$$M = N_1 \dotplus \dots \dotplus N_p \dotplus M_{p+1} \dotplus \dots \dotplus M_{p+q}$$

is in $J(A,H)$. $\square$

PROOF OF THEOREM 3.2. To prove Theorem 3.2 we can by localization consider three separate cases: $\sigma(A) \cap i\mathbb{R} = \emptyset$, $\sigma(A) = \{0\}$ and $\sigma(A) = \{\pm ib\}$. First assume A has no pure imaginary eigenvalues. Using a transformation $(A,H) \to (S^{-1}AS, S^{T}HS)$ one can replace (A,H) by

$$A' = \begin{pmatrix} K & 0 \\ 0 & -K^{T} \end{pmatrix} \qquad H' = \begin{pmatrix} 0 & I_{\frac{n}{2}} \\ -I_{\frac{n}{2}} & 0 \end{pmatrix},$$

where K has all its eigenvalues in the right half plane. It suffices to prove the theorem for (A',H') instead of (A,H). Let M'_+, resp. M'_-, denote the spectral subspace of A' corresponding to the right, resp. left, half plane. So let $N \subset M'_+$ be an arbitrary A'-invariant subspace in M'_+. Then $N = \left\{\begin{pmatrix} x \\ 0 \end{pmatrix} \mid x \in N_+\right\}$ for some K-invariant subspace N_+. Put

$$M = N \dotplus (H'N)^{\perp} \cap M'_- = \left\{\begin{pmatrix} x \\ y \end{pmatrix} \mid x \in N_+, y \in N_+^{\perp}\right\}$$

Clearly M is A'-invariant and lagrangian, and $M \cap M_+ = N$. Conversely, let $M_1 \in J(A',H')$ be a subspace with $M_1 \cap M'_+ = N$. Then clearly $M_1 \cap M'_-$ must be contained in $(H'N)^{\perp} \cap M'_-$. Indeed

$$M_1 = H'^{-1}M_1^{\perp} \subset H'^{-1}N^{\perp},$$

so

$$M_1 \cap M'_- \subset H'^{-1}N^{\perp} \cap M'_- = (H'N)^{\perp} \cap M'_-.$$

However, $\dim M_1 \cap M'_- = \frac{n}{2} - \dim N_+ = \dim N_+^{\perp} = \dim (H'N)^{\perp} \cap M'_-$. So $M_1 = M$. We have shown that for each $N \subset M'_+$ which is A'-invariant there is a unique $M \in J(A,H)$ with $M \cap M'_+ = N$.

Next, suppose $\sigma(A) \subset i\mathbb{R}$ and that the conditions of the theorem are fulfilled. We have to show that $J(A,H)$ consists of only one element. Clearly, by Theorem 3.1, $J(A,H) \neq \emptyset$. Suppose $M_1, M_2 \in J(A,H)$ and $M_1 \neq M_2$. Then also $J^{c}(iA,iH)$ contains at least two elements. However, Proposition 3.3 implies that if (A,H) satisfies the conditions of theorem then all partial multiplicities of iA at real eigenvalues are even, and for each real eigenvalue of iA the signs in the sign characteristic of (iA,iH) corresponding to this eigenvalue are all the same. By Theorem 2.2 in [RR3] (see also Section 1.2) we have that $J^{c}(iA,iH)$ has only one element. Hence $J(A,H)$ consists of only one element.

In case the conditions of the theorem are not satisfied we have to show that $J(A,H)$ contains at least two elements. We start with the case $\sigma(A) = \{0\}$. Let (A,H) be given as in (3.1). Suppose first there is an odd partial multiplicity corresponding to zero, i.e. $q > 0$. Then (3.5) gives one element in J(A,H), and another one is provided by

$$M' = N_1 + \dots N_p + M'_{p+1} + \dots + M'_{p+q}.$$

where N_i is as in (3.6) and

$$M_i' = \operatorname{span}\{e_{i2n_i+2},\dots,e_{i4n_i+2}\}, \quad i = p+1,\dots,p+q.$$

So we can suppose q = 0, and for sake of simplicity we may assume $(-1)^{n_1}\kappa_1 \neq (-1)^{n_2}\kappa_2$, and $n_1 \geq n_2$. Consider $A' = J_{2n_1}(0) \oplus J_{2n_2}(0)$, $H' = \kappa_1 F_{2n_1} \oplus \kappa_2 F_{2n_2}$. Then, put

$$M' = \operatorname{span}\{e_{1,2n_1},\dots,e_{1,n_1+n_2+1},e_{1,n_1+n_2}+e_{2,2n_2},\dots,e_{1,n_1-n_2+1}+e_{2,1}\}$$

(In case $n_1 = n_2$, this starts with $e_{1,2n_1}+e_{2,2n_1}$.) This subspace is easily seen to be A'-invariant. Further, the span of $e_{1,2n_1},\dots,e_{1,n_1+n_2+1}$ is H-orthogonal to itself and to the span of the other basisvectors for M', and

$$\langle He_{1,n_1+n_2-k}+e_{2,2n_2-k}, e_{1,n_1+n_2-j}+e_{2,2n_2-j}\rangle$$

$$= \begin{cases} 0 & \text{if } 2n_1+2n_2-k-j \neq 2n_1+1 \text{ (or, equivalently:} \\ & 2n_2-k+2n_2-j \neq 2n_2+1) \\ (-1)^{n_2-k}\{(-1)^{n_1}\kappa_1+(-1)^{n_2}\kappa_2\} = 0 & \end{cases}$$

(The latter follows from $(-1)^{n_1}\kappa_1 \neq (-1)^{n_2}\kappa_2$, so one of these numbers is +1, the other -1.) Hence M' is in J(A',H'). Then $M' \dotplus N_3 \dotplus \dots \dotplus N_p$ is in J(A,H) and is clearly different from $M = N_1 \dotplus N_2 \dotplus N_3 \dotplus \dots \dotplus N_p$.

Now consider the case $\sigma(A) = \{\pm bi\}$. Let (A,H) be given by (3.3). We may suppose that either n_1,n_2 are both odd and $\kappa_1 \neq \kappa_2$ or n_1,n_2 are both even and $\kappa_1 \neq \kappa_2$. For sake of simplicity we shall only consider the case p = 2 since the general case can be reduced to this, as in the previous paragraph. First suppose n_1 and n_2 are odd, and let $n_i = 2p_i+1$, suppose also $n_1 \geq n_2$. Put

$$M(\varepsilon) = \operatorname{span}\{e_{1,2n_1},\dots,e_{1,n_1+n_2+1},e_{1,n_1+n_2}+\varepsilon e_{2,2n_2}, \dots,e_{1,n_1-n_2+1}+\varepsilon e_{2,1}\} \quad \varepsilon = \pm 1.$$

Then $M(\varepsilon)$ is easily seen to be A-invariant. Next we use

$$\langle He_{i,k},e_{i,j}\rangle = \begin{cases} 0 & \text{if } k+j \neq 2n_i+1 \\ (-1)^{[\frac{1}{2}j]+p_i}\kappa_i & \text{if } k+j = 2n_i+1 \end{cases}.$$

So one sees that the span of $e_{1,2n_1},\dots,e_{1,n_1+n_2+1}$ is H-orthogonal both to itself and to the span of the remainder basis vectors of $M(\varepsilon)$. Further

$$\langle He_{1,n_1+n_2-k}+\varepsilon e_{2,2n_2-k}, e_{1,n_1+n_2-j}+\varepsilon e_{2,2n_2-j}\rangle$$

$$= \langle He_{1,n_1+n_2-k}, e_{1,n_1+n_2-j}\rangle + \langle He_{2,2n_2-k}, e_{2,2n_2-j}\rangle$$

$$= \begin{cases} 0 \text{ if } 2n_1+2n_2-k-j \neq 2n_1+1, \text{ or, equivalently,} \\ \qquad 2n_2+2n_2-k-j \neq 2n_2+1 \\ (-1)^{[\frac{1}{2}(n_1+n_2-j)]+p_1}\kappa_1 + (-1)^{[n_2-\frac{1}{2}j]+p_2}\kappa_2. \end{cases}$$

Now $[\frac{1}{2}(n_1+n_2-j)]+p_1 = 2p_1+p_2+1+[-\frac{1}{2}j]$ and $[n_2-\frac{1}{2}j]+p_2 = 3p_2+1+[-\frac{1}{2}j]$. So

$$(-1)^{[\frac{1}{2}(n_1+n_2-j)]+p_1}\kappa_1 + (-1)^{[n_2-\frac{1}{2}j]+p_2}\kappa_2 = (-1)^{p_2+1+[-\frac{1}{2}j]}(\kappa_1+\kappa_2) = 0,$$

since $\kappa_1+\kappa_2 = 0$. As $M(1) \neq M(-1)$ we have shown that $J(A,H)$ contains at least two elements.

Now suppose n_1 and n_2 are even. Again we assume $n_1 \geq n_2$. Let $n_i = 2p_i$. We take $M(\varepsilon)$, $\varepsilon = \pm 1$ to be the same as in the previous case, so that $M(\varepsilon)$ is certainly A-invariant. In this case we have

$$\langle He_{i,k}, e_{i,j}\rangle = \begin{cases} 0 \text{ if } k+j \neq 2n_i+2 \text{ and } k,j \text{ even} \\ 0 \text{ if } k+j \neq 2n_i \text{ and } k,j \text{ odd.} \end{cases}$$

Further,

$$\langle He_{i,2n_i-j}, e_{i,j}\rangle = (-1)^{[\frac{1}{2}j]+p_i}\kappa_i, \qquad j \text{ odd}$$

and

$$\langle He_{i,2n_i-j}, e_{i,j+2}\rangle = (-1)^{\frac{1}{2}j+p_i}\kappa_i, \qquad j \text{ even.}$$

Again, one checks that $\text{span}\{e_{1,2n_1},\dots,e_{1,n_1+n_2+1}\}$ is H-orthogonal to the whole $M(\varepsilon)$ and further

$$\langle He_{1,n_1+n_2-k}+\varepsilon e_{2,2n_2-k}, e_{1,n_1+n_2-j}+\varepsilon e_{2,2n_2-j}\rangle$$

$$= \langle He_{1,n_1+n_2-k}, e_{1,n_1+n_2-j}\rangle + \langle He_{2,2n_2-k}, e_{2,2n_2-j}\rangle$$

$$= \begin{cases} 0 \text{ if } 2n_1+2n_2-k-j \neq 2n_1+2,\ k,j \text{ even, or equivalently} \\ \quad 4n_2-k-j \neq 2n_2+2,\ k,j \text{ even} \\ 0 \text{ if } 2n_1+2n_2-k-j \neq 2n_1,\ k,j \text{ odd, or equivalently} \\ \quad 4n_2-k-j \neq 2n_2,\ k,j \text{ odd.} \end{cases}$$

and finally, if $2n_1+2n_2-k-j = 2n_1+2$, k,j even:

$$\langle He_{1,n_1+n_2-k}, e_{1,n_1+n_2-j}\rangle + \langle He_{2,2n_2-k}, e_{2,2n_2-j}\rangle$$
$$= (-1)^{p_2-\frac{1}{2}j-1}\kappa_1 + (-1)^{n_2-\frac{1}{2}j-1+p_2}\kappa_2 = (-1)^{p_2-\frac{1}{2}j-1}\{\kappa_1+\kappa_2\} = 0,$$

and if $2n_1 + 2n_2 - k - j = 2n_1$, k,j odd, this expression equals

$$(-1)^{[\frac{1}{2}(n_1+n_2-j)]+p_1}\kappa_1 + (-1)^{[n_2-\frac{1}{2}j]+p_2}\kappa_2 = (-1)^{p_2+[-\frac{1}{2}j]}\{\kappa_1+\kappa_2\} = 0.$$

So $M(\varepsilon)$ is actually Lagrangian for $\varepsilon = \pm 1$. As in the previous case we conclude that J(A,H) contains at least two elements. This concludes the proof. □

3.3. Stability of invariant Lagrangian subspaces.

First we shall study conditional and unconditional stability of Lagrangian subspaces. The following theorem gives necessary and sufficient conditions for existence of such subspaces, and provides also a description.

THEOREM 3.4. *Suppose* $(A,H) \in L_n(-1,-1)$.

(i) *There exists a conditionally stable invariant Lagrangian subspace if and only if the partial multiplicities of* A *at each pure imaginary eigenvalue* ib *are even, and the signs in the sign characteristic of* (iA,iH) *corresponding to* ib *are all the same.*

(ii) *There exists an unconditionally stable invariant Lagrangian subspace if and only if* $\sigma(A) \cap i\mathbb{R} = \emptyset$.

(iii) *In case* (i) *or* (ii) *holds, a subspace* $M \in J(A,H)$ *is (un)conditionally stable if and only if the following hold:*

(a) $M \cap R(A,\lambda)$ *is either* (0) *or* $R(A,\lambda)$ *whenever* $0 \neq \lambda$ *is a real eigenvalue of* A *with* $\dim \operatorname{Ker}(A-\lambda) > 1$, *in this case* $M \cap R(A,-\lambda)$ *is* $R(A,-\lambda)$ *or* (0), *respectively,*

(b) $M \cap R(A,\lambda)$ *is an arbitrary even dimensional* A*-invariant subspace of* $R(A,\lambda)$ *whenever* $0 \neq \lambda$ *has geometric multiplicity one and the algebraic multiplicity of* λ *is even. In this case*

$$M \cap R(A,-\lambda) = [H(M \cap R(A,\lambda))]^{\perp} \cap R(A,-\lambda), \tag{3.11}$$

(c) $M \cap R(A,\lambda)$ *is an arbitrary* A*-invariant subspace of* $R(A,\lambda)$ *whenever* λ *has geometric multiplicity one and the algebraic multiplicity of* λ *is odd. Again, in this case* $M \cap R(A,-\lambda)$ *is given by* (3.11),

(d) $M \cap R(A,a \pm ib)$ *is either* (0) *or* $R(A,a\pm ib)$ *whenever* $a \pm ib$, $a \neq 0$, *are eigenvalues of* A *with geometric multiplicity at least two, in this case* $M \cap R(A,-a\pm ib)$ *is* $R(A,-a\pm ib)$ *or* (0), *respectively,*

(e) $M \cap R(A, a \pm ib)$ *is an arbitrary A-invariant subspace of* $R(A, a \pm ib)$ *whenever* $a \pm ib$, $a \neq 0$, *are eigenvalues of A with geometric multiplicity one. In this case*

$$M \cap R(A, -a \pm ib) = [H(M \cap R(A, a \pm ib))]^{\perp} \cap R(A, -a \pm ib). \tag{3.12}$$

PROOF. By Theorem 2.2 we have to consider only four separate cases: $\sigma(A) = \{a \pm ib, -a \pm ib\}$ $a > 0$, $b > 0$, $\sigma(A) = \{\lambda, -\lambda\}$, $\lambda > 0$, $\sigma(A) = \{\pm ib\}$, $b > 0$ and $\sigma(A) = \{0\}$. In effect, we shall treat the first two cases together. Suppose A has no pure imaginary eigenvalues. Then we may assume that (A,H) is in fact given by

$$A = \begin{pmatrix} K & 0 \\ 0 & -K^T \end{pmatrix} \qquad H = \begin{pmatrix} 0 & I_{\frac{n}{2}} \\ -I_{\frac{n}{2}} & 0 \end{pmatrix},$$

where all eigenvalues of K are in the open right half plane (see the proof of Theorem 3.2), and any $M \in J(A,H)$ has the form

$$M = \left\{ \begin{pmatrix} x \\ y \end{pmatrix} \mid x \in N_+, y \in N_+^{\perp} \right\}$$

for some K-invariant subspace N_+. We shall show that M is unconditionally stable if and only if N_+ is stable as a K-invariant subspace. Indeed suppose M is unconditionally stable. Consider a perturbation $\tilde{K}$ of K and put $\tilde{A} = \tilde{K} \oplus -\tilde{K}^T$. Then $(H, \tilde{A}) \in L_n(-1,-1)$. Since $\tilde{A}$ has no pure imaginary eigenvalues for $\tilde{K}$ sufficiently close to K we have $J(\tilde{A}, H) \neq \emptyset$ for $\|\tilde{K} - K\|$ sufficiently small. Since M is unconditionally stable in $J(A,H)$ there exists an $\tilde{M} \in J(\tilde{A}, H)$ close to M. Then

$$\tilde{M} = \left\{ \begin{pmatrix} x \\ y \end{pmatrix} \mid x \in \tilde{N}_+, y \in \tilde{N}_+^{\perp} \right\},$$

where $\tilde{N}_+$ is $\tilde{K}$-invariant. Then clearly $\mathrm{gap}(N_+, \tilde{N}_+) \leq \mathrm{gap}(M, \tilde{M})$, so N_+ is stable as a K-invariant subspace. Conversely, assume N_+ is stable as a K-invariant subspace. Then $N = \{\begin{pmatrix} x \\ 0 \end{pmatrix} \mid x \in N_+\}$ is stable as an A-invariant subspace. Let $(H, \tilde{A}) \in L_n(-1,-1)$ with $\tilde{A}$ close to A and such that $\sigma(\tilde{A}) \cap i\mathbb{R} = \emptyset$. Then there is an A-invariant subspace $\tilde{N}$ close to N. Clearly $\tilde{N}$ has to be contained in the spectral subspace of $\tilde{A}$ corresponding to the right half plane. Then $\tilde{M} := \tilde{N} \dotplus (H\tilde{N})^{\perp} \cap \tilde{M}_-$, where $\tilde{M}_-$ is the spectral subspace of $\tilde{A}$ corresponding to the open left half plane, is $\tilde{A}$-invariant and Lagrangian. Clearly, since $\mathrm{gap}(N, \tilde{N})$ is as small as we wish, so too is $\mathrm{gap}(M, \tilde{M})$. Hence M is stable in $J(A,H)$.

Combining this result with [BGK] Theorem 9.9 (see also Theorem 1.9), we obtain that if $\sigma(A) \cap i\mathbb{R} = \emptyset$ there exists an unconditionally stable (and hence also stable) subspace in $J(A,H)$, and we also obtain the description as provided in (iii).

Next, we consider the cases where $\sigma(A) \subset i\mathbb{R}$. Note that if the conditions under (i) are satisfied, then by Theorem 3.2 $J(A,H)$ consists only of one element, say M. By Theorem 2.4 M is conditionally stable, even conditionally *strongly* stable. It remains to show the converse statement in (i), and the fact that M is not unconditionally stable in case (i) holds, thereby establishing (ii).

Suppose first $\sigma(A) = \{0\}$ and let (A,H) be given as in (3.1). Let $M \in J(A,H)$ be conditionally stable. First we show that

$$M = N_1 \dotplus \dots \dotplus N_p \dotplus M_{p+1} \dotplus \dots \dotplus M_{p+q} \tag{3.13}$$

where N_i is $J_{2n_i}(0)$-invariant and Lagrangian with respect to $\kappa_i F_{2n_i}$, and M_{p+i} is $J_{2n_{p+i}+1}(0) \oplus -(J_{2n_{p+i}+1}(0))^T$-invariant and Lagrangian with respect to

$$\begin{pmatrix} 0 & I_{2n_{p+i}+1} \\ -I_{2n_{p+i}+1} & \end{pmatrix}.$$

Indeed, let $\varepsilon_1,\dots,\varepsilon_{p+q}$ be different real numbers, and put

$$A(\varepsilon) = J_1'(\varepsilon_1) \oplus \dots \oplus J_p'(\varepsilon_p) \oplus J_{2n_{p+1}+1}(\varepsilon_{p+1}) \oplus$$
$$\oplus -(J_{2n_{p+1}+1}(\varepsilon_{p+1}))^T \oplus \dots \oplus J_{2n_{p+q}+1}(\varepsilon_{p+q}) \oplus -(J_{2n_{p+q}+1}(\varepsilon_{p+q}))^T,$$

where $J_i'(\varepsilon_i)$ is given by

$$J_i'(\varepsilon_i) = \begin{pmatrix} -\varepsilon_i & & & & & & \\ 1 & \varepsilon_i & & & & & \\ & \ddots & -\varepsilon_i & & & & \\ & & \ddots & \ddots & & & \\ & & & \ddots & \ddots & & \\ & & & & \ddots & -\varepsilon_i & \\ & & & & & 1 & \varepsilon_i \end{pmatrix} \qquad (2n_i \times 2n_i).$$

Then $(A(\varepsilon),H) \in L_n(-1,-1)$, and every subspace $M' \in J(A(\varepsilon),H)$ (which is not empty since $\sigma(A(\varepsilon)) \cap i\mathbb{R} = \emptyset$) has the form

$$M' = N_1' \dotplus \dots \dotplus N_p' \dotplus M_{p+1}' \dotplus \dots \dotplus M_{p+q}',$$

where N_i' is in $J(J_i'(\varepsilon_i),\kappa_i F_{2n_i})$ and

$$M'_{p+i} \in J\left(J_{2n_{p+i}+1}(\varepsilon_{p+i}) \oplus -(J_{2n_{p+i}+1}(\varepsilon_{p+i}))^T, \begin{pmatrix} 0 & I_{2n_{p+i}+1} \\ -I_{2n_{p+i}+1} & 0 \end{pmatrix}\right)$$

Since M is stable in J(A,H) taking limits $\varepsilon_i \to 0$ yields (3.13).

Now we show that there are actually no odd partial multiplicities. Because of what we have shown in the previous paragraph we may as well suppose $A = J_{2k+1}(0) \oplus -J_{2k+1}(0)^T$, and

$$H = \begin{pmatrix} 0 & I_{2k+1} \\ -I_{2k+1} & 0 \end{pmatrix}$$

(see Proposition 2.5). Put $Z = \mathrm{diag}(0,\ldots,0,\varepsilon)$ of size $(2k+1)\times(2k+1)$, and $Y = \mathrm{diag}(\varepsilon,0,\ldots,0)$ of the same size. First consider the perturbation

$$A(\varepsilon) = \begin{bmatrix} J_{2k+1}(0) & 0 \\ Z & -J_{2k+1}(0)^T \end{bmatrix}.$$

Then $(A(\varepsilon),H) \in L_n(-1,-1)$, $\dim \mathrm{Ker}\, A(\varepsilon) = 1$, so $J(A(\varepsilon),H)$ consists of only one element, being $(0) \dotplus \mathbb{R}^{2k+1}$. Letting $\varepsilon \to 0$ we see that if $M \in J(A,H)$ is conditionally stable then $M = (0) \dotplus \mathbb{R}^{2k+1}$. Next, perturb A as follows

$$\tilde{A}(\varepsilon) = \begin{bmatrix} J_{2k+1}(0) & Y \\ 0 & -J_{2k+1}(0)^T \end{bmatrix}.$$

Applying the same arguments we see that M has to be $\mathbb{R}^{2k+1} \dotplus (0)$, which is a contradiction. So there are no blocks of odd order in the Jordan canonical form of A.

Next we show that the sign $(-1)^{n_i}\kappa_i$ are all the same. Suppose not. Again, for sake of simplicity assume (cf. Proposition 2.5)

$$A = J_{2n_1}(0) \oplus J_{2n_2}(0), \qquad H = \kappa_1 F_{2n_1} \oplus \kappa_2 F_{2n_2}$$

and $(-1)^{n_1}\kappa_1 \neq (-1)^{n_2}\kappa_2$. We have already seen from (3.13) that if M is conditionally stable then $M = (0)_{n_1} \oplus \mathbb{R}^{n_1} \oplus (0)_{n_2} \oplus \mathbb{R}^{n_2}$. Assume $n_1 \geq n_2$. Let Z be the $2n_2 \times 2n_2$ matrix with ε in the upper right corner and zeros elsewhere, and put

$$A(\varepsilon) = A + \begin{pmatrix} 0_{n_1-n_2} & 0 & 0 & 0 \\ 0 & -Z & 0 & Z \\ 0 & 0 & 0_{n_1-n_2} & 0 \\ 0 & -Z & 0 & Z \end{pmatrix}$$

Decomposing H as

$$H = \begin{pmatrix} & & -H_1^T & \\ & \tilde{\kappa}_1 F_{2n_2} & & \\ H_1 & & & \\ & & & \kappa_2 F_{2n_2} \end{pmatrix},$$

where $\tilde{\kappa}_1 = (-1)^{n_1-n_2}\kappa_1$ and H_1 is given by

$$H_1 = \kappa_1 \begin{pmatrix} & & & (-1)^{n_1-n_2} \\ & & \cdot^{\cdot^{\cdot}} & \\ & 1 & & \\ -1 & & & \end{pmatrix}$$

one easily checks that $(A(\varepsilon),H) \in L_n(-1,-1)$. Moreover, $\sigma(A(\varepsilon)) = \{0\}$ and $\dim \operatorname{Ker} A(\varepsilon) = 1$. So there is a unique element in $J(A(\varepsilon),H)$, which in fact is given by

$$M' = \operatorname{span}\{e_{1,2n_1},\dots,e_{1,n_1+n_2+1}, e_{1,n_1+n_2}+e_{2,2n_2},\dots,$$
$$\dots,e_{1,n_1-n_2+1}+e_{2,1}\}.$$

Since M' does not depend on ε, we have, by letting $\varepsilon \to 0$, that $M = M'$. This contradicts $M = (0)_{n_1} \dotplus \mathbb{R}^{n_1} \dotplus (0)_{n_2} \dotplus \mathbb{R}^{n_2}$. So (i) is proved for the case $\sigma(A) = \{0\}$.

Finally we show (ii) for the case $\sigma(A) = \{0\}$. If there is an unconditionally stable element in $J(A,H)$ this element is certainly conditionally stable, so the conditions of (i) hold for (A,H). Consider the following perturbation on the block $J_{2n_1}(0)$ in A: $A(\varepsilon) = A + Z$, where Z has zeros everywhere except on the n_1,n_1+1 entry, where it has $-\varepsilon^2$. Then $\sigma(A(\varepsilon)) = \{\pm i\varepsilon\} \cup \{0\}$ if $n > 2$ and $\sigma(A(\varepsilon)) = \{\pm i\varepsilon\}$ if $n = 2$ (here n is the size of A). In any case the partial multiplicity of $A(\varepsilon)$ at $i\varepsilon$ is one, so $J(A(\varepsilon),H) = \emptyset$. Hence there cannot exist an unconditionally stable element in $J(A,H)$.

Next, we shall consider the case $\sigma(A) = \{\pm bi\}$. Let (A,H) be given by

$$A = J_{n_1}\begin{pmatrix} 0 & b \\ -b & 0 \end{pmatrix} \oplus \dots \oplus J_{n_p}\begin{pmatrix} 0 & b \\ -b & 0 \end{pmatrix} \oplus J_{n_{p+1}}\begin{pmatrix} 0 & b \\ -b & 0 \end{pmatrix} \oplus \dots \oplus J_{n_{p+2q}}\begin{pmatrix} 0 & b \\ -b & 0 \end{pmatrix}$$

and H as in (3.3), where $n_1,\dots,n_p$ are even and $n_{p+1},\dots,n_{p+2q}$ are odd, and κ_{p+i} is +1 for i odd, $\kappa_{p+i} = -1$ for i even. Let $\varepsilon_1,\dots,\varepsilon_{p+q}$ be different real numbers, and put

$$A(\varepsilon) = J_{n_1}\begin{pmatrix} 0 & b+\varepsilon_1 \\ -b-\varepsilon_1 & 0 \end{pmatrix} \oplus \dots \oplus J_{n_p}\begin{pmatrix} 0 & b+\varepsilon_p \\ -b-\varepsilon_p & 0 \end{pmatrix} \oplus$$

$$\oplus J_{n_{p+1}}\begin{pmatrix} 0 & b+\varepsilon_{p+1} \\ -b-\varepsilon_{p+1} & 0 \end{pmatrix} \oplus J_{n_{p+2}}\begin{pmatrix} 0 & b+\varepsilon_{p+1} \\ -b-\varepsilon_{p+1} & 0 \end{pmatrix} \oplus \dots \oplus$$

$$\oplus J_{n_{p+2q-1}}\begin{pmatrix} 0 & b+\varepsilon_{p+q} \\ -b-\varepsilon_{p+q} & 0 \end{pmatrix} \quad J_{n_{p+2q}}\begin{pmatrix} 0 & b+\varepsilon_{p+q} \\ -b-\varepsilon_{p+q} & 0 \end{pmatrix} .$$

Then $(A(\varepsilon),H) \in L_n(-1,-1)$, $J(A(\varepsilon),H) \neq \emptyset$, and every element in $J(A(\varepsilon),H)$ has the form

$$M' = N_1' \dotplus \dots \dotplus N_p' \dotplus M_{p+1}' \dotplus \dots \dotplus M_{p+q}'$$

where N_i' is $J_{n_i}\begin{pmatrix} 0 & b \\ -b & 0 \end{pmatrix}$-invariant and Lagrangian with respect to the i'th block in H, and M_{p+i}' is $J_{n_{p+2i-1}}\begin{pmatrix} 0 & b \\ -b & 0 \end{pmatrix} \oplus J_{n_{p+2i}}\begin{pmatrix} 0 & b \\ -b & 0 \end{pmatrix}$-invariant and Lagrangian with respect to the direct sum of the p+2i-1'th and the p+2i'th block in H. Letting $\varepsilon_i \to 0$ for all i one sees that M itself has this form.

Now we shall show that there are no odd partial multiplicities of A corresponding to zero. For sake of simplicity we can assume (by Proposition 2.5)

$$A = J_{n_1}\begin{pmatrix} 0 & b \\ -b & 0 \end{pmatrix} \oplus J_{n_2}\begin{pmatrix} 0 & b \\ -b & 0 \end{pmatrix},$$

where n_1,n_2 are odd, $n_i = 2p_i+1$ and $n_1 \geq n_2$. Let Z be the $2n_2 \times 2n_2$ matrix with zeros everywhere except for the places $1,2n_2$ and $2,2n_2-1$ where it is ε. Consider for $\mu = \pm 1$ the perturbation

$$A(\mu,\varepsilon) = A + \begin{pmatrix} 0_{n_1-n_2} & & & 0 \\ & -Z & & \mu Z \\ & & 0_{n_1-n_2} & 0 \\ 0 & -(-1)^{p_1-p_2}\mu Z & 0 & (-1)^{p_1-p_2}Z \end{pmatrix} .$$

Since H can be blockdecomposed as

$$H = \begin{pmatrix} & & H_1 & \\ & (-1)^{p_1-p_2}\kappa_1 H_2 & & \\ -H_1^T & & & \\ & & & \kappa_2 H_2 \end{pmatrix}$$

and $\kappa_1 \neq \kappa_2$ one easily checks that $(A(\mu,\varepsilon),H) \in L_n(-1,-1)$. Further, one checks that $\sigma(A(\mu,\varepsilon)) = \{\pm bi\}$ and $\dim \operatorname{Ker}(A(\mu,\varepsilon) \pm bi) = 1$. So $J(A(\mu,\varepsilon),H)$ consists of only one element, which is given by

$$M(\mu) = \{e_{1,2n_1},\dots,e_{1,n_1+n_2+1},e_{1,n_1+n_2}+\mu e_{2,2n_2},$$
$$\dots,e_{1,n_1-n_2+1}+\mu e_{2,1}\}.$$

Letting $\varepsilon \to 0$ one sees that M has to be equal both to M(1) and M(-1). However $M(1) \neq M(-1)$, so in fact there is no stable element in J(A,H).

To prove (i) it remains to show that in case the signs κ_i are not all the same there is no conditionally stable subspace in J(A,H). Again, we can assume

$$A = J_{n_1}\begin{pmatrix} 0 & b \\ -b & 0 \end{pmatrix} \oplus J_{n_2}\begin{pmatrix} 0 & b \\ -b & 0 \end{pmatrix}$$

where $n_i = 2p_i$, $n_1 \geq n_2$ and $\kappa_1 \neq \kappa_2$. Consider the same perturbation $A(\mu,\varepsilon)$ as above. Precisely the same argument yields that also in this case there is no conditionally stable subspace.

Finally we show (ii) by proving that there exists no unconditionally stable subspace in this case. Suppose there exists an unconditionally stable subspace, then there certainly exists a conditionally stable subspace. So let A be as in (3.3) with all n_i's even and all κ_i's equal. Perturb $J_{n_1}\begin{pmatrix} 0 & b \\ -b & 0 \end{pmatrix}$ such that in the middle 4×4 block of the perturbation we have the following matrix

$$\begin{pmatrix} 0 & b & -\varepsilon^2 & 0 \\ -b & 0 & 0 & -\varepsilon^2 \\ 1 & 0 & 0 & b \\ 0 & 1 & -b & 0 \end{pmatrix}.$$

A little computation shows that in this case $\pm i(b-\varepsilon)$ and $\pm i(b+\varepsilon)$ are eigenvalues of the perturbed matrix with partial multiplicities one. Letting $\varepsilon \to 0$ one sees that there is no unconditionally stable subspace. □

Note that because of Lemma 1.13 and Theorem 2.4 every (un)conditionally stable subspace is also (un)conditionally strongly stable. So we have established in the preceding proof also the following statement.

THEOREM 3.5. *Suppose* $(A,H) \in L_n(-1,-1)$. *Then there exists an (un)conditionally stable invariant Lagrangian subspace if and only if there exists an (un)conditionally strongly stable invariant Lagrangian subspace, and these two classes of subspaces coincide.*

Lipschitz stability of Lagrangian invariant subspaces for a pair $(A,H) \in L_n(-1,-1)$ is considered next.

THEOREM 3.6. *Let* $(A,H) \in L_n(-1,-1)$. *Then the following are equivalent:*

(i) *there exists an unconditionally strongly Lipschitz stable subspace in* $J(A,H)$,

(ii) *there exists a conditionally strongly Lipschitz stable subspace in* $J(A,H)$,

(iii) *there exists a conditionally Lipschitz stable subspace in* $J(A,H)$,

(iv) $\sigma(A) \cap i\mathbb{R} = \emptyset$.

In this case $M \in J(A,H)$ *is Lipschitz stable in* $J(A,H)$ *if and only if* M *is a spectral subspace.*

PROOF. Clearly (i) $\Rightarrow$ (ii) $\Rightarrow$ (iii). To prove (iv) $\Rightarrow$ (i) just let $M \in J(A,H)$ be the spectral subspace of A corresponding to the open right half plane. Then M is obviously unconditionally strongly Lipschitz stable.

To prove (iii) $\Rightarrow$ (iv) suppose $ib \in \sigma(A)$, $b \in \mathbb{R}$, and suppose $M \in J(A,H)$ is conditionally Lipschitz stable. Then certainly M is conditionally stable and hence we may assume that either (A,H) is of the form (3.3), with n_i even and all κ_i's the same, or (A,H) is of the form (3.1) with $q = 0$ and all κ_i's the same. In the former case consider the perturbation

$$A(\alpha) = \left(J_{n_1}\begin{pmatrix} 0 & b \\ -b & 0\end{pmatrix} + \begin{pmatrix} 0 & \cdots & 0 & \alpha & 0 \\ \vdots & & & 0 & \alpha \\ \vdots & & & & 0 \\ \vdots & & & & \vdots \\ 0 & \cdots & \cdots & \cdots & 0 \end{pmatrix}\right) \oplus \bigoplus_{i=2}^{p} J_{n_i}\begin{pmatrix} 0 & b \\ -b & 0\end{pmatrix}.$$

Clearly $(A(\alpha),H) \in L_n(-1,-1)$. We will let $\alpha > 0$ if $n_1 = 2+4m$ for some m and $\alpha < 0$ if $n_1 = 4m$ for some m. One checks that in that case the first block in $A(\alpha)$ does not have pure imaginary eigenvalues. So $J(A(\alpha),H) \neq \emptyset$. Now apply an argument used in the proof of Theorem 15.9.7 in [GLR1] (see also Theorem 15.5.1 in [GLR1]), to see that M cannot be Lipschitz stable as for all elements $M(\alpha)$ in $J(A(\alpha),H)$ one has $\mathrm{gap}(M(\alpha),M) > C \cdot \alpha^{2/n_1}$, whereas $\|A - A(\alpha)\| = \alpha$. (Here we use that M is nontrivial.) In the latter case, i.e. $b = 0$, the arguments are analogous, considering the perturbation

$$A(\alpha) = \left(J_{2n_1}(0) + \begin{pmatrix} 0 & \cdots & 0 & \alpha \\ \vdots & & & 0 \\ \vdots & & & \vdots \\ 0 & \cdots & \cdots & 0 \end{pmatrix}\right) \oplus \bigoplus_{i=2}^{p} J_{2n_i}(0),$$

where $\alpha > 0$ if $n_1 = 2+4m$ and $\alpha < 0$ if $n_1 = 4m$. Again $(A(\alpha),H) \in L_n(-1,-1)$ and $J(A(\alpha),H) \neq \emptyset$. From here on the argument is as in the previous case.

Finally we prove the last statement of the theorem. If M is a spectral subspace then M is clearly Lipschitz stable in J(A,H) (compare Theorem 15.9.7 in [GLR1]). Conversely, if M is Lipschitz stable it is certainly unconditionally stable, so by Theorem 3.4 we only have to consider the cases $\sigma(A) = \{\pm(a \pm ib)\}$ $a \neq 0$, $b \neq 0$, $\dim \operatorname{Ker}(A \pm (a \pm ib)) = 1$, and $\sigma(A) = \{-\lambda, \lambda\}$, $\lambda \in \mathbb{R}$, $\dim \operatorname{Ker}(A \pm \lambda) = 1$, i.e. the cases (3.4) and (3.2), respectively, with $p = 1$. In the former case consider the perturbation

$$A(\alpha) = (J_n\begin{pmatrix} a & b \\ -b & a \end{pmatrix} + Z) \oplus (-J_n\begin{pmatrix} a & b \\ -b & a \end{pmatrix} - Z)^T$$

where

$$Z = \begin{bmatrix} 0 & \cdots & 0 & \alpha & 0 \\ \vdots & & & 0 & \alpha \\ \vdots & & & & 0 \\ \vdots & & & & \vdots \\ 0 & \cdots & & \cdots & 0 \end{bmatrix}.$$

Applying the argument in the proof of Theorem 15.9.7 of [GLR1] we mentioned before one sees that a subspace $M \in J(A,H)$ cannot be Lipschitz stable unless it is spectral. In the latter case the argument is analogous, considering the perturbation

$$A(\alpha) = \left(J_n(a) + \begin{bmatrix} 0 & \cdots & 0 & \alpha \\ \vdots & & & 0 \\ \vdots & & & \vdots \\ 0 & \cdots & & C \end{bmatrix}\right) \oplus \left(-J_n(a)^T + \begin{bmatrix} 0 & \cdots & & 0 \\ \vdots & & & \vdots \\ 0 & & & \vdots \\ -\alpha & 0 & \cdots & 0 \end{bmatrix}\right).$$

□

We next consider the question whether or not a stable invariant Lagrangian subspace can be approximated with *stable* invariant Lagrangian subspaces.

THEOREM 3.7. *Let* $(A,H) \in L_n(-1,-1)$ *and suppose* $M \in J(A,H)$ *is (un)conditionally stable. Then for each* $\varepsilon > 0$ *there exists* $\delta > 0$ *such that for every pair* $(A',H') \in L_n(-1,-1)$ *with* $\|A - A'\| + \|H - H'\| < \delta$ *there is a (un)conditionally stable* $M' \in J(A',H')$ *with* $\operatorname{gap}(M,M') < \varepsilon$.

PROOF. Since $(iA,iH) \in L_n^c$ and this pair satisfies the *sign condition* we can apply Theorem 6.5 in [RR3] to see that if $\|A - A'\| + \|H - H'\|$ is small enough then (iA',iH') also satisfied the sign condition and hence there exists an (un)conditionally stable element in J(A,H). It remains to show that this subspace can be chosen close to M. For this we only have to consider the cases $\sigma(A) \cap i\mathbb{R} = \emptyset$ and $\sigma(A) \subset i\mathbb{R}$. In the former case every subspace $M' \in J(A',H')$ with $\operatorname{gap}(M,M')$ small enough is (un)conditionally stable (compare Corollary 4.3 in [RR4]). In the latter case we can take for M' the unique element in

$J(A',H')$ with $\mathrm{Re}\,\sigma(A'|_{M'}) \le 0$. Since M is strongly stable by Theorem 3.5, we can make $\mathrm{gap}(M,M')$ arbitrary small by taking $\|A - A'\| + \|H - H'\|$ small. Clearly also M' is (un)conditionally stable by Theorem 3.4. □

It is not true that any subspace $M' \in J(A',H')$ is conditionally stable provided $\|A - A'\| + \|H - H'\| + \mathrm{gap}(M,M')$ is small enough. This is seen by the following example. Let $\eta \ge 0$ and take

$$A(\eta) = \begin{pmatrix} \eta & 1 & 0 & 0 \\ 0 & -\eta & 0 & 0 \\ 0 & 0 & -\eta & 1 \\ 0 & 0 & 0 & \eta \end{pmatrix} \qquad H = \begin{pmatrix} 0 & 1 & 0 & 0 \\ -1 & 0 & 0 & 0 \\ 0 & 0 & 0 & 1 \\ 0 & 0 & -1 & 0 \end{pmatrix}$$

Then $(A(\eta),H) \in L_4(-1,-1)$. Put $M = \mathrm{span}\{e_1,e_3\}$ then $M \in J(A(\eta),H)$ for all η, and for $\eta = 0$ this M is conditionally stable. However for $\eta \ne 0$ this M is not conditionally stable as is seen from Theorem 3.4. As was already remarked in the proof of Theorem 3.7 in the case of unconditional stability (and thus in case $\sigma(A) \cap i\mathbb{R} = \emptyset$) every subspace $M' \in J(A',H')$ is unconditionally stable provided $\|A - A'\| + \|H - H'\| + \mathrm{gap}(M,M')$ is small enough.

3.4. Isolatedness of invariant Lagrangian subspaces.

In this section we study the connection between isolatedness and stability of elements in $J(A,H)$. Our main result is described in the following theorem.

THEOREM 3.8. *Let $(A,H) \in L_n(-1,-1)$ and suppose there is a conditionally stable subspace in $J(A,H)$. Then a subspace $M \in J(A,H)$ is isolated in $J(A,H)$ (with respect to the gap topology) if and only if the following hold*

(a) *$M \cap R(A,\lambda)$ is either (0) or $R(A,\lambda)$ whenever $0 \ne \lambda$ is a real eigenvalue of A of geometric multiplicity larger than one,*

(b) *$M \cap R(A,\lambda)$ is an arbitrary A-invariant subspace of $R(A,\lambda)$ whenever $0 \ne \lambda$ has geometric multiplicity one,*

(c) *$M \cap R(A,a \pm ib)$ is either (0) or $R(A,a \pm ib)$ whenever $a \pm ib$, $a \ne 0$, are eigenvalues of A with geometric multiplicity larger than one,*

(d) *$M \cap R(A,a \pm ib)$ is an arbitrary A-invariant subspace of $R(A,a \pm ib)$ whenever $a \pm ib$, $a \ne 0$, are eigenvalues of A with geometric multiplicity one.*

In case (a) and (b) we have

$$M \cap R(A,-\lambda) = [H(M \cap R(A,\lambda))]^{\perp} \cap R(A,-\lambda),$$

in case (c) and (d) we have

$$M \cap R(A,-a \pm ib) = [H(M \cap R(A,a \pm ib))]^{\perp} \cap R(A,-a \pm ib).$$

PROOF. By Theorem 2.2 we have to consider only the cases $\sigma(A) \subset i\mathbb{R}$ or $\sigma(A) \cap i\mathbb{R} = \emptyset$. In the first case J(A,H) consists of only one element, and there is nothing to prove. In the second case we may assume, as in the proofs of Theorems 3.2 and 3.4, that (A,H) is given by

$$A = \begin{pmatrix} K & 0 \\ 0 & -K^T \end{pmatrix}, \qquad H = \begin{pmatrix} 0 & I_{\frac{n}{2}} \\ -I_{\frac{n}{2}} & 0 \end{pmatrix}$$

where $\operatorname{Re}\sigma(K) > 0$, and

$$M = \left\{ \begin{pmatrix} x \\ y \end{pmatrix} \mid x \in N_+, y \in N_+^{\perp} \right\}$$

for some K-invariant subspace N_+. An argument similar to the one used to show that M is conditionally stable if and only if N_+ is stable as a K-invariant subspace (see the proof of Theorem 3.4), now shows that M is isolated in J(A,H) if and only if N_+ is isolated as a K-invariant subspace. It remains to apply Theorem 9.9 in [BGK]. □

Note that the theorem shows that stable subspaces are isolated in J(A,H), but not conversely. That is, a subspace isolated in J(A,H) need not be stable. Note also that the theorem requires the existence of a conditionally stable element of J(A,H) in advance. Concerning this point we conjecture that *the existence of an isolated element in* J(A,H) *implies the existence of a stable element in* J(A,H), i.e. implies that the condition (i) of Theorem 3.4 holds.

The following examples show that at least for small size matrices this conjecture holds.

EXAMPLES 1). Let $A = \begin{pmatrix} 0 & 0 \\ 0 & 0 \end{pmatrix}$, $H = \begin{pmatrix} 0 & 1 \\ -1 & 0 \end{pmatrix}$. Then every A-invariant subspace of dimension one is in J(A,H), so there is no isolated one.

2) Let

$$A = \begin{pmatrix} 0 & 0 & 0 & 0 \\ 1 & 0 & 0 & 0 \\ 0 & 0 & 0 & 0 \\ 0 & 0 & 1 & 0 \end{pmatrix}, \qquad H = \begin{pmatrix} 0 & 1 & 0 & 0 \\ -1 & 0 & 0 & 0 \\ 0 & 0 & 0 & -1 \\ 0 & 0 & 1 & 0 \end{pmatrix}.$$

Any element in J(A,H) is either Ker A or of the form

$$\operatorname{span}\left\{ \begin{pmatrix} 0 \\ 1 \\ 0 \\ 1 \end{pmatrix}, \begin{pmatrix} 1 \\ \gamma \\ 1 \\ \delta \end{pmatrix} \right\} \qquad \gamma, \delta \in \mathbb{R},$$

or of the form

$$\operatorname{span}\left\{\begin{pmatrix}0\\1\\0\\-1\end{pmatrix},\begin{pmatrix}1\\\gamma\\-1\\\delta\end{pmatrix}\right\},\qquad \gamma,\delta\in\mathbb{R}.$$

We shall show that Ker A is not isolated, thereby clearly showing there are no isolated elements in J(A,H). Take $\gamma = \delta = \frac{1}{\alpha}$, then

$$\operatorname{span}\left\{\begin{pmatrix}0\\1\\0\\1\end{pmatrix},\begin{pmatrix}1\\1/\alpha\\1\\1/\alpha\end{pmatrix}\right\}\to \operatorname{Ker} A. \qquad \square$$

REFERENCES

[AFS] C. Apostol, C. Foias and N. Salinas: On stable invariant subspaces, Integral Equations and Operator Theory 8 (1985), 721-750.

[BGK] H. Bart, I. Gohberg and M.A. Kaashoek: Minimal factorization of matrix and operator functions, Birkhäuser Verlag, Basel, 1979.

[CD] S. Campbell and J. Daughtry: The stable solutions of quadratic matrix equations, Proc. Amer. Math. Soc. 74 (1979), 19-23.

[DPWZ] D.Z. Djokovic, J. Potera, P. Winternitz and H. Zassenhaus: Normal forms of elements of classical real and complex Lie and Jordan algebras, Journal of Math. Physics 24 (1983), 1363-1374.

[GLR1] I. Gohberg, P. Lancaster and L. Rodman: Invariant subspaces of matrices with applications, J. Wiley, New York etc, 1986.

[GLR2] I. Gohberg, P. Lancaster and L. Rodman: Matrix polynomials, Academic Press, New York, 1982.

[GLR3] I. Gohberg, P. Lancaster and L. Rodman: Matrices and indefinite scalar products, Birkhäuser Verlag, Basel, 1983.

[GLR4] I. Gohberg, P. Lancaster and L. Rodman: Spectral analysis of self-adjoint matrix polynomials, Ann. of Math. 112 (1980), 34-71.

[GLR5] I. Gohberg, P. Lancaster and L. Rodman: Perturbations of H-selfadjoint matrices, with applications to differential equations, Integral Equations and Operator Theory 5 (1982), 718-757.

[GR] I. Gohberg and L. Rodman: On the distance between lattices of invariant subspaces, Linear Algebra. Appl. 76 (1986), 85-120.

[J] N. Jacobson: Lectures in Abstract Algebra II. Linear Algebra, Van Nostrand, Princeton etc, 1953.

[KMR] M.A. Kaashoek, C.V.M. van der Mee and L. Rodman: Analytic operator functions with compact spectrum II. Spectral pairs and factorization, Integral Equations and Operator Theory 5 (1982), 791-827.

[LT] P. Lancaster and M. Tismenetsky: Theory of matrices with applications, Academic Press, New York, 1985.

[RR1] A.C.M. Ran and L. Rodman: Stable real invariant semidefinite subspaces and stable factorizations of symmetric rational matrix functions, Linear and Multilinear Algebra, 1987, to appear.

[RR2] A.C.M. Ran and L. Rodman: Stability of invariant maximal semidefinite subspaces II. Applications: Selfadjoint rational matrix functions, algebraic Riccati equations, Linear Algebra Appl. 63 (1984), 133-173.

[RR3] A.C.M. Ran and L. Rodman: Stability of invariant maximal semidefinite subspaces I, Linear Algebra Appl. 62 (1984), 51-86.

[RR4] A.C.M. Ran and L. Rodman: Stability of neutral invariant subspaces in indefinite inner products and stable symmetric factorizations, Integral Equations and Operator Theory 6 (1983), 536-571.

[U] F. Uhlig: A canonical form for a pair of real symmetric matrices that generate a nonsingular pencil, Linear Algebra Appl. 14 (1976), 189-209.

André C.M. Ran
Faculteit Wiskunde en Informatica
Vrije Universiteit
Amsterdam
The Netherlands

Leiba Rodman
Department of Mathematics
Arizona State University
Tempe, Arizona
USA

Operator Theory:
Advances and Applications, Vol. 32

PRODUCTS OF KERNEL FUNCTIONS AND MODULE TENSOR PRODUCTS

Dedicated to the memory of Constantin Apostol

Norberto Salinas

0. INTRODUCTION

In the present paper, we relate the functional Hilbert space generated by the product of two kernel functions on a bounded domain $\Omega \subset \mathbb{C}^n$, to the module tensor product of the spaces generated by the factor kernel functions over the algebra of holomorphic functions on neighborhoods of $\overline{\Omega}$. The point of our results is that the first space is generated in a purely analytic manner, while the second is constructed in a purely algebraic fashion.

The definition of products of kernels was introduced in [3], but explicit examples of products of kernel functions were given in [4], and also in [2] where powers of the Szego kernel on the unit disk were considered. However, the need of understanding the structure of functional Hilbert spaces generated by products of kernel functions first arose in [4], where a very elegant solution to a conjecture posed in [7] was given. This conjecture was also solved, in the negative, independently and simultaneously in [6] by a direct computational method.

The above mentioned conjecture can be formulated, using the language of kernel functions (see section 1 for the appropriate definitions) as follows. Let K and K' be two generalized Bergman kernels on the unit disk $\mathbb{D}$, and assume that $\overline{\mathbb{D}}$ is a spectral set for the canonical models associated with K and K', respectively. Let $k: \mathbb{D} \to \mathbb{R}$ and $k': \mathbb{D} \to \mathbb{R}$ be the curvatures of K and K', respectively (we recall that $k(z) = -[\partial\bar{\partial}K(.,.)](z,z)$, $z \in \Omega$). If the canonical models associated with K and K' are similar, then

$$\lim_{z \to \partial\mathbb{D}} k(z)/k'(z) = 1.$$

The counter example given in [4], was the following. Let $K(z,w) = 1/(1 - z\overline{w})$ (the

Szego kernel), and $L(z,w) = -\log(1 - z\overline{w})$ (the Dirichlet kernel), z,w on $\mathbb{D}$. Consider the product $k'(z,w) = K(z,w)L(z,w)$. Then K' is a generalized Bergman kernel on $\mathbb{D}$, whose canonical model is a contraction. Further, the canonical models associated with K and K' are not similar, because K'/K = L is unbounded on $\mathbb{D}$ (see [8], Remark 4.7), but it is easy to check that if $z \to \partial\mathbb{D}$ then

$$k'(z)/k(z) = 1 - (1 - |z|^2)^2[\partial,\bar{\partial} \log L(.,.)](z,z) \to 1.$$

The organization of the paper is as follows. In section 1, we present all the definitions and facts needed in later sections. Along the way, we produce an extension of the analytic functional calculus for canonical models of sharp kernel functions (see Theorem 1.5). Section 2 is devoted to a discussion of the direct sum and tensor product of kernel functions, as well as their pointwise sums and products, and we prove some structural theorems concerning these operations. Finally, in section 3, we introduce the concept of cyclicity with respect to a pair of spaces $(\mathcal{H},\mathcal{M})$ (where $\mathcal{M} \subset \mathcal{H}$), and use this notion (see Theorem 3.5) to derive our main result, Corollary 3.6. This result asserts that, under very natural assumptions, the polynomial Hilbert module generated by the product of two kernel functions K and K' is isomorphic to the module tensor product of the spaces generated by K and K' over the polynomial algebra.

1. PRELIMINARIES

Although in the last section of these notes we restrict attention to certain specific kernels, the theory developed in the next two sections may be applied to more general kernel functions. Thus, in the present section, we recall some of the properties of kernel functions that will be needed in our discussions.

It may be worth pointing out (see Remark 1.6) that there are certain interesting pathological situations where our theory is applicable and where the behavior of the kernel functions involved is quite different from the classical kernels.

As in [1] and [8], we shall work with a Hilbert space $\mathcal{K}$ of holomorphic functions on a bounded domain Ω in $\mathbb{C}^n$, with values in a fixed separable Hilbert space ℓ^2_m), which is the space of all square summable complex sequences $\{\xi_i: 0 < i < m+1\}$, where $1 \leq m \leq \infty$.

We assume that the evaluation map $E_w: \mathcal{K} \to \ell^2_m$ at any point w of Ω is bounded and surjective. We also assume that the space $\mathcal{K}$ is invariant under multiplication by the standard coordinate functions $Z_j: \Omega \to \mathbb{C}$, $1 \leq j \leq n$.

The above assumptions are equivalent (see [1], Section 1, or [8], section 3) to the existence of an operator valued function K: $\Omega\times\Omega \to \mathcal{L}(\ell^2_m)$, enjoying the following properties:

I) $K(z,w)^* = K(w,z)$, and $K(w,w)$ is positive and invertible for all z and w in Ω.

II) K is holomorphic in the first variable and antiholomorphic in the second variable (i.e., K is a sesquianalytic function).

III) The Hilbert space generated by K (see [8], Remark 3.3) coincides with $\mathcal{K}$, and K is the reproducing kernel for $\mathcal{K}$, i.e., for a given w in Ω, for every $f \in \mathcal{K}$ and for all $\xi \in \ell^2_m$, we have

$$(f(w),\xi) = \langle f,K(.,w)\xi \rangle_{\mathcal{K}}.$$

IV) The n-tuple $Z = (z_1,\ldots,Z_n)$ of coordinate functions on Ω induces the n-tuple T_Z of multiplication operators on $\mathcal{K}$, so that each component of T_Z is bounded on $\mathcal{K}$.

Henceforth, a function K: $\Omega\times\Omega \to \mathcal{L}(\ell^2_m)$, satisfying properties I) through IV) will be called a kernel function on Ω, and the n-tuple T_Z will be called the canonical model associated with K (see [1] and [8]). The cardinal number m will be called the rank of K, i.e., m = Rank(K). Although all the results of this paper that deal with an arbitrary rank are valid for any cardinal number, we shall assume, for simplicity sake, that $m \leq \aleph_0$, so that the Hilbert spaces involved in our discussions are all separable.

REMARK 1.1. Let K be a kernel function on Ω, and let $\Gamma: \Omega \to \mathcal{L}(\ell^2_m,\mathcal{K})$ be the antiholomorphic function given by

(1) $$\Gamma(\omega)\xi = K(.,\omega)\xi,\ \omega \in \Omega,\ \xi \in \ell^2_m.$$

It follows that $\Gamma(w)$ coincides with the adjoint $(E_w)^*$ of the evaluation map at w. We also have:

(2) $$(T_Z)^*\Gamma(\omega) = \overline{\omega}\Gamma(\omega),\ \omega \in \Omega.$$

In other words, $\{\Gamma(\omega): \omega \in \Omega\}$ is a set of expanding "eigenfunctions" for $(T_Z)^*$. Moreover, let P(w) be the (orthogonal) projection onto $\mathrm{Ran}(\Gamma(w))$, or, equivalently,

(3) $$P(\omega) = \Gamma(\omega)[\Gamma(\omega)^*\Gamma(\omega)]^{-1}\Gamma(\omega)^*,\ \omega \in \Omega.$$

From [14], it follows that P represents an antiholomorphic hermitian vector bundle, and that Γ is a global antiholomorphic frame for P. We recall that Theorem 3.8 of [14] states that a real analytic projection valued map P on Ω represents an antiholomorphic hermitian vector bundle if and only if

$$(4) \qquad [\bar{\partial}P(w)]P(w) = \bar{\partial}P(w),$$

for every w in Ω, where $\bar{\partial}$ is the standard $\bar{\partial}$-differential operator from (0,0) forms into (0,1) forms with coefficients in the algebra of $\mathcal{L}(\mathcal{K})$-valued smooth functions on Ω. Indeed, since Γ is antiholomorphic, we have:

$$(5) \qquad \bar{\partial}P(\omega) = \{\bar{\partial}[\Gamma(\omega)(\Gamma(\omega)^*\Gamma(\omega))^{-1}]\}\Gamma(\omega)^*.$$

If we multiply (5) on the right by (3), we obtain (4), as desired. Conversely, given a Hilbert space $\mathcal{H}$, if there exist an operator T in $\mathcal{L}(\mathcal{H})$ and an antiholomorphic function $\Gamma\colon \Omega \to \mathcal{L}(\ell^2_m, \mathcal{H})$, such that $\Gamma(w)^*\Gamma(w)$ is invertible for every w in Ω, and such that (2) holds with T in place of T_Z, and the set $\{\mathrm{Ran}[\Gamma(\omega)]\colon \omega \in \Omega\}$ is dense in $\mathcal{H}$, then $\mathcal{H}$ is a functional Hilbert space of ℓ^2_m valued holomorphic functions on Ω, and it has a reproducing kernel K satisfying conditions I) through IV), with $\mathcal{H}$ in place of $\mathcal{K}$ (see [8]). In fact, $K(z,w) = \Gamma(z)^*\Gamma(w)$, z,w in Ω.

Given a kernel function K on Ω, we shall call the antiholomorphic function Γ defined by (1) the canonical section associated with K.

REMARK 1.2. The simplest and most important examples of kernel functions are those with a diagonal expansion about a given point, say the origin, namely, a function of the form:

$$K(z,w) = \sum A_\alpha z^\alpha \bar{w}^\alpha,$$

where the sum runs over all multi-indices $\alpha \in \mathbb{Z}^n_+$, $z^\alpha = z_1^{a_1},\ldots,z_n^{\alpha_n}$, and A_α is a positive invertible operator on ℓ^2_m for every $\alpha \in \mathbb{Z}^n_+$. Using the formalism of [8], Remark 4.3, in order to generate the Hilbert space $\mathcal{K}$ out of the kernel function K, we first construct the linear space $\mathcal{K}_0$ of all functions on Ω of the form:

$$f(z) = \sum_{0\le\alpha\le\gamma} \bar{\partial}^\alpha K(z,0)\xi_\alpha/\alpha! = \sum_{0\le\alpha\le\gamma} A_\alpha\xi_\alpha z^\alpha.$$

If $g(z) = \sum_{0 \le \beta \le \delta} A_\alpha \eta_\alpha z^\alpha$ is another function of the same type, we provide $\mathcal{K}_0$ with the inner-product

$$\langle f,g \rangle_0 = \sum_{0 \le \alpha \le \gamma, 0 \le \beta \le \delta} \langle \partial^\alpha \bar{\partial}^\beta K(0,0)\xi_\alpha, \eta_\beta \rangle = \sum_{0 \le \alpha \le \gamma} \langle \xi_\alpha, A_\alpha, \eta_\alpha \rangle.$$

The space $\mathcal{K}$ is obtained by completing $\mathcal{K}_0$ with respect to $\langle .,. \rangle_0$. It follows that $\mathcal{K}$ can be identified with a weighted sequence space with operator weights in $\mathcal{L}(\ell^2_m)$. The jth component T_{Z_j} of the canonical model is identified with the forward shift of weight sequence $W_{\alpha,j} = (A_{\alpha+\varepsilon j})^{-1/2}(A_\alpha)^{1/2}$, where εj is the multi-index all of whose components are zero except the jth one which is one. Condition IV) simply means that $\sup\|W_{\alpha,j}\| < \infty$, where the supremun is taken over all $\alpha \in \mathbb{Z}^n_+$, and $0 \le j \le n$. Via this identification, the canonical section Γ associated with K is given by

$$\Gamma(\omega)\xi = \{\xi\bar{\omega}^\alpha\},\ \omega \in \Omega,\ \xi \in \ell^2_m.$$

The above identification is carried out by the function Γ^*, as follows:

$$\Gamma^*(\omega)\{\xi_\alpha\} = \sum_\alpha A_\alpha \xi_\alpha\, z^\alpha.$$

Let Ω_0 be the (logarithmically convex complete Reinhardt) domain where the series representing the kernel function K around zero converges absolutely in norm. Then the function associated with a sequence $\{\xi_\alpha\}$, i.e. $f(z) = \sum_\alpha A_\alpha \xi_\alpha z^\alpha$ also converges absolutely in norm and uniformly on compacta of Ω_0. Furthermore, it is easy to see that the following strengthening of (1) holds, in this case, for all w in Ω_0.

V) $$\begin{aligned} \mathrm{Ker}(T_Z - w)^* &(= \pi_i\, \mathrm{Ker}(T_{Z_i} - w_i)^*) \\ &= \mathrm{Ran}[\Gamma(w)]. \end{aligned}$$

The Szego kernel, the Bergman kernel and the Dirichlet kernel on Reinhardt domain (see [1], Section 4) are very special examples of kernel functions of the above type. As we shall see in Theorems 1.4 and 1.5 below, condition V) is a very important property for a kernel function to have. Following [1], we introduce the next definition.

DEFINITION 1.3. A kernel function on Ω is called sharp if it satisfies property V) for all points w in Ω.

The following theorem was proved in [8], but used more extensively in [1].

THEOREM 1.4. *Let* K *and* K' *be two sharp kernel functions on a bounded domain* Ω *in* $\mathbb{C}^n$ *and let* $\mathcal{K}$ *and* $\mathcal{K}'$ *denote the Hilbert spaces they generate. Also, let* Γ *and* Γ' *be the canonical sections associated with* K *and* K'. *Then the following statements are equivalent.*

a) There exists a unitary transformation $U: \mathcal{K}' \to \mathcal{K}$ *that intertwines the corresponding canonical models on* $\mathcal{K}$ *and* $\mathcal{K}'$.

b) There exists a unitary transformation $U: \mathcal{K}' \to \mathcal{K}$, *such that* $U^*\Gamma(w)U = \Gamma'(w)$, *for every* w *in* Ω.

If in addition, K *and* K' *are normalized around some point* z *in* Ω, *i.e., there exists a neighborhood* Ω_z *of* z *such that* $K(z,w) = 1 = K'(z,w)$ *for every* w *in* Ω_z, *then the above statements a) and b) are equivalent to the following one:*

c) There exists a unitary operator V *on* ℓ^2_m *such that* $V^*K(z,w)V = K'(z,w)$, *for every* w *in* Ω_z.

THEOREM 1.5. *Let* K *be a sharp kernel function on a bounded domain* Ω *in* $\mathbb{C}^n$, *and let* $\mathcal{K}$ *be the Hilbert space generated by* K. *Let* Γ *be the canonical section associated to* K, *and let* $(T_Z)'$ *(respectively,* $(T_Z)''$ *) denote the commutant (resp. double commutant) of the canonical model* T_Z *associated with* K. *Then*

a) There exists an injective (algebraic) unital homomorphism F_K *from* $(T_Z)'$ *into the algebra of all holomorphic* $\mathcal{L}(\ell^2_m)$ *valued functions on* Ω, *such that:*

$$S^*\Gamma(\omega) = \Gamma(\omega)(F_K S)^*(\omega),\ \omega \in \Omega \text{ and for every } S \text{ in } (T_Z)'. \tag{6}$$

b) Let $\mathcal{A}_K$ *be the intersection of* $\mathrm{Ran}(F_K)$ *with the algebra of holomorphic functions of the form* fI_m, *where* $f: \Omega \to \mathbb{C}$, *and* I_m *is the identity on* ℓ^2_m, $m = \mathrm{Rank}(K)$. *Then the algebra* $\mathcal{A}(T_Z) = (F_K)^{-1}\mathcal{A}_K$ *is a weakly closed subalgebra of* $(T_Z)''$, *and the restriction of the map* F_K *to* $\mathcal{A}(T_Z)$ *is norm decreasing and extends the analytic functional calculus of the commuting n-*

tuple T_Z *(see [16]). In particular,* $\mathcal{A}_K$ *will be identified with a subalgebra of the Banach algebra* $H^\infty(\Omega)$ *of all bounded complex valued holomorphic functions on* Ω.

PROOF. The statement in a) is essentially contained in [8], Theorem 3.7. For the proof of b), we first point out that $\mathcal{A}(T_Z)$ is obviously contained in $(T_Z)''$ and that $\mathcal{A}(T_Z)$ clearly contains the analytic functional calculus of T_Z. From (6), it follows that if $S \in \mathcal{A}(T_Z)$, then

$$(F_K S)(w)K(w,w)(F_K S)^*(w) = \Gamma(w)^* SS^* \Gamma(w) \leq \|S\|^2 K(w,w).$$

Since $F_K S$ is actually a scalar valued function, and $K(w,w)$ is positive and invertible for all w in Ω, it follows that F_K is norm decreasing. We next prove that the map $S \to (F_K S)(w)$ from $(T_Z)'$ into $\mathcal{L}(\ell^2_m)$ is continuous for every w in Ω, when both the domain and range spaces are given the weak operator topology. Let $\{S_\nu\} \subset \mathcal{A}(T_Z)$, and assume that $\{S_\nu\}$ converges weakly to an operator S on $\mathcal{K}$. Then S is in $(T_Z)'$, and, by (6), for every $\xi \in \ell^2_m$, and every f in $\mathcal{K}$, we have:

$$\begin{aligned}\langle\Gamma(w)(F_K S)^*(w)\xi, f\rangle_{\mathcal{K}} &= (S^*\Gamma(w)\xi, f)_{\mathcal{K}} = \\ \lim_{\upsilon} \langle\Gamma(w)\xi, S_\nu f\rangle_{\mathcal{K}} &= \lim_{\upsilon} \langle\Gamma(F_K S_\nu)^*(w)\xi, f\rangle_{\mathcal{K}}\end{aligned}$$

Since $K(w,w)$ is invertible, we can find, for an arbitrary given η in ℓ^2_m, a (unique) ζ in ℓ^2_m such that $\eta = K(w,w)\zeta$. So, we let $f = \Gamma(w)\zeta$ in the above chain of identities, and we obtain:

$$\langle(F_K S)^*(w)\xi,\eta\rangle = \langle(F_K S)^*(w)\xi, K(w,w)\zeta\rangle = \lim_{\upsilon} \langle(F_K S_\nu)^*(w)\xi,\eta\rangle.$$

This establishes our claim. Thus, it follows that $\mathcal{A}(T_Z)$ is weakly closed, and the proof of the theorem is complete.

We now need to introduce some more terminology, in order to discuss another important class of kernel functions. Given a kernel function K on Ω and the space $\mathcal{K}$ generated by K, we shall associate to the n-tuple Z of coordinate functions, two bounded linear transformations, $D_Z\colon \mathcal{K} \to \mathcal{K} \otimes \mathbb{C}^n$, and $D^Z\colon \mathcal{K} \otimes \mathbb{C}^n \to \mathcal{K}$, as follows:

$$D_Z f = (T_{Z_1} f, T_{Z_2} f, \ldots, T_{Z_n} f),$$

$$D^Z(f_1,f_2,\ldots,f_n) = \sum_i T_{Z_i} f_i.$$

For every w in Ω, we denote by Z - w, the n-tuple $Z - w = (Z_1 - w_1,\ldots,Z_n - w_n)$.

REMARK 1.6. a) Let K be a kernel function on Ω, and let Γ be the canonical section associated with K. It can be easily seen that property V) holds (i.e., K is sharp) if and only if $\mathrm{Ran}(D^{Z-w})$ is dense in $\mathrm{Ran}[\Gamma(w)]^{\perp}$, $w \in \Omega$. Equivalently, K is a generalized Bergman kernel function on Ω if and only if condition V) holds, and, in addition, the following property is satisfied:

VI) $\mathrm{Ran}(D^{Z-w})$ is closed for every w in wwVaa.

It may be worth noting that, except in very special cases, the theory developed in [7] and [8] is also applicable to sharp kernel functions which are not generalized Bergman kernels in any subdomain. For example, let $\mathbb{D}$ be the open unit disk in $\mathbb{C}$, and let $\Omega = \mathbb{D}^n$ be the unit polydisk in $\mathbb{C}^n$. Let K_0 be a kernel function on $\mathbb{D}$ with a diagonal expansion of the form:

$$K_0(z,w) = \sum_r a_r z\bar{w}^r,$$

which is uniformly and absolutely convergent on $\mathbb{D}\times\mathbb{D}$, and the positive sequence $\{a_r\}$ is chosen so that the radius of the compression spectrum $r_1(T_Z)$ of the canonical model associated with K_0 be zero, but the spectral radius $r(T_Z) = 1$ (see [1], Remark 2.2). Let K be the function on $\mathbb{D}^n$, given by:

$$K(z_1,\ldots,z_n,w_1,\ldots,w_n) = \prod_i K_0(z_i,w_i).$$

If $\mathcal{K}_0$ denotes the space generated by K_0, then the space $\mathcal{K}$ generated by K is Hilbert space isomorphic to the n-fold tensor product of $\mathcal{K}_0$ and K can be identified with the n-fold tensor product of K_0 (see Lemma 2.4 below). It is easy to see that the kernel function K fulfills the requirements of the desired example.

b) All the well known classical kernel functions, such as the Szego kernel, the Bergman kernel, and the Dirichlet kernel, (see [1]) are sharp kernel functions. However, although we suspect that such kernel functions are also generalized Bergman kernels, this more delicate property has only been checked for Bergman kernels on pseudoconvex domains and in some other special situations (see [13]).

2. OPERATIONS WITH KERNEL FUNCTIONS

We already saw in the introduction that pointwise products of kernel functions arise naturally (see also the examples at the end of Remarks 1.6, and 2.8) and the study of geometrical properties of the hermitian holomorphic vector bundles associated with operators in the class of Cowen and Douglas. In this section we present some general facts concerning products and tensor products of kernel functions, and, in the next section, we exhibit a purely algebraic approach to identifying the Hilbert spaces generated by such products. We begin with a discussion on sums and direct sums of kernel functions which also leads to the construction of new kernels from old ones.

THEOREM 2.1. *Let* K *and* K' *be kernel functions on (a bounded domain)* $\Omega \subset \mathbb{C}^n$, Rank(K) = m, Rank(K') = m', *and let* $\mathcal{K}$ *and be the Hilbert spaces generated by* K *and* K', *respectively.*

a) The Hilbert space $\mathcal{K}\oplus\mathcal{K}'$ *is generated by the pointwise direct sum* K⊕K' *of the kernel functions* K *and* K', *so that* Rank(K⊕K') = m + m'. *Further,* K⊕K' *is sharp (resp. generalized Bergman kernel) whenever* K *and* K' *are sharp (resp. generalized Bergman kernel).*

b) Assume that Rank(K) = m = m' = Rank(K') *and let* $\mathcal{H}$ *be the subspace of* $\mathcal{K}\oplus\mathcal{K}'$ *given by* $\mathcal{H} = \text{clspan}\{[\Gamma(\omega) \oplus \Gamma'(\omega)]\xi : \omega \in \Omega, \xi \in \ell^2_m\}$. *Then* $\mathcal{H}$ *can be identified with the space generated by the kernel function consisting of the pointwise sum* K+K' *of* K *and* K'. *Furthermore, if* K *and* K' *are sharp (resp. generalized Bergman kernels) then* K+K' *is also sharp (resp. generalized Bergman kernel).*

PROOF. The proof of a) follows immediately from properties I) through VI). In order to prove b), we first let $M \in \mathcal{L}(\ell^2_p, \ell^2_m)$ be a m×p matrix, $p \leq m$, such that Rank(M) = p, and let $L = M^*KM$. Then it easily follows that L is a kernel function, Rank(L) = p, and L is sharp (respectively, a generalized Bergman kernel) if K has the same property. Notice that the canonical section associated with L is ΓM, where Γ is the canonical section associated with K. Statement b) now follows readily from the last remarks, by considering $\mathcal{K}\oplus\mathcal{K}'$, and taking M so that $M^* = (I_m \; I_m)$.

REMARKS 2.2. a) One way of obtaining interesting looking kernels is by perturbing the classical kernels. We shall see, now, that certain kinds of perturbations are quite manageable. Let K be a sharp kernel on Ω, and let $\mathcal{K}$ be the space generated by K. Further, let Γ be the canonical section associated with K, and let $\mathcal{M}$ be a subspace of $\mathcal{K}$ which is invariant under T_Z. If P: $\mathcal{K} \to \mathcal{M}$ denotes the projection onto $\mathcal{M}$, then the function $K_{\mathcal{M}}(\zeta,\omega) = \Gamma(\zeta)^*P\Gamma(\omega)$, $\zeta,\omega \in \Omega$, is a kernel function on the complement Ω' of those points w where the restriction of the evaluation map $E_{w\,|\,\mathcal{M}}$ is not onto. Indeed, $K_{\mathcal{M}}$ is the reproducing kernel of $\mathcal{M}$. It is an interesting and important problem to determine under what conditions $K_{\mathcal{M}}$ is sharp. Sufficient conditions were given in [1]. For instance, if $X = \Omega\backslash\Omega'$ is the set of zeroes of $\mathcal{M}$ and it is finite, then an argument similar to the one used in the proof of [1] Lemma 2.7 shows that $K_{\mathcal{M}}$ is sharp on Ω'. Now, let $K' = K + 3K_{\mathcal{M}}$. If $K_{\mathcal{M}}$ is sharp on Ω', it then follows, from Theorem 2.1, that K' is sharp on Ω'. However, the sharpness of $K_{\mathcal{M}}$ is not really needed for K' to be sharp. In fact, we claim that if $\mathcal{M}$ is an invariant subspace of T_Z on $\mathcal{K}$ such that $\Omega' \neq 0$, then K' is sharp on Ω, and the canonical model associated with K' is similar to T_Z on $\mathcal{K}$. Indeed, we first observe that the function $\Gamma': \Omega \to \mathcal{L}(\ell^2_m, \mathcal{K})$ given by $\Gamma' = (1 + p)\Gamma$ satisfies $\Gamma'(z)^*\Gamma'(w) = K'(z,w)$, and the set span{Ran[$\Gamma'(z)$]: $z \in \Omega'$} is dense in $\mathcal{K}$. It follows that the space $\mathcal{H}$ of Theorem 2.1 b), constructed from K and $3K_{\mathcal{M}}$, is Hilbert space isomorphic (via a transformation that intertwines the canonical models) to $\mathcal{K}$ endowed with the inner product induced by K'. Also, note that Γ' is a global section of K' in the space $\mathcal{K}$. Moreover, by (1), we see that

$$(1 + P)(T_Z)^*\Gamma(w) = \overline{w}(1 + P)\Gamma(w) = (T_Z)^*(1 + P)\Gamma(w),$$

for every w in Ω'. Since span{Ran[$\Gamma(\omega)$]: $\omega \in \Omega'$} is dense in $\mathcal{K}$, we conclude that $(1 + P)T_Z = T_Z(1 + P)$ and the similarity statement is established. Since K is a kernel function on Ω already, we see that the same is true for K'. We also deduce that Ker[$(T_Z - w)^*$] in $\mathcal{K} = \mathcal{K}_{K'}$ is the image under 1+P of Ker[$(T_Z - w)^*$] in $\mathcal{K} = \mathcal{K}_K$, for every w in Ω. Since $\Gamma'(\omega) = (1 + P)\Gamma(\omega)$, $\omega \in \Omega$ is a global section of K' in $\mathcal{K}$, we conclude that K' is sharp on Ω, as desired.

b) Notice that the property of kernel functions discussed in Remark 1.2 (i.e., having a diagonal expansion about a given point) is invariant under unitary transformations that intertwine the canonical models (see Theorem 1.4). One can easily construct kernel functions that do not have a diagonal expansion about zero, but whose canonical models are still manageable. Examples of such kernels were first found in [4], where the case m = n = 1 was discussed. More generally, let K be a kernel function on Ω with a diagonal expansion about $0 \in \Omega$. We also choose $\mathcal{M}$ as in a) above, so that zero is in the set of zeros X of $\mathcal{M}$. Now, we let K': $\Omega\times\Omega \to \mathcal{L}(\ell^2_m)$ be defined as in a). We claim that, with a further suitable choice of $\mathcal{M}$, the space $\mathcal{K}'$ generated by K' cannot be isomorphic, via a unitary transformation that intertwines the corresponding canonical models, to any functional Hilbert space generated by a kernel function on Ω_0 that has a diagonal expansion around zero. In order to see this, we first choose Ω_0 as at the end of Remark 1.2, so that the restriction of K to $\Omega_0\times\Omega_0$ is sharp, and assume, by way of contradiction, that there is such a unitary transformation U from $\mathcal{K}'$ to a functional Hilbert space $\mathcal{K}''$ on $\mathbb{Z}$ with a reproducing kernel which is a kernel function and that has a diagonal expansion about zero. Since $K'(0,z) = K'(0,0) = K'(z,0)$ for every z in Ω_0, because $0 \in X$, we can assume, via a change of scales (see [8], Remark 3.8) that $K'(0,z) = 1 = K''(0,z)$, for all z in Ω_0. But, then, from Theorem 1.4, there exists a unitary operator V on ℓ^2_m, such that $V^*K''(z,w)V = K'(z,w)$, $z,w \in \Omega_0$. This implies that

$$K_{\mathcal{M}}(z,w) = (1/3)V^*K''(z,w)V - K(z,w), \quad z,w \in \Omega_0$$

also has a diagonal expansion around zero. But, if $\mathcal{M}$ is chosen so that it is not a canonical subspace of $\mathcal{K}$ (see [1]), then $K_{\mathcal{M}}$ cannot have a diagonal expansion around zero, reaching the desired contradiction. (For instance, if we take

$$\mathcal{M} = \{f \in \mathcal{K}: f(0) = 0 = \sum_{|\alpha|=1} \partial^\alpha f(0)\},$$

then $\mathcal{M}$ satisfies the desired requirements.) From the discussion of a) above, we know, however, that the canonical models of K and K' are similar.

Now we turn our attention to products and tensor products of kernel functions.

DEFINITION 2.3. Let K and K' be two kernel functions on Ω. We denote by $K\otimes K'$ the tensor product of K and K', namely,

(7) $$K\otimes K'(z,w,y,x) = K(z,y) \otimes K'(w,x), \quad w,x,y,z \in \Omega.$$

We also denote by KK' the pointwise tensor product of K and K', namely:

(8) $$KK'(z,w) = K(z,w) \otimes K'(z,w), \quad z,w \text{ in } \Omega.$$

Note that while $K\otimes K'$ is a function on $(\Omega\times\Omega) \times (\Omega\times\Omega)$, KK' is a function on $\Omega\times\Omega$. The next lemma is elementary and its proof is similar to that of Theorem 2.1a).

LEMMA 2.4. *Let* K *and* K' *be two kernel functions on* Ω, *with* $\mathrm{Rank}(K) = m$ *and* $\mathrm{Rank}(K') = m'$. *Let* $\mathcal{K}$ *and* $\mathcal{K}'$ *be the spaces generated by* K *and* K', *respectively and let* Γ *and* Γ' *be the canonical sections associated with* K *and* K'.

a) $K \otimes K'$ *is a kernel function, whose rank is* mm', *and it generates the space* $\mathcal{K} \otimes \mathcal{K}'$, *which coincides with the closure of the set*

$$\mathrm{span}\{\Gamma(\zeta) \otimes \Gamma'(\omega)(\xi \otimes \eta): \zeta, \omega \in \Omega, \xi \in \ell^2_m, \eta \in \ell^2_{m'}\}.$$

The canonical model associated with $K \otimes K'$ *is the 2n-tuple consisting of the pair*

$$(T_Z \otimes (1_{\mathcal{K}'}), (1_{\mathcal{K}}) \otimes T_Z).$$

b) KK' *is a kernel function, whose rank is* mm', *and it generates a Hilbert space that can be identified with the subspace*

$$\mathcal{L} = \mathrm{clspan}\{\Gamma(\omega) \otimes \Gamma'(\omega)(\xi \otimes \eta): \omega \in \Omega, \xi \in \ell^2_m, \eta \in \ell^2_{m'}\} \tag{9}$$

of $\mathcal{K} \otimes \mathcal{K}'$. *Moreover,* $\mathcal{L}$ is *a hyper-invariant subspace of the adjoint of the canonical model associated with* $K \otimes K'$, *(i.e., it is invariant under the commutant of such a 2n-tuple), and both components of its restriction to* $\mathcal{L}$ *coincide with the adjoint of the canonical model associated with* KK', *so that we have:*

$$(T_Z \otimes 1_{\mathcal{K}'})^*|_{\mathcal{L}} = (1_{\mathcal{K}} \otimes T_Z)^*|_{\mathcal{L}} = ()^*|_{\mathcal{L}}. \tag{10}$$

The following lemma allows us to compute the kernels of Definition 2.3, in many important situations.

LEMMA 2.5. *Let* K *and* K' *be kernel functions on* Ω, *assume that* $0 \in \Omega$, *and suppose that* K *and* K' *have diagonal expansion in a neighborhood* Ω_0 *of zero, of the form:*

$$K(\zeta,\omega) = \sum_{\alpha} A_\alpha \zeta^\alpha \bar{\omega}^\alpha,$$

$$K'(\zeta,\omega) = \sum_{\alpha} A'_\alpha \zeta^\alpha \bar{\omega}^\alpha,$$

$\zeta,\omega \in \Omega_0$. *Then* K⊗K' *and* KK' *have also diagonal expansions about zero in* Ω_0, *where the coefficients* $\{B_\alpha\}$ *can be computed by the standard convolution formula*

$$B_\alpha = \sum_{\beta\leq\alpha} A_\beta \otimes A'_{\alpha-\beta}, \tag{11}$$

where $\beta\leq\alpha$ *means* $\beta_j\leq\alpha_j$, $1 \leq j \leq n$.

PROOF. It is an immediate consequence of the fact that the coefficients of a product of two convergent power series are computed by a convolution formula such as (11).

The following theorem is the main result of this section.

THEOREM 2.6. *Let* K *and* K' *be kernel functions on* Ω. *If* K *and* K' *are sharp (respectively, generalized Bergman kernels), then so are* K ⊗ K' *and* KK'.

PROOF. Let $\mathcal{K}$ and $\mathcal{K}'$ be the spaces generated by K and K', respectively, and let $P(w)$: $\mathcal{K} \to \operatorname{Ker}(T_Z - w)^*$ and $P'(w')$: $\mathcal{K} \to \operatorname{Ker}(T_Z - w')^*$ be the corresponding (orthogonal) projections. Then the projections, in $\mathcal{L}(\mathcal{K}\otimes\mathcal{K}')$, $P(w) \otimes (1_{\mathcal{K}'})$ and $(1_{\mathcal{K}}) \otimes P'(w')$ commute, and hence their product $P(w) \otimes P'(w')$ is the projection onto the intersection of the corresponding ranges. Therefore, we have:

$$\begin{aligned}\operatorname{Ker}[(T_Z - w)^* \otimes (1_{\mathcal{K}'})] \cap \operatorname{Ker}[(1_{\mathcal{K}}) \otimes (T_Z - w')^*] &= \\ = [\operatorname{Ker}(T_Z - w)^*] \otimes [\operatorname{Ker}(T_Z - w')^*],\end{aligned} \tag{12}$$

for every $w,w' \in \Omega$. Since the canonical model on $\mathcal{K}\otimes\mathcal{K}'$ is the 2n-tuple consisting of the pair of n-tuples $(T_Z \otimes (1_{\mathcal{K}'}), (1_{\mathcal{K}}) \otimes T_Z)$, we see that the canonical section associated with K ⊗ K' is given by $\Gamma(\omega) \otimes \Gamma'(\omega')$, $\omega,\omega' \in \Omega$. From (12), it readily follows that conditions V) and VI) hold for K ⊗ K', whenever they hold for both K and K'. Now, let $\mathcal{L}$ be as in (9) of Lemma 2.4. Since it is an invariant subspace for both $(T_Z - w)^* \otimes (1_{\mathcal{K}'})$ and $(1_{\mathcal{K}}) \otimes (T_Z - w)^*$, we see, from (10), that if K and K' are sharp, then

$$\begin{aligned}\operatorname{Ker}[((T_Z - w) \otimes 1_{\mathcal{K}'})^*|_{\mathcal{L}}] &= \operatorname{Ker}[(1_{\mathcal{K}}\otimes (T_Z - w))^*|_{\mathcal{L}}] \subset \\ &\subset \operatorname{Ker}[(T_Z - w) \otimes 1_{\mathcal{K}'}]^* \cap \operatorname{Ker}[1_{\mathcal{K}}\otimes (T_Z - w)^*] = \\ &= \operatorname{Ran}(\Gamma(w)) \otimes \operatorname{Ran}(\Gamma'(w)),\end{aligned}$$

for every w in Ω, and hence KK' is sharp. Now, assume that K and K' are generalized Bergman kernels. Since we know that KK' is then a sharp kernel, from what we already proved, in order to show that KK' is a generalized Bergman kernel, it suffices to prove that condition VI) holds, i.e. that the range of $D_Z - w: \mathcal{L} \otimes \mathbb{C}^n \to \mathcal{L}$ is closed for every w in Ω (see Remark 1.6). An immediate consequence of the definition (see Remark 1.6 and [1]) implies that the diagonal transformations $D_{(Z-w)^*}$ acting on $\mathcal{K}$ and on $\mathcal{K}'$ have closed range, for every w in Ω. Therefore, from the statement of Lemma 2.7 (proved below), we deduce that the diagonal transformation induced by the 2n-tuple consisting of the pair

$$(13) \qquad [(T_Z - w) \otimes 1_{\mathcal{K}}]^*, [1_{\mathcal{K}} \otimes (T_Z - w)]^*,$$

has also closed range. Again, from Lemma 2.7 and the first part of the present proof, it follows that the restriction to $\mathcal{L}$ of the diagonal transformation induced by the 2n-tuple in (13) has closed range. But, from (10), we see that the last diagonal transformation consists of a pair of the same diagonal transformation namely $D_{(Z-w)^*}: \mathcal{L} \to \mathcal{L} \otimes \mathbb{C}^n$, so $D^{Z-w}: \mathcal{L} \otimes \mathbb{C}^n \to \mathcal{L}$ has closed range, as desired.

LEMMA 2.7. *Let $\mathcal{H}, \mathcal{H}', \mathcal{K}, \mathcal{K}'$ be Hilbert spaces, and let* $S \in \mathcal{L}(\mathcal{H},\mathcal{H}')$ *and* $T \in \mathcal{L}(\mathcal{K},\mathcal{K}')$. *Assume that both* S *and* T *have closed ranges.*

a) If $\mathcal{M} \subset \mathcal{K}$ *is an invariant subspace under* T, *and* $\mathrm{Ker}(T) \subset \mathcal{M}$, *then the restriction of* T *to* $\mathcal{M}$, *i.e.,* $T|_{\mathcal{M}}$ *has closed range.*

b) Let $S' \in \mathcal{L}[(\mathcal{H} \otimes \mathcal{K}), (\mathcal{H}' \otimes \mathcal{K})]$, $T' \in \mathcal{L}[(\mathcal{H} \otimes \mathcal{K}), (\mathcal{H} \otimes \mathcal{K}')]$ *be given by* $S' = S \otimes (1_{\mathcal{K}})$, $T' = (1_{\mathcal{K}}) \otimes T$. *Further, let* $D: (\mathcal{H} \otimes \mathcal{K}) \to (\mathcal{H}' \otimes \mathcal{K}) \oplus (\mathcal{H} \otimes \mathcal{K}')$, *be the diagonal linear transformation defined by* $Dx = (S'x, T'x)$, $x \in \mathcal{H} \otimes \mathcal{K}$. *Then* D *has closed range and*

$$(14) \qquad \mathrm{Ker}(D) = \mathrm{Ker}(S') \otimes \mathrm{Ker}(T').$$

PROOF. Let q be the (orthogonal) projection onto Ker(T). Then T has closed range if and only if there exists a constant c > 0 such that

$$(15) \qquad \|Tx\| = \|T(1 - Q)x\| \geq \|(1 - Q)x\|,\ x \in \mathcal{K}.$$

For the proof of a), we simply observe that, since $\mathrm{Ker}(T) \subset \mathcal{M}$, (15) is obviously satisfied by $t|_{\mathcal{M}}$. In order to prove b), we first point out that (14) is proved with an argument identical to one used in the proof of Theorem 2.6. Indeed, let P be the projection onto Ker(S). Thus the projections onto $\mathrm{Ker}(S \otimes 1_{\mathcal{K}})$ and $\mathrm{Ker}(1_{\mathcal{H}} \otimes T)$ are $P \otimes 1_{\mathcal{K}}$ and $1_{\mathcal{H}} \otimes Q$, respectively. Thus, the projection onto Ker(S',T') is $P \otimes Q$. In order to show that D has closed range, we prove that (15) holds for D. Let

U: $\mathcal{H}' \to \mathcal{H}$, V: $\mathcal{K}' \to \mathcal{K}$ be unitary transformations, and let U': $(\mathcal{H}' \otimes \mathcal{K}) \to (\mathcal{H} \otimes \mathcal{K})$, V': $(\mathcal{H}' \otimes \mathcal{K}') \to (\mathcal{H}' \otimes \mathcal{K})$ be unitary transformations defined from U and V, in a similar fashion as S' and T' were defined from S and T. Then U'S' = (US) $\otimes 1_{\mathcal{K}}$ and V'T' = $1_{\mathcal{H}} \otimes$ (VT). (In the rest of this proof, we drop the subindices $\mathcal{H}$ and $\mathcal{K}$ of the identity operator 1, for brevity sake.) It follows, from (15), that for every z in $\mathcal{H} \otimes \mathcal{K}$, we have:

$$\begin{aligned}\|T'z\| &= \|T'[1 - (1 \otimes Q)]z\| = \|T'[1 \otimes (1 - Q)]z\| = \\ &= \|V'T'[1 \otimes (1 - Q)]z\| = \|[1 \otimes (VT)][1 \otimes (1 - Q)]z\| = \\ &= \|1 \otimes [VT(1 - Q)]z\| \geq c\|[1 \otimes (1 - Q)]z\|.\end{aligned}$$

Since Ran(S) is closed, there exists a constant b > 0, such that (15) is valid, with T and c replaced by S and b. Reasoning as above, we also have:

$$\|S'z\| \geq b\|[(1 - P) \otimes 1]z\|.$$

Let d = min(b,c). Putting the above estimates together, we obtain:

$$\begin{aligned}\|Dz\|^2 &= \|S'z\|^2 + \|T'z\|^2 \geq b^2\|[(1 - P) \otimes 1]z\|^2 + c^2\|[1 \otimes (1 - Q)]z\|^2 \geq \\ &\geq b^2\|[(1 - P) \otimes Q]z\|^2 + c^2\|[1 \otimes (1 - Q)]z\|^2 \geq \\ &\geq d^2\|\{[(1 - P) \otimes Q] + [1 \otimes (1 - Q)]\}z\|^2 = d^2\|[1 - (P \otimes Q)]z\|^2.\end{aligned}$$

We conclude that D satisfies (15), as desired.

REMARK 2.8. Let K and K' be two kernel functions on Ω, and let $\mathcal{K}$, $\mathcal{K}'$ and $\mathcal{L}$ be as in Theorem 2.6. There is a natural map δ: $\mathcal{K} \otimes \mathcal{K}' \to \mathcal{L}$, given by δf(w) = f(w,w), f ∈ $\mathcal{K} \otimes \mathcal{K}'$ and w in Ω. Recall that the elements of $\mathcal{K} \otimes \mathcal{K}'$ are functions on Ω×Ω, so that δ is the restriction map to the diagonal of Ω×Ω. The map δ has already been studied in various special cases, (see [12] and [15]). Notice that when Rank(K) = 1 = Rank(K'), then K and K' are complex valued, and KK' is just the pointwise product of K and K'. It is well known that if Ω = $\mathbb{D}$ (the open unit disk) and K = K' is the Szego kernel, then KK' is the Bergman kernel, $\mathcal{K} \otimes \mathcal{K}'$ is the Hardy space on the bidisk $\mathbb{D} \times \mathbb{D}$ and δ is the restriction map onto the Bergman space $\mathcal{L}$ (see [12] page 53).

3. HILBERT MODULE TENSOR PRODUCTS

Our basic references for this section are [9] and [18]. Given a (complex, unital) algebra $\mathcal{A}$, a (left) Hilbert module over $\mathcal{A}$ or an $\mathcal{A}$-Hilbert module is a Hilbert space $\mathcal{H}$ together with an action (or representation) of $\mathcal{A}$ as an algebra of bounded operators on $\mathcal{H}$, i.e., (a,x) → ax, a ∈ $\mathcal{A}$, x ∈ $\mathcal{H}$. If $\mathcal{A}$ is a Banach algebra, we shall automatically assume that this action is continuous in the norm topologies of $\mathcal{A}$ and $\mathcal{L}(\mathcal{H})$.

Our primordial example of an $\mathcal{A}$-Hilbert module will be the space $\mathcal{K}$ generated by a kernel function K: $\Omega\times\Omega \to \mathcal{L}(\ell^2_m)$, where, as in the previous sections, Ω is a bounded domain in $\mathbb{C}^n$. Let F_K: $\mathcal{A}(T_Z) \to \mathcal{A}_K$ be the homomorphism produced by Theorem 1.5. Further, let $\mathcal{A}$ be a subalgebra of $\mathcal{A}_K$. Then $\mathcal{K}$ is an $\mathcal{A}$-Hilbert module, via the action $(F_K)^{-1}|_{\mathcal{A}}$ (note that the action of $\mathcal{A}$ is faithful in this case). In [8], Theorem 5.2, sufficient conditions were given for $\mathcal{A}_K$ to be maximal, i.e., for $\mathcal{A}_K$ to coincide with $H^\infty(\Omega)$. Since, in this case, F_K becomes a topological isomorphism, the action of a subalgebra $\mathcal{A}$ will be continuous whenever $\mathcal{A}$ has the induced topology from $H^\infty(\Omega)$.

Each ideal $\mathcal{I}$ in an (abelian) algebra $\mathcal{A}$, induces in a given $\mathcal{A}$-Hilbert module $\mathcal{H}$, an $\mathcal{A}$-Hilbert submodule, by taking the closure of $\mathcal{I}\mathcal{H}$. The quotient space $\mathcal{H}/(\overline{\mathcal{I}\mathcal{H}})$ is then an $\mathcal{A}/\mathcal{I}$-Hilbert module because it is $\mathcal{A}/\mathcal{I}$-isomorphic to the submodule $(\mathcal{I}\mathcal{H})^\perp$ in $\mathcal{H}$ (which is an $\mathcal{A}/\mathcal{I}$-Hilbert module, via the compression map).

Going back to our primordial example, the following special submodules are of interest. Let V be an analytic subvariety of Ω (see [10]), and let $\mathcal{I}$ be the ideal in $\mathcal{A} \subset \mathcal{A}_K$, consisting of those functions in $\mathcal{A}$ that vanish on V. Then, there are two submodules of $\mathcal{K}$ naturally associated with V:

(16) $\mathcal{L}_V = (\overline{\mathcal{I}\mathcal{K}})$, and

(17) $\mathcal{M}_V = \{f \in \mathcal{K}: f(z) = 0,\ z \in V\}$.

The following notion provides a useful sufficient criterion for the submodules in (16) and (17) to be the same (see Theorem 3.3).

DEFINITION 3.1. Given an $\mathcal{A}$-Hilbert module $\mathcal{H}$ and a subspace $\mathcal{M}$ of $\mathcal{H}$, the pair $(\mathcal{H},\mathcal{M})$ is said to be $\mathcal{A}$-m-cyclic if there are m vectors, and no less than m vectors, $x_1,\ldots,x_m$ in $\mathcal{H}$ such that, denoting by φ the map φ: $A \otimes \mathbb{C}^n \to \mathcal{H}$ given by $\varphi(f_1,\ldots,f_m) = \sum f_i x_i$, the following property holds:

(*) Given any y in $\mathcal{H}$, there exists a sequence $\{y_\nu\} \subset \mathcal{H}$ that tends to y, and whose projection on $\mathcal{M}$ is also in $\mathrm{Ran}(\varphi)$.

REMARKS 3.2. a) The above definition is a generalization of the notion of $\mathcal{A}$-m-cyclicity (just take $\mathcal{M}$ to be trivial). Also, since $\mathcal{A}$ is unital, $(\mathcal{H},\mathcal{M})$ is $\mathcal{A}$-m-cyclic if and only if $(\mathcal{H},\mathcal{M}^{\perp})$ is $\mathcal{A}$-m-cyclic.

b) An important sufficient condition for the pair $(\mathcal{H},\mathcal{M})$ to be $\mathcal{A}$-m-cyclic is that Ran(φ) be dense (i.e., $\mathcal{H}$ is $\mathcal{A}$-cyclic), and that the projection from $\mathcal{H}$ onto $\mathcal{M}$ leaves Ran(φ) invariant. In remarks 3.4, we exhibit a large class of examples of $\mathcal{A}$-m-cyclic pairs in which this property holds, and we also give examples where it doesn't necessarily hold.

c) Consider a sharp kernel function K on Ω with Rank(K) = m < ∞ and with a diagonal expansion around $0 \in \Omega$. Let $\mathcal{K}$ be the space generated by K. For each multi-index α, let $\mathcal{M}_\alpha = \{f \in \mathcal{K}: \partial^\beta f(0) = 0, \beta \leq \alpha\}$. Then the pair $(\mathcal{K}, \mathcal{M}_\alpha)$ is $\mathcal{A}$-m-cyclic, where $\mathcal{A} \subset \mathcal{A}_K$. Indeed, assume, via a change of scale, that $K(0,z) = 1_m$, $z \in \Omega$. Let $\{e_j: 1 \leq j \leq m\}$ be the standard basis of ℓ^2_m. Then, because of Remark 3.2 b), the vectors $x_i = K(.,0)e_i = e_i$, $1 \leq i \leq m$, implement the polynomial-m-cyclicity of the above pair, and hence, it also implements the $\mathcal{A}$-m-cyclicity of the same pair, where $\mathcal{A}$ is any subalgebra of $\mathcal{A}_K$ containing the coordinate functions. Notice that there should be at least m multicyclic vectors because Rank(K) = m (see the proof of Theorem 3.3, below). We also observe that if $\mathcal{I}^\alpha$ is the ideal in $\mathcal{A}$ given by $\mathcal{I}^\alpha = \{f \in \mathcal{A}: \partial^\beta f(0) = 0, \beta \leq \alpha\}$, then the closure of span $\mathcal{I}^\alpha\mathcal{K}$ coincides with $\mathcal{M}_\alpha$.

THEOREM 3.3. *Let* K *be a kernel function on* Ω, m = Ran(K) < ∞. *Also, let* $\mathcal{A}$ *be a subalgebra of* $\mathcal{A}_K$, *and let* $\mathcal{K}$ *be the space generated by* K. *Given* $X \subset \Omega$, *let* $\mathcal{M}_X = \{x \in \mathcal{K}: x(\omega) = 0, \omega \in X\}$, *and let* $\mathcal{L}_X$ *be the closure of* span$\{fx: f(\omega) = 0, \omega \in X, f \in \mathcal{A}, x \in \mathcal{K}\}$. *If the pair* $(\mathcal{K},\mathcal{M}_X)$ *is* $\mathcal{A}$*-m-cyclic, then* $\mathcal{M}_X = \mathcal{L}_X$.

PROOF. Let $x_1,\dots,x_m$ be m vectors implementing the $\mathcal{A}$-m-cyclicity of the pair $(\mathcal{K},\mathcal{M}_X)$, and let $F: \Omega \to \mathcal{L}(\ell^2_m)$ be the mapping whose value at $z \in \Omega$ is the operator represented on the standard basis of ℓ^2_m by the m×m matrix whose i-th column is the vector $x_i(z)$. It follows that F(z) is invertible, because the evaluation map $E_z: \mathcal{K} \to \ell^2_m$ at z is onto, and Ran(φ) is dense in $\mathcal{K}$ by (*), so that Rank[F(z)] = m for every $z \in \Omega$. Let $y \in \mathcal{M}_X$. By assumption, there exists a sequence $\{y_\nu\} \subset \mathrm{Ran}(\varphi) \cap \mathcal{M}_X$ that tends to y. Thus, $y_\nu = \sum f_{i\nu} x_i$, where $f_{i\nu} \in \mathcal{A}$ for all ν and all i =1,...,m. But, since F(z) is invertible and $y_\nu(z) = 0$, $z \in X$, we deduce that $f_{i\nu}(z) = 0$, $z \in X$, for all ν and all i, so that $\{y_\nu\} \subset \mathcal{L}_X$, as desired.

REMARK 3.4. a) There are many situations where the assumptions of Theorem 3.3 are satisfied. Indeed, as in Remark 3.2 c), let K be a sharp kernel function on Ω with a diagonal expansion around $0 \in \Omega$. Let $X_h = \{\omega \in \Omega: \omega_j = 0, 1 \le j \le h\}$, and let $\mathcal{K}$ be the space generated by K. Just as in Remark 3.2 c), it follows that the pair $(\mathcal{K}, \mathcal{M}_{X_h})$ is $\mathcal{A}$-m-cyclic, where $\mathcal{A}$ is any unital subalgebra of $\mathcal{A}_K$ that contains the coordinate functions, and hence, by Theorem 3.3, we see that $\mathcal{M}_{X_h} = \mathcal{L}_{X_h}$.

b) Let $\Omega \subset \mathbb{C}^n$ be a star-like domain, i.e., $r\Omega \subset \Omega$ for every $r \in [0,1]$, and let K be a sharp kernel function on Ω with Rank(K) = m < ∞. Assume that $\sigma(T_Z) = \overline{\Omega}$, and let $\mathcal{A}$ be the algebra of complex-valued holomorphic functions whose domains of definition contain $\overline{\Omega}$. Further, assume that for every function f in the space $\mathcal{K}$ generated by K, the function f_r defined by $f_r(z) = f(rz)$, $z \in \Omega$, is in $\mathcal{K}$ for all $r \in [0,1]$, and that $\lim_{r \to 1^-} f_r = f$. Also, let $X \subset \Omega$ be a star-like set, and let $\mathcal{M}_X$ be as in the statement of Theorem 3.3. Then the pair $w(\mathcal{K}, \mathcal{M}_X)$ is $\mathcal{A}$-m-cyclic. Indeed, let Γ be the canonical section associated with K, and notice that for every $\xi \in \ell^2_m$, the constant function $K(0,0)\xi = [\Gamma(0)\xi]_0$ is in $\mathcal{K}$, by assumption (take r = 0). Using a change of scales, if necessary, we can assume, without loss of generality, that $K(0,0) = 1_m$. Let x_i, $0 < i < m+1$, be the vectors defined just as in Remark 3.2 c). Given $f \in \mathcal{K}$, we can write $f = \sum f_i x_i$, where f_i is the i-th component of f. It follows that $f_{i,r}$ is in $\mathcal{A}$, for $0 < i < m+1$, $r \in (0,1)$. Also, since $\sigma(T_Z) = \overline{\Omega}$, we see that $\mathcal{A} \subset \mathcal{A}_K$. Let $\varphi: \mathcal{A} \otimes \mathbb{C}^m \to \mathcal{K}$ be as in Definition 3.1, and let $P: \mathcal{K} \to \mathcal{M}_X$ be the (orthogonal) projection map. Since $Pf_r = (Pf)_r$ for every $f \in \mathcal{K}$ we deduce that, given f in $\mathcal{K}'$ we have: $f_r = \sum f_{i,r} x_i \in \mathrm{Ran}(\varphi)$, and $pf_r \in \mathrm{Ran}(\varphi)$, $\mathbb{R} \in (0,1)$. Since $f_r \to f$ by assumption, the pair $(\mathcal{K}, \mathcal{M}_X)$ is $\mathcal{A}$-m-cyclic, and, hence, by Theorem 3.3, we conclude that $\mathcal{M}_X = \mathcal{L}_X$. Notice that $\mathrm{Ran}(\varphi)$ may not be invariant under P, i.e., the condition of Remark 3.2 b) may not be satisfied, in this case. We shall see another important application of Theorem 3.3, when we consider the tensor product of functional Hilbert modules over the polynomial algebra.

Let $\mathcal{A}$ be a unital algebra, and let $\mathcal{H}$ and $\mathcal{H}'$ be two $\mathcal{A}$-Hilbert modules. Via the (algebraic) tensor product action, $\mathcal{H} \otimes \mathcal{H}'$ has a natural structure of $\mathcal{A} \otimes \mathcal{A}$-Hilbert module. From this module, we produce an $\mathcal{A}$-Hilbert module by proceeding with the usual algebraic construction. We obtain this $\mathcal{A}$-module, by identifying the action of $a\otimes 1$ and $1\otimes a$. Thus, let $\mathcal{I}$ be the ideal in $\mathcal{A}\otimes\mathcal{A}$ generated by the elements of the form $a\otimes 1 - 1\otimes a$, $a \in \mathcal{A}$. It follows that $\mathcal{A}\otimes\mathcal{A}/\mathcal{I}$ is isomorphic to $\mathcal{A}$ via the maps $\rho: \mathcal{A}\otimes\mathcal{A}/\mathcal{I} \to \mathcal{A}$, $\sigma: \mathcal{A} \to \mathcal{A}\otimes\mathcal{A}/\mathcal{I}$ given by:

$$\rho(\sum a_i \otimes b_i + \mathcal{I}) = \sum a_i b_i,$$

$\sigma_a = a\otimes 1 + \mathcal{I}$. Therefore, the $(\mathcal{A}\otimes\mathcal{A}/\mathcal{I})$-module is $(\mathcal{I}\,\mathcal{H}\otimes\mathcal{H}')^{\perp}$ is a subspace of $\mathcal{H}\otimes\mathcal{H}'$ that is actually an $\mathcal{A}$-Hilbert module, via the above isomorphism. Such an $\mathcal{A}$-Hilbert module will be denoted by $\mathcal{H}\otimes_{\mathcal{A}}\mathcal{H}'$, and it will be called the module tensor product of $\mathcal{H}$ and $\mathcal{H}'$ over $\mathcal{A}$.

If $\mathcal{A}$ is an algebra of holomorphic functions on $\Omega \subset \mathbb{C}^n$, then $\mathcal{A}\otimes\mathcal{A}$ can be regarded as an algebra of holomorphic functions on $\Omega\times\Omega$. The map δ: $\mathcal{A}\otimes\mathcal{A} \to \mathcal{A}$ defined by

$$\delta(\sum a_i \otimes b_i) = \sum a_i b_i$$

can be identified with the restriction map to the diagonal of $\Omega\times\Omega$, so that $\mathrm{Ker}(\delta) = \mathcal{I}$, where $\mathcal{I}$ is as above. If $\mathcal{A}$ is a Banach algebra then δ can be extended to a map π on the completion of $\mathcal{A}\otimes\mathcal{A}$ with respect to any suitable tensor product norm, but $\mathrm{Ker}(\pi)$ (i.e., the ideal of functions vanishing on the diagonal) may not be the closure of $\mathcal{I}$ in such a norm.

The following theorem is one of the main results of this paper.

THEOREM 3.5. *Let* K *and* K' *be sharp kernel functions on* Ω, $\mathrm{Rank}(K) = m < \infty$, $\mathrm{Rank}(K') = m' < \infty$, *and let* $\mathcal{A} \subset \mathcal{A}_K \cap \mathcal{A}_{K'}$ *be a unital subalgebra. Further, let* $\mathcal{L}$ *be the space generated by* KK', *and assume that the pair* $(\mathcal{K}\otimes\mathcal{K}', \mathcal{L})$ *is* $\mathcal{A}\otimes\mathcal{A}$*-mm'-cyclic. Then* $\mathcal{L}$ *is an* $\mathcal{A}$*-Hilbert module which is* $\mathcal{A}$*-isomorphic to* $\mathcal{K}\otimes_{\mathcal{A}}\mathcal{K}'$.

PROOF. Let X be the diagonal of $\Omega\times\Omega$ and let $\mathcal{M}_X$, $\mathcal{L}_X$ be the submodules of $\mathcal{K}\otimes\mathcal{K}'$ defined as in Theorem 3.3. Then, by Remark 2.8, $\mathcal{M}_X = \mathcal{L}^{\perp}$. Moreover, by Theorem 3.3, we deduce that

(18) $$\mathcal{L}_X^{\perp} = \mathcal{L}.$$

Notice that the ideal in (the algebraic tensor product) $\mathcal{A}\otimes\mathcal{A}$ consisting of those functions that vanish on X coincides with the ideal $\mathcal{I}$ generated by functions of the form $f \otimes 1 - 1 \otimes f$, $f \in \mathcal{A}$. Thus, by definition, (18) means precisely that $\mathcal{L} = \mathcal{K} \otimes_{\mathcal{A}} \mathcal{K}'$, as desired.

COROLLARY 3.6. *Let* K *and* K' *be sharp kernel functions on* $\Omega \subset \mathbb{C}^n$, *with finite ranks, and assume that both* K *and* K' *have diagonal expansions around* $0 \in \Omega$. *Let* $\mathcal{K}$ *and* $\mathcal{K}'$ *be the spaces generated by* K *and* K', *and let* $\mathcal{P}$ *be the polynomial algebra in n indeterminates. Then the* $\mathcal{P}$*-Hilbert module* $\mathcal{K} \otimes_{\mathcal{P}} \mathcal{K}'$ *is* $\mathcal{P}$*-isomorphic to the space generated by* KK'.

PROOF. We can assume, without loss of generality, that $K(0,0) = 1_m$ and $K'(0,0) = 1_{m'}$, where $m = \text{Rank}(K)$, $m' = \text{Rank}(K')$. Let $\mathcal{L}$ be the space generated by KK'. By Lemma 2.4, $\text{Rank}(KK') = mm' = \text{Rank}(K \otimes K')$. Recall, from Remark 1.2, that $\mathcal{K}$ is isomorphic to a weighted sequence space of elements $\{\xi_\alpha\} \subset \ell^2_m$, such that

$$\sum \langle A_\alpha \xi_\alpha, \xi_\alpha \rangle < \infty,$$

and analogously for $\mathcal{K}'$. Therefore, $\mathcal{K} \otimes \mathcal{K}'$ is isomorphic to the (double indexed) weighted sequence space with elements

$$\{\xi_{\alpha\beta}\} \subset \ell^2_m \otimes \ell^2_{m'} = \ell^2_{mm'},$$

such that

$$\sum \langle \xi_{\alpha\beta}, (A_\alpha \otimes A'_\beta)\xi_{\alpha\beta} \rangle < \infty.$$

Let $\{B_\gamma\} \subset \mathcal{L}(\ell^2_{mm'})$, be as in (11). It follows that $\mathcal{L}$ is isomorphic to a weighted sequence space with elements $\{\eta_\gamma\} \subset \ell^2_{mm'}$, such that

$$\sum \langle \eta_\gamma, B_\gamma \eta_\gamma \rangle < \infty.$$

The projection Q: $\mathcal{K} \otimes \mathcal{K}' \to \mathcal{L}$ is given by

$$\{(Q\{\xi_{\alpha\beta}\})_{\lambda\mu}\} = \{ \sum_{\alpha+\beta=\lambda+\mu} \xi_{\alpha\beta} / [\prod_{j=1}^{n} (\alpha_j + \beta_j + 1)]\}. \tag{19}$$

In the representation of $\mathcal{L} = \text{Ran}(Q)$ given by (19), the sequence space is double indexed because it is obtained from the fact that $\mathcal{L}$ is a subspace of $\mathcal{K}\otimes\mathcal{K}'$, but notice that the λ,μ terms of $Q\{\xi_{\alpha\beta}\}$ are the same for $\gamma = \lambda + \mu$. We also observe that if the sequence $\{\xi_{\alpha\beta}\}$ in $\mathcal{K}\otimes\mathcal{K}'$ has only finitely many non-zero terms i.e., they are the coefficients of a polynomial function, then the same property holds for the sequence $Q\{\xi_{\alpha\beta}\}$. Let φ: $(\mathcal{A}\otimes\mathcal{A})^{mm'} \to \mathcal{P}\otimes\mathcal{P}'$ be the extension of the map given by

$$\varphi: (f_i \otimes f'_h) = \sum (f_i \otimes f'_h)(x_i \otimes x'_h),$$

where $x_i = K(0,0)e_i$, $x'_h = K'(0,0)e_h$, $0 < i < m+1$, $0 < h < m'+1$ are the vectors defined as in Remark 3.2 c). Just as in Remark 3.2 c) and Remark 3.4 a), Ran(φ) is dense in $\mathcal{K}\otimes\mathcal{K}'$. Moreover, Ran($\varphi$) is invariant under Q. Therefore, the corollary follows from Remark 3.2 b), Theorem 3.3, and Theorem 3.5.

REMARK 3.7. Even for m = m' = n = 1, Corollary 3.6 has an interesting consequence. Notice that in this case, the corresponding canonical model T_Z determines completely the modular structure of $\mathcal{K}$, $\mathcal{K}'$, and $\mathcal{K}\otimes_{\mathcal{P}}\mathcal{K}'$. From Corollary 3.6 we see that if the canonical models on $\mathcal{K}$ and $\mathcal{K}'$ are weighted shifts with weight sequences $\{w_i\}$ and $\{w'_i\}$, respectively, then the canonical model on $\mathcal{K}\otimes\mathcal{K}'$ is also a weighted shift whose weight sequence $\{z_i\}$ can be explicitly computed from $\{w_i\}$ and $\{w'_i\}$. Indeed, $w_i = \sqrt{(A_i/A_{i+1})}$, $w'_i = \sqrt{(A'_i/A'_{i+1})}$, where A_i and A'_i are the coefficients of the corresponding kernel functions, and $z_i = \sqrt{(B_i/B_{i+1})}$, where $\{B_i\}$ is given by a convolution formula such as (11). This fact was first noticed in [4], where a direct computational argument was used.

b) It would be of interest to determine what properties are inherited by the canonical model on $\mathcal{K}\otimes_{\mathcal{P}}\mathcal{K}'$, from the canonical models on $\mathcal{K}$ and $\mathcal{K}'$. Even for m = m' = n = 1, this problem is far from resolved. For instance, if T_Z is hyponormal on $\mathcal{K}$ and on $\mathcal{K}'$, is the canonical model on $\mathcal{K}\otimes_{\mathcal{P}}\mathcal{K}'$ also hyponormal? What about subnormality? For weighted shifts the first question has an affirmative answer (see [5]), while the second remains open, and seems to be rather difficult. Its reformulation in terms of moments of measures (see [16]) appears to be highly non-trivial.

REFERENCES

1. Agrawal, Om and Salinas, Norberto: Sharp kernels and canonical subspaces, Amer. J. of Math., 109, 1987.

2. Agler, Jim: Hyper-contractions and subnormality, J. of Operator Theory, 13, 1985, 203-217.

3. Aronszajn, Naham: Theory of reproducing kernels, Transac. Amer. Math. Soc., 68, 1950, 337-404.

4. Badri, Matooq: On perturbations and products of generalized Bergman kernels, Ph.D. Dissertation, University of Kansas, May 1986.

5. Badri, Matooq and Szeptycki, Pawel: Cauchy products of positive sequences, to appear in the Proceedings of 1987 G.P.O.T.S., Rocky Mountain J. of Math.

6. Clark, Douglas and Misra, Gadar: The curvature function and similarity of operators, Matematicki Vesnik, 37, 1985, 21-32.

7. Cowen, Michael and Douglas, Ronald: Complex geometry and operator theory, Acta Math. 141, 1978, 187-261.

8. Curto, Raul and Salinas, Norberto: Generalized Bergman kernels and the Cowen-Douglas theory, Amer. J. of Math., 106, 1984, 447-488.

9. Douglas, Ronald: Hilbert modules over function algebras, preprint, 1984.

10. Gunning, Robert and Rossi, Hugo: Analytic functions of several complex variables, Prentice Hall, New Jersey, 1965.

11. Krantz, Steve: Function theory of several complex variables, John Wiley and Sons, New York, 1982.

12. Rudin, Walter: Function theory in polydisks, Benjamin, New York, 1969.

13. Salinas, Norberto: Toeplitz operators of bounded pseudoconvex domains, preprint.

14. Salinas, Norberto: The Grassman manifold of a C^*-algebra and Hermitian holomorphic bundles, to appear in the Proceedings of the XI International Conference in Operator Theory, Bucharest, Romania, 1986.

15. Shapiro, Joel: Mackey topologies, reproducing kernels and diagonal maps on the Hardy and Bergman spaces, Duke Math. J. 43, 1976, 187-202.

16. Taylor, Joe: The analytic functional calculus of several commuting operators, Acta Math. 125, 1970, 1-38.

17. Taylor, Joe: A general framework for a multi-operator functional calculus, Advances in Math., 9, 1972, 183-225.

Department of Mathematics
University of Kansas
Lawrence, Kansas 66045-2142

Operator Theory:
Advances and Applications, Vol. 32

SPECTRAL CAPACITIES IN QUOTIENT FRÉCHET SPACES

Florian-Horia Vasilescu

Dedicated to the memory of Constantin Apostol

The definition of a spectral capacity has been proved to be a fruitful one in the theory of spectral decompositions of linear operators. Most of its standard properties still hold in the context of quotient Fréchet spaces.

1. INTRODUCTION

One of the most important concepts introduced by C. Apostol in the theory of spectral decompositions of linear operators is that of *spectral capacity* ([2], Definition 2.1). He defined it as a map E from the family of all closed subsets of the complex plane $\mathbf{C}$, with values closed linear subspaces of a given Banach space X, satisfying the following conditions:

(1.1)
(i) $E(\emptyset) = \{0\}$, $E(\mathbf{C}) = X$;

(ii) $E(\bigcap_{m=1}^{\infty} F_m) = \bigcap_{m=1}^{\infty} E(F_m)$ for every sequence of closed subsets $\{F_m\}_{m=1}^{\infty}$ of $\mathbf{C}$;

(iii) $X = E(\overline{G}_1) + \ldots + E(\overline{G}_n)$, where $\{G_j\}_{j=1}^{n}$ is an arbitrary finite open cover of $\mathbf{C}$.

C. Apostol showed, in particular, that every decomposable operator T on X (in the sense of C. Foiaş [8]) has a spectral capacity with the property

(1.2) $TE(F) \subset E(F)$ and $\sigma(T,E(F)) \subset F$ for all closed sets $F \subset \mathbf{C}$.

(We denote here by $\sigma(T,Y)$ the *spectrum* of T when acting on Y, where Y is a subspace of X invariant under T.)

This concept of Apostol's has been adapted to various situations, assuming changes of both the domain of definition and the range, assigned to one or several operators; nevertheless, all these versions essentially preserved the requirements (1.1) and (1.2) (see [1], [10], [16], [17], [22], etc.).

Let X be a Fréchet space (in this work all Fréchet spaces are assumed to be locally convex) and let Lat(X) denote the family of all *Fréchet subspaces* ([21]) of X (i.e. those linear subspaces Y of X that have a Fréchet space structure of their own which makes the inclusion $Y \subset X$ continuous). As the notation suggests, Lat (X) is a lattice with respect to the sum and intersection of subspaces (see, for instance, [18], Lemma 2.1). A *quotient Fréchet space* ([21]) is a linear space of the form X/Y, where X is a Fréchet space and $Y \in \mathrm{Lat}(X)$.

Let X_1/Y_1, X_2/Y_2 be quotient Fréchet spaces and let $T : X_1/Y_1 \to X_2/Y_2$ be a linear map. We define the set

$$G_o(T) = \{(x_1,x_2) \in X_1 \times X_2;\ x_2 \in T(x_1 + Y_1)\},$$

which is obviously a linear space. We say that T is a *linear operator* (or simply *operator*) if $G_o(T) \in \mathrm{Lat}(X_1 \times X_2)$. (Here we adopt the terminology from [19]. A linear operator is called in [18] a *morphism* in the category of quotient Fréchet spaces; it is equivalent to the homonymic concept introduced in [21] in a different manner.) The family of all linear operators from X_1/Y_1 into X_2/Y_2 is a linear space under the usual operations ([18]), denoted by $L(X_1/Y_1,X_2/Y_2)$. When $X_1/Y_1 = X/Y = X_2/Y_2$, we set $L(X/Y) = L(X/Y,X/Y)$, which is, in this case, an algebra ([18]). A *strict operator* $T : X_1/Y_1 \to X_2/Y_2$ is a linear map which is induced by a linear and continuous operator $T_o : X_1 \to X_2$ satisfying $T_o(Y_1) \subset Y_2$; in other words $T(x + Y_1) = T_ox + Y_2$ for each $x \in X_1$. A strict operator is a linear operator ([18]) (Strict operators are called *strict morphisms* in [18] or [21]).

Let X/Y be a fixed quotient Fréchet space. A linear manifold $D = D_o/Y \subset X/Y$ will be called a *(quotient Fréchet) subspace* of X/Y if $D_o \in \mathrm{Lat}(X)$. The family of all subspaces of X/Y, which is easily seen to be a lattice with respect to the sum and intersection of subspaces, will be also denoted by Lat (X/Y). Direct and inverse images of subspaces via linear operators are subspaces too ([22]; see also [18], Lemma 2.1).

Let $P(X/Y)$ be the family of those linear operators T that are defined on subspaces $D = D(T) \in \mathrm{Lat}(X/Y)$, with values in X/Y. Then D(T), which is called the *domain of definition* of T, has the form $D(T) = D_o(T)/Y$, with $D_o(T) \in \mathrm{Lat}(X)$. This class of operators, which has been studied in [22], is a natural extension of the family of closed operators (and even of the larger family of those linear maps between two Fréchet spaces, whose graph is a Fréchet subspace, which originates in [4]).

If $T \in P(X/T)$, we may also define its iterates. Namely, let $D(T^2) = T^{-1}(D(T))$. Then $D(T^2) \in \mathrm{Lat}(X/Y)$ and $T^2 : D(T^2) \to X/Y$, given by $T^2\xi = T(T\xi)$ $(\xi \in D(T^2))$, is a member of $P(X/Y)$. In general, if T^n has been defined $(n \geq 1)$, we set $D(T^{n+1}) = T^{-1}(D(T^n)) \subset D(T)$. Then $D(T^{n+1}) \in \mathrm{Lat}(X/Y)$ and $T^{n+1}\xi = T(T^n\xi)$ $(\xi \in D(T^{n+1}))$. Note also that $D(T^{n+1}) \in \mathrm{Lat}(D(T^n))$.

Let $\mathbf{C}_\infty = \mathbf{C} \cup \{\infty\}$ be the Riemann sphere and let $U \subset \mathbf{C}_\infty$ be open. We denote by

$\mathcal{O}(U,X)$ the Fréchet space of all holomorphic X-valued functions on U. Let ${}_0\mathcal{O}(U,X)$ be equal to $\mathcal{O}(U,X)$ if $\infty \notin U$ and equal to $\{f \in \mathcal{O}(U,X);\ f(\infty) = 0\}$ if $\infty \in U$. It is known that the assignment $U \to \mathcal{O}(U,X)/\mathcal{O}(U,Y)$ ($U \subset \mathbf{C}_\infty$ open) is an analytic sheaf $\mathcal{O}_{X/Y}$ on $\mathbf{C}_\infty$ whose space of global sections $\Gamma(U,\mathcal{O}_{X/Y})$ on U is given by

$$\Gamma(U,\mathcal{O}_{X/Y}) = \mathcal{O}(U,X)/\mathcal{O}(U,Y).$$

The space $\Gamma(U,\mathcal{O}_{X/Y})$ will be denoted by $\mathcal{O}(U,X/Y)$. The assignment $U \to {}_0\mathcal{O}(U,X)/{}_0\mathcal{O}(U,Y)$ is a subsheaf ${}_0\mathcal{O}_{X/Y}$ of $\mathcal{O}_{X/Y}$ whose space of global sections on U equals ${}_0\mathcal{O}(U,X)/{}_0\mathcal{O}(U,Y)$ and will be denoted by ${}_0\mathcal{O}(U,X/Y)$. Note that both X/Y and ${}_0\mathcal{O}(U,X/Y)$ are (isomorphic to) subspaces of $\mathcal{O}(U,X/Y)$ (see [18], Section 3, for some details).

Let $T \in P(X/Y)$ and let $U \subset \mathbf{C}_\infty$ be open. Then T induces a linear operator

$$T_U : \mathcal{O}(U,D(T)) \to \mathcal{O}(U,X/Y)$$

which extends T and maps the subspace ${}_0\mathcal{O}(U,D(T))$ into ${}_0\mathcal{O}(U,X/Y)$ (see [18], Section 2). In other words, $T_U \in P(\mathcal{O}(U,X/Y))$ and $D(T_U) = \mathcal{O}(U,D(T))$.

Let ζ be the coordinate function on $\mathbf{C}$. Then ζ induces by multiplication a linear operator

$$\zeta_U : {}_0\mathcal{O}(U,D(T)) \to \mathcal{O}(U,D(T)).$$

Therefore we have a linear operator

$$\zeta_U - T_U : {}_0\mathcal{O}(U,D(T)) \to \mathcal{O}(U,X/Y) \tag{1.3}$$

for every open $U \subset \mathbf{C}_\infty$. (It is, in fact, a sheaf morphism.)

The *resolvent set* $\rho(T,X/Y)$ of $T \in P(X/Y)$ is the largest open set $V \subset \mathbf{C}_\infty$ such that the linear operator (1.3) is bijective for every open $U \subset V$. The complement $\sigma(T,X/Y)$ of $\rho(T,X/Y)$ in $\mathbf{C}_\infty$ (which is a nonempty closed set) is called the *spectrum* of T. (These concepts have been defined in [18] for $T \in L(X/Y)$ and extended in [22] for $T \in P(X/Y)$.)

A subspace $Z \in \mathrm{Lat}(X/Y)$ is said to be *invariant* under $T \in P(X/Y)$ if $T(Z \cap D(T)) \subset Z$. We denote by $T|Z$ the linear map $T : Z \cap D(T) \to Z$ and call it the *restriction* of T to Z. It is easily seen that $T|Z \in P(Z)$. The family of all invariant subspaces of T will be designated by Inv(T). The spectrum of the operator $T|Z$ will be denoted by $\sigma(T,Z)$.

Let $T \in P(X/Y)$ and let $Z = Z_0/Y \in \mathrm{Inv}(T)$. Then T induces a linear operator $\hat{T} \in P(X/Z_0)$ with $D(\hat{T}) = (D_0(T) + Z_0)/Z_0$, given by $\hat{T}(x + Z_0) = y + Z_0$ ($(x,y) \in G_0(T)$).

From now on by *spectral capacity* we mean a map E defined on the closed subsets of $\mathbf{C}_\infty$, with values in Lat(X/Y), such that (1.1) is fulfilled (with $\mathbf{C}$ replaced by $\mathbf{C}_\infty$ and X by X/Y). A linear operator $T \in P(X/Y)$ is said to be *decomposable* if there exists a spectral capacity E with values in Inv(T) such that $\sigma(T,E(F)) \subset F$ for all closed $F \subset \mathbf{C}_\infty$ (which is essentially (1.2)). Such a spectral capacity is said to be *attached* to T.

This concept of decomposable operator extends the homonymic one due to C. Foiaş (see [8] or [3]), via the characterization from [9] (see also [16], [17], [22] for other extensions). Spectral capacities of this type and decomposable operators from $L(X/Y)$ have been studied in [22]. Unlike [22], we consider in this work "unbounded" decomposable operators, that is, operators from $P(X/Y)$. Our main concern is to recapture, in the present setting, some of the properties of (unbounded) decomposable operators in Fréchet spaces (see [16], Chapter IV). At the same time, we try to prove that the framework of quotient Fréchet spaces allows the development of a sufficiently sophisticated theory of spectral decompositions, in which the contributions of C. Apostol play a central rôle.

2. THE SPECTRUM OF A CLOSED OPERATOR

In this section we shall present a characterization of the resolvent set of a closed operator in a Fréchet space (and therefore of its spectrum) in terms of spaces of holomorphic vector-valued functions (see [18], Proposition 1.2 for continuous operators). Although elementary, it provides, in our opinion, the necessary explanation for the definition of the spectrum of a linear operator in a quotient Fréchet space, as given in the Introduction.

Let X be a fixed Fréchet space (which can be regarded as the quotient Fréchet space X/{0}), and let $C(X)$ be the family of all closed linear operators, defined on linear subspaces of X, with values in X. It is known ([18], Lemma 2.5) that $L(X)(= L(X/\{0\})$ is precisely the algebra of all linear and continuous operators on X. If $T \in C(X)$, the algebraic isomorphism between D(T) and G(T) (i.e. the graph of T) shows that $D(T) \in$ Lat(X). Moreover, the operator $T : D(T) \to X$ becomes continuous when D(T) is endowed with this topology.

Following [20] (see also [16]), a point $z_0 \in \mathbf{C}_\infty$ is said to be *regular* for $T \in C(X)$ if z_0 has a neighbourhood V_0 in $\mathbf{C}_\infty$ such that

(1) $(z - T)^{-1} \in L(X)$ for every $z \in V_0 \cap \mathbf{C}$;

(2) the set $\{(z - T)^{-1}x;\ z \in V_0 \cap \mathbf{C}\}$ is bounded in X for each $x \in X$.

The set of all regular points for T, which is obviously an open set in $\mathbf{C}_\infty$, will be denoted by $\rho_W(T,X)$. We shall prove in this section that $\rho_W(T,X) = \rho(T,X)$, where the latter has been defined in the Introduction.

2.1. LEMMA. *Let* $T \in C(X)$ *and let* $z_0 \in \rho_W(T,X) \cap \mathbf{C}$. *Then there exists an open set* $V \subset \mathbf{C}$, $z_0 \in V$, *such that the operator* $\zeta - T : O(V,D(T)) \to O(V,X)$ *is bijective.*

PROOF. Let $V \subset \mathbf{C}$ be an open neighbourhood of z_0 such that $(z - T)^{-1} \in L(X)$ for every $z \in V$ and the set $\{(z - T)^{-1}x;\ z \in V\}$ is bounded in X for every $x \in X$. Then the operator $\zeta - T$ is obviously injective on $O(V,D(T))$. Let us prove that it is surjective too.

Let $g \in O(V,X)$ and set $f(z) = (z - T)^{-1}g(z)$ for $z \in V$. We first show that the function $f : V \to D(T)$ is continuous. For, let $w_0 \in V$ be fixed and let $W \subset V$ be a compact neighbourhood of w_0. It follows from the uniform boundedness principle that the family $\{(z - T)^{-1};\ z \in V\}$ is equally continuous. In particular, the set

$$\{(w - T)^{-1}(w_0 - T)^{-1}g(w);\ w \in W\} \tag{2.1}$$

is bounded in X. Since the set

$$\{T(w - T)^{-1}(w_0 - T)^{-1}g(W);\ w \in W\} \subset$$
$$\subset W \cdot \{(w - T)^{-1}(w_0 - T)^{-1}g(w);\ w \in W\} - \{(w_0 - T)^{-1}g(w);\ w \in W\}$$

is also bounded in X, it results that (2.1) is actually bounded in D(T). Therefore

$$\lim_{w\to w_0} (w_0 - w)(w - T)^{-1}(w_0 - T)^{-1}g(w) = 0 \tag{2.2}$$

in D(T). Using the continuity of g, we infer that

$$\lim_{w\to w_0} T(w_0 - T)^{-1}(g(w) - g(w_0)) =$$
$$= \lim_{w\to w_0} w_0(w_0 - T)^{-1}(g(w) - g(w_0)) - \lim_{w\to w_0} (g(w) - g(w_0)) = 0.$$

Therefore

$$\lim_{w\to w_0} (w_0 - T)^{-1}(g(w) - g(w_0)) = 0 \tag{2.3}$$

in D(T). From (2.2) and (2.3) we deduce that

$$\lim_{w\to w_0} (f(w) - f(w_0)) = \lim_{w\to w_0} (w_0 - w)(w - T)^{-1}(w_0 - T)^{-1}g(w) +$$
$$+ \lim_{w\to w_0} (w_0 - T)^{-1}(g(w) - g(w_0)) = 0$$

in D(T), which proves the continuity of the function f at w_0. Since $w_0 \in V$ is arbitrary, $f : V \to D(T)$ is continuous.

Note also that

$$\lim_{w\to w_o} ((w - w_o)^{-1}(w_o - T)^{-1}(g(w) - g(w_o)) - (w_o - T)^{-1}f(w)) =$$

$$= (w_o - T)^{-1}g'(w_o) - (w_o - T)^{-1}f(w_o)$$

in X, and that

$$\lim_{w\to w_o} T((w - w_o)^{-1}(w_o - T)^{-1}(g(w) - g(w_o)) - (w_o - T)^{-1}f(w)) =$$

$$= w_o(w_o - T)^{-1}g'(w_o) - g'(w_o) + f(w_o) - w_o(w_o - T)^{-1}f(w_o)$$

in X, where $g'(w_o) = (dg/dw)(w_o)$. Hence the limit

$$\lim_{w\to w_o} (w - w_o)^{-1}(f(w) - f(w_o))$$

exists in D(T). This shows that the function $f : V \to D(T)$ is differentiable at every point $w_o \in V$. By the (vector version of the classical) Looman-Menchoff theorem, it follows that $f \in O(V,D(T))$.

2.2. LEMMA. *Let* $T \in C(X)$ *and let* $V \subset \mathbf{C}_\infty$ *be open. If* $\zeta - T : {}_oO(V,D(T)) \to \to O(V,X)$ *is bijective, then* $V \subset \rho_W(T,X)$.

PROOF. For every $z \in V \cap \mathbf{C}$ we define the linear map

$$T_z x = ((\zeta - T)^{-1}x)(z), \ x \in X,$$

where x is regarded as as constant function from $O(V,X)$. Since $(\zeta - T)^{-1} : O(V,X) \to \to {}_oO(V,D(T))$ is continuous (by the closed graph theorem), it is clear that $T_z : X \to D(T)$ is continuous. Notice that

$$(z - T)T_z x = (z - T)((\zeta - T)^{-1}x)(z) = ((\zeta - T)(\zeta - T)^{-1}x)(z) = x$$

for each $x \in X$, and

$$T_z(z - T)y = ((\zeta - T)^{-1}(z - T)y)(z) = ((z - T)(\zeta - T)^{-1}y)(z) =$$

$$= ((\zeta - T)(\zeta - T)^{-1}y)(z) = y$$

for every $y \in D(T)$. This shows that $T_z = (z - T)^{-1}$ for all $z \in V \cap \mathbf{C}$.

Let $z_o \in V$ be fixed and let $V_o \subset V$ be a compact neighbourhood of z_o. Then the set

$$\{(z - T)^{-1}x;\ x \in V_o \cap \mathbf{C}\} \subset ((\zeta - T)^{-1}x)(V_o)$$

is bounded in D(T), and therefore in X. Thus $z_o \in \rho_W(T,X)$, and so $V \subset \rho_W(T,X)$.

2.3. LEMMA. *Let* $T \in C(X)$ *and assume that* $\infty \in \rho_W(T,X)$. *Then there exists an open set* $V_\infty \subset \mathbf{C}_\infty$, $\infty \in V_\infty$, *such the operator* $\zeta - T : {}_0O(V_\infty, D(T)) \to O(V_\infty, X)$ *is bijective.*

PROOF. Since $\infty \in \rho_W(T,X)$, there exists an open neighbourhood V_∞ of ∞ such that $(z - T)^{-1} \in L(X)$ for all $z \in V_\infty \cap \mathbf{C}$ and the set $\{(z - T)^{-1}x; z \in V_\infty \cap \mathbf{C}\}$ is bounded for each $x \in X$. Then the operator $\zeta - T$ is clearly injective on ${}_0O(V_\infty, D(T))$. We shall prove that $\zeta - T$ is onto $O(V_\infty, X)$.

Let $g \in O(V_\infty, X)$ and set $f(z) = (z - T)^{-1}g(z)$ for $z \in V_\infty \cap \mathbf{C}$. Since every point $z \in V_\infty \cap \mathbf{C}$ is regular for T, it follows by Lemma 2.1 that $f \in O(V_\infty \cap \mathbf{C}, D(T))$. We shall prove that f is analytic and null at infinity.

Let $W_\infty \subset V_\infty$ be a compact neighbourhood of ∞. Then the set

$$f(W_\infty \cap \mathbf{C}) = \{(z - T)^{-1}g(z); z \in W_\infty \cap \mathbf{C}\}$$

is bounded in X (by the uniform boundedness principle). This shows that $f \in O(W_\infty, X)$ (see the proof of Corollary II.4.14 in [16]). Then, from the equation

$$f(z) = z^{-1}Tf(z) + z^{-1}g(z),\ z \in V_\infty \cap \mathbf{C},\ z \neq 0, \tag{2.4}$$

it follows that $\lim_{z\to\infty} z^{-1}Tf(z)$ exists in X. Since $\lim_{z\to\infty} z^{-1}f(z) = 0$ and T is closed, we must have $\lim_{z\to\infty} z^{-1}Tf(z) = 0$. Using this fact, we obtain, again from (2.4), that $\lim_{z\to\infty} f(z) = 0$. Therefore

$$\lim_{z\to\infty} Tf(z) = \lim_{z\to\infty} (zf(z) - g(z)) \tag{2.5}$$

exists in X. Then we have as above that $\lim_{z\to\infty} Tf(z) = 0$. Consequently both f(z) and $Tf(z) = zf(z) - g(z)$ are analytic in V_∞ and null at ∞. Hence $f \in {}_0O(V_\infty, D(T))$.

The next consequence of Lemma 2.3 is known (see, for instance, [16], Lemma III.3.5) but its proof is seemingly new.

2.4. COROLLARY. *Let* $T \in C(X)$ *be such that* $\infty \in \rho_W(T,X)$. *Then* $T \in L(X)$.

PROOF. We take in the previous proof $g = x$, where $x \in X$ is fixed. Then from (2.5) we infer that $\lim_{z\to\infty} zf(z) = x$. Since $\lim_{z\to\infty} zTf(z)$ exists in X and the operator T is closed, we deduce that $x \in D(T)$. Hence $D(T) = X$, and so $T \in L(X)$.

2.5. THEOREM. *Let* $T \in C(X)$ *and let* $V \subset \mathbf{C}_\infty$ *be open. We have* $V \subset \rho_W(T,X)$ *if and only if the operator*

$$\zeta - T : {}_0O(V,D(T)) \to O(V,X)$$

is bijective.

PROOF. Let $V \subset \rho_W(T,X)$. From Lemmas 2.1 and 2.3 we derive the existence of an open cover $\{V_j\}_{j\in J}$ of V such that $\zeta - T : {}_0O(V_j,D(T)) \to O(V_j,X)$ is bijective for all $j \in J$. Let $g \in O(V,X)$ be given. Then for every $j \in J$ we can find $f_j \in {}_0O(V_j,D(T))$ such that $(\zeta - T)f_j = g|V_j$. Since $(\zeta - T)(f_j - f_k) = 0$ on $V_j \cap V_k$, it follows $f_j = f_k$ on $V_j \cap V_k$. Therefore there is a function $f \in {}_0O(V,D(T))$ such that $f|V_j = f_j$ for all $j \in J$. This shows that $\zeta - T$ is onto $O(V,X)$. As $\zeta - T$ is obviously injective, it must be bijective.

Conversely, the assertion follows from Lemma 2.2.

2.6. COROLLARY. *The set $\rho_W(T,X)$ is the largest open set $V \subset \mathbf{C}_\infty$ with the property that $\zeta - T : {}_0O(V,D(T)) \to O(V,X)$ is bijective. Therefore $\rho_W(T,X) = \rho(T,X)$.*

2.7. REMARK. The family $P(X)(= P(X/\{0\})$ is strictly larger than the family $C(X)$. Indeed, if $Y \in \mathrm{Lat}(X)$ is not closed in X, then the inclusion $i : Y \to X$ is in $P(X)$ but not in $C(X)$. Nevertheless, it is the class $C(X)$ which is the most interesting from the spectral point of view. Specifically, if $T \in P(X)$ and $\rho(T,X) \neq \emptyset$, then $T \in C(X)$.

3. NATURAL SPECTRAL CAPACITIES

The uniqueness of the spectral capacity attached to a decomposable operator, first proved by C. Foiaş [9] (see also [10], [16], [17], [22] for some extensions) makes it a very useful concept in the study of spectral decompositions of linear operators. In this section we shall prove a version of this uniqueness result in our more general setting. Some other properties, extensions of statements from [2], [3], [7], [13], [14], [16], [17], [22], will be also presented.

We shall rely heavily upon the work [18]. We shall also use some assertions from [19] and [22] (generally accompanied by an outline of the proof).

3.1. REMARK. Let

$$0 \to X_1/Y_1 \xrightarrow{S} X_2/Y_2 \xrightarrow{T} X_3/Y_3 \to 0$$

be an exact complex of quotient Fréchet spaces. Then for every open set $U \subset \mathbf{C}_\infty$ the complex

$$0 \to O(U,X_1/Y_1) \xrightarrow{S_U} O(U,X_2/Y_2) \xrightarrow{T_U} O(U,X_3/Y_3) \to 0$$

is also exact. This assertion is proved in [19]. For the convenience of the reader we shall sketch its proof.

If $S \in L(X_1/Y_1,X_2/Y_2)$ is arbitrary, $N(S) = N_0(S)/Y_1$ is the null-space of S, $R(S) = R_0(S)/Y_2$ is the range of S and $U \subset \mathbf{C}_\infty$ is open, then we have the equalities $N_0(S_U) =$

$= \mathcal{O}(U,N_o(S))$ and $R_o(S_U) = \mathcal{O}(U,R_o(S))$. These equalities follow from the elementary properties of tensor products with nuclear spaces (see, for instance, [5]). Then, if $T \in L(X_2/Y_2, X_3/Y_3)$ and $R(S) \subset N(T)$, we have the equality

$$\mathcal{O}(U,N_o(T)/R_o(S)) = N_o(T_U)/R_o(S_U)\,,$$

from which we derive the desired exactness.

From now on X/Y will be a fixed quotient Fréchet space.

3.2. LEMMA. *Let* $T \in P(X/Y)$, *let* $Z = Z_o/Y \in \mathrm{Inv}(T)$ *and let* $\hat{T} \in P(X/Z_o)$ *be the operator induced by* T. *Then the union of any two of the sets* $\sigma(T,X/Y)$, $\sigma(T,Z)$ *and* $\sigma(\hat{T},X/Z_o)$ *contains the third.*

PROOF. For every open $U \subset \mathbf{C}_\infty$ the following diagram

$$\begin{array}{ccccccccc} 0 \to & {}_o\mathcal{O}(U,D(T|Z)) & \xrightarrow{i_U} & {}_o\mathcal{O}(U,D(T)) & \xrightarrow{k_U} & {}_o\mathcal{O}(U,D(\hat{T})) & \to 0 \\ & \downarrow \zeta_U - (T|Z)_U & & \downarrow \zeta_U - T_U & & \downarrow \zeta_U - \hat{T}_U & \\ 0 \to & \mathcal{O}(U,Z) & \xrightarrow{i_U} & \mathcal{O}(U,X/Y) & \xrightarrow{k_U} & \mathcal{O}(U,X/Z_o) & \to 0 \end{array} \tag{3.1}$$

is commutative, where $i : Z \to X/Y$ is the inclusion and $k : X/Y \to X/Z_o$ is the canonical map. The commutativity of (3.1) follows from the results of [18] (see especially Theorem 2.9). Moreover, the rows of (3.1) are exact, by Remark 3.1. Therefore if any two of the columns of (3.1) are exact (i.e. the corresponding operators are bijective) then the third is exact as well, whence we derive our assertion.

If $T \in P(X/Y)$ and $\xi \in X/Y$, we denote by $\delta_T(\xi)$ the set of those points $z \in \mathbf{C}_\infty$ for which there is an open set V containing z and a section $\phi \in {}_o\mathcal{O}(V,D(T))$ such that $(\zeta_V - T_V)\phi = \xi$ (where ξ is regarded as a section in $\mathcal{O}(U,X/Y)$). $\delta_T(\xi)$ is an open set which is called the *local resolvent (set)* of T at ξ. The set $\gamma_T(\xi) = \mathbf{C}_\infty \setminus \delta_T(\xi)$ is called the *local spectrum* of T at ξ (see [3], [16], [22] for some stages of these concepts).

A linear operator $T \in P(X/Y)$ is said to have the *single valued extension property* (briefly SVEP) if the operator

$$\zeta_U - T_U : {}_o\mathcal{O}(U,D(T)) \to \mathcal{O}(U,X/Y)$$

is injective for every open $U \subset \mathbf{C}_\infty$. In this case, for each $\xi \in X/Y$ there exists a uniquely determined section $\xi_T \in {}_o\mathcal{O}(W,D(T))$ such that $(\zeta_W - T_W)\xi_T = \xi$, where $W = \delta_T(\xi)$ (see the above references).

3.3. LEMMA. *Let* $T \in P(X/Y)$. *The local spectrum has the following properties:*

(1) $\gamma_T(0) = \emptyset$;

(2) $\gamma_T(\xi + \gamma) \subset \gamma_T(\xi) \cup \gamma_T(\eta)$ *for all* $\xi, \eta \in X/Y$;

(3) $\gamma_T(z\xi) = \gamma_T(\xi)$ *for all* $\xi \in X/Y$ *and* $z \in \mathbf{C} \setminus \{0\}$;

(4) *if* $\xi \in D(T)$ *and* $\phi \in {}_oO(V,D(T))$ *satisfies* $(\zeta_V - T_V)\phi = \xi$ *for some open* $V \subset \delta_T(\xi)$, *then* $\phi \in {}_oO(V,D(T^2))$.

PROOF. Properties (1), (2) and (3) are simple exercises.

Let us prove (4). We have:

$$T_V\phi = \phi + \zeta_V\phi \in O(V,D(T)) = D(T_V)\,.$$

In other words,

$$\phi \in D((T_V)^2) = D((T^2)_V) = O(V,D(T^2)).$$

On the other hand, $\phi \in P(X/Y)$, and so $\phi \in {}_oO(V,D(T^2))$.

Let $T \in P(X/Y)$, let $W \subset \mathbf{C}_\infty$ be open and let $F = \mathbf{C}_\infty \setminus W$. We define the linear manifold

$$(3.1) \qquad {}_oO_c(W,D(T)) = \{\phi \in {}_oO(W,D(T)),\ (\zeta_W - T_W)\phi \in X/Y\},$$

which is a subspace of ${}_oO(W,D(T))$. Then the image

$$(3.2) \qquad E_T^o(F) = (\zeta_W - T_W)({}_oO_c(W,D(T)))$$

is a subspace of X/Y. If $F = \mathbf{C}_\infty$, we set $E_T^o(F) = X/Y$. Since T_W extends T, from the equation

$$T(\zeta_W - T_W)\phi = T_W(\zeta_W - T_W)\phi = (\zeta_W - T_W)T_W\phi,$$

valid for every $\phi \in {}_oO(W,D(T^2))$, it follows that T_W maps ${}_oO(W,D(T^2))$ into ${}_oO_c(W,D(T))$.

3.4. LEMMA. *Let* $T \in P(X/Y)$ *and let* $W \subset \mathbf{C}_\infty$ *be open. Then for every open* $V \subset W$ *the operator*

$$\zeta_V - T_{WV} : {}_oO(V,{}_oO_c(W,D(T^2))) \to O(V,{}_oO_c(W,D(T)))$$

is bijective.

PROOF. There exists a linear and continuous operator

$$\tau_o : O(V,O(W,X)) \to {}_oO(V,{}_oO(W,X))$$

which is given by the equation

$$(3.3) \qquad (z - w)(\tau_o f)(z,w) = f(z,w) - f(z,z), \qquad z \in \mathbf{C} \cap V, \qquad w \in \mathbf{C} \cap W.$$

We shall also use the linear and continuous operator

$$\delta_o : \mathcal{O}(V,\mathcal{O}(W,X)) \to \mathcal{O}(V,X)$$

given by $(\delta_o f)(z) = f(z,z)$ $(z \in V)$. Let τ (resp. δ) be the strict operator induced by τ_o (resp. δ_o) from $\mathcal{O}(V,\mathcal{O}(W,X/Y))$ into ${}_o\mathcal{O}(V,{}_o\mathcal{O}(W,X/Y))$ (resp. from $\mathcal{O}(V, \mathcal{O}(W,X/Y))$ into $\mathcal{O}(V,X/Y)$). Then from (3.3) we derive easily the equality

$$(3.4) \qquad (\zeta_V - \zeta_W)\tau\phi = \phi - \delta\phi, \ \phi \in \mathcal{O}(V,\mathcal{O}(W,X/Y))$$

(where we use some obvious identifications). We shall show that the operator τ induces a map

$$(3.5) \qquad \tau : \mathcal{O}(V,{}_o\mathcal{O}_c(W,D(T))) \to {}_o\mathcal{O}(V,{}_o\mathcal{O}_c(W,D(T^2))),$$

which provides an inverse for $\zeta_V - T_{WV}$.

If $\phi \in \mathcal{O}(V,{}_o\mathcal{O}_c(W,D(T)))$, then $\zeta_W\phi$ is a section in $\mathcal{O}(V,\mathcal{O}(W,D(T)))$ and we may write the equalities

$$(3.6) \qquad \begin{aligned}&(\zeta_V - \zeta_W)(\zeta_W - T_{WV})\tau\phi = (\zeta_W - T_{WV})(\zeta_V - \zeta_W)\tau\phi = \\ &= (\zeta_W - T_{WV})(\phi - \delta\phi) = (\zeta_W - T_{WV})\phi + (\zeta_V - \zeta_W)\delta\phi - (\zeta_V - T_{WV})\delta\phi.\end{aligned}$$

Let us prove that

$$(\zeta_W - T_{WV})\phi = (\zeta_V - T_{WV})\delta\phi.$$

Indeed, it is clear that $\delta\zeta_W\phi = \zeta_V\delta\phi$. We also have $\delta T_{WV}\phi = T_V\delta\phi$ since if $(f,g) \in G_o(T_{WV})$, then $\delta_o(f,g) = (\delta_o f,\delta_o g) \in G_o(T_V)$. Therefore

$$\delta(\zeta_W - T_{WV})\phi = (\zeta_V - T_V)\delta\phi.$$

On the other hand, the restriction of δ to $\mathcal{O}(V,E_T^o(F))$ is just the identity, where $F = \mathbf{C}_\infty \setminus W$. Hence

$$(\zeta_W - T_{WV})\phi = \delta(\zeta_W - T_{WV})\phi = (\zeta_V - T_V)\delta\phi.$$

If we return to (3.6), we get

$$(\zeta_V - \zeta_W)(\zeta_W - T_{WV})\tau\phi = (\zeta_V - \zeta_W)\delta\phi.$$

Since the map $\zeta_V - \zeta_W$ is injective (which follows from the fact that if the function $(z - w)f(z,w)$ belongs to $\mathcal{O}(V,\mathcal{O}(W,Y))$, then f itself must be in $\mathcal{O}(V,\mathcal{O}(W,Y))$), we obtain

$$(\zeta_W - T_{WV})\tau\phi = \delta\phi.$$

Thus

$$(\zeta_V - T_{WV})\tau\phi = (\zeta_V - \zeta_W)\tau\phi + (\zeta_W - T_{WV})\tau\phi = \phi,$$

by (3.4). This shows that τ is a right inverse of $\zeta_V - T_{WV}$. Moreover

$$T_{WV}\tau\phi = \zeta_V\tau\phi - \phi \in O(V,O(W,D(T)),$$

and therefore $\tau\phi \in {}_oO(V,{}_oO_c(W,D(T^2)))$.

Note also that

$$\tau(\zeta_V - T_{WV})\phi = (\zeta_V - T_{WV})\tau\phi = \phi$$

(since obviously $\tau\zeta_V = \zeta_V\tau$ and if $(f,g) \in G_o(T_{WV})$, then $\tau_o(f,g) = (\tau_o f,\tau_o g) \in G_o(T_{WV})$). Consequently (3.5) must be the desired inverse.

3.5. COROLLARY. *Let* $T \in P(X/Y)$, *let* $W \subset \mathbf{C}_\infty$ *be open and let* $F = \mathbf{C}_\infty \setminus W$. *If* $\xi \in E^o_T(F)$ *and* $\phi \in {}_oO_c(W,D(T))$ *satisfies* $(\zeta_W - T_W)\phi = \xi$, *then* $\varepsilon_w\phi \in E^o_T(F)$ *for every* $w \in W$, *where* $\varepsilon_w : O(W,X/Y) \to X/Y$ *is the strict operator induced by the evalution at the point* w.

PROOF. It follows from Lemma 3.4 that we can find a section $\psi \in {}_oO(V,{}_oO_c(W,D(T^2)))$ such that $(\zeta_V - T_{WV})\psi = \phi$, where $V \subset W$ is an arbitrary open set. Then we have

$$(\zeta_V - T_V)\varepsilon_{w,V}\psi = \varepsilon_{w,V}(\zeta_V - T_{WV})\psi = \varepsilon_{w,V}\phi = \varepsilon_w\phi,$$

and $\varepsilon_{w,V}\psi \in {}_oO(V,D(T))$. Therefore, for $V = W$, we infer that $\gamma_T(\varepsilon_w\phi) \subset F$.

As a matter of fact, we actually have

$$\gamma_T(\varepsilon_w\phi) = \gamma_T(\xi)$$

for every $w \in \mathbf{C} \cap W$, which can be shown by similar arguments. We omit the details (see [16], Proposition IV.3.4).

3.6. LEMMA. *Let* $T \in P(X/Y)$ *have the* SVEP. *For every closed* $F \subset \mathbf{C}_\infty$ *we set*

$$E_T(F) = \{\xi \in X/Y;\ \gamma_T(\xi) \subset F\}\ .$$

Then $E_T(F) = E^o_T(F) \in \mathrm{Inv}(T)$.

PROOF. If $\xi \in E_T(F)$, then $\xi = (\zeta_W - T_W)\phi$, where $\phi = \xi_T | W \in {}_oO(W,D(T))$ and $W = \mathbf{C}_\infty \setminus F$. Hence $\xi \in E^o_T(F)$. The inclusion $E^o_T(F) \subset E_T(F)$ is obvious.

That $E_T(F) \in \mathrm{Inv}(T)$ follows from the equality

$$(\zeta_W - T_W)T_W\phi = T(\zeta_W - T_W)\phi\ ,$$

valid for every $\phi \in {}_oO_c(W,D(T^2))$ (which has been already noticed).

The next result extends an assertion which originates in [14] (see also [13], [7], [22]).

3.7. THEOREM. *Let* $T \in \mathcal{P}(X/Y)$ *have the* SVEP. *Then for every closed* $F \subset \mathbf{C}_\infty$ *one has the inclusion*

$$\sigma(T,\mathcal{E}_T(F)) \subset F \cap \sigma(T,X/Y) .$$

PROOF. If $F = \mathbf{C}_\infty$ the assertion is obvious, so that we may assume $F \neq \mathbf{C}_\infty$. Let $W = \mathbf{C}_\infty \setminus F$ and let $V \subset W$ be open. Then the diagram

$$\begin{array}{ccc} \mathcal{E}_T(F) \cap D(T) & \xrightarrow{\;T\;} & \mathcal{E}_T(F) \\ \uparrow \zeta_W - T_W & & \uparrow \zeta_W - T_W \\ {}_0\mathcal{O}_c(W,D(T^2)) & \xrightarrow{\;T_W\;} & {}_0\mathcal{O}_c(W,D(T)) \end{array}$$

is easily seen to be commutative. Using the functors ${}_0\mathcal{O}(V,\cdot\,)$ and $\mathcal{O}(V,\cdot\,)$, we obtain the commutative diagram

$$(3.7)\qquad \begin{array}{ccc} {}_0\mathcal{O}(V,\mathcal{E}_T(F) \cap D(T)) & \xrightarrow{\;\zeta_V - T_V\;} & \mathcal{O}(V,\mathcal{E}_T(F)) \\ \uparrow \zeta_W - T_{WV} & & \uparrow \zeta_W - T_{WV} \\ {}_0\mathcal{O}(V,{}_0\mathcal{O}_c(W,D(T^2))) & \xrightarrow{\;\zeta_V - T_{WV}\;} & \mathcal{O}(V,{}_0\mathcal{O}_c(W,D(T))) \end{array}$$

(see [18] for some details).

We have to prove that the operator

$$(3.8)\qquad \zeta_V - T_V : {}_0\mathcal{O}(V,\mathcal{E}_T(F) \cap D(T)) \to \mathcal{O}(V,\mathcal{E}_T(F))$$

is bijective. The space $\mathcal{E}_T(F)$ is isomorphic to the space ${}_0\mathcal{O}_c(W,D(T))$, since T has the SVEP. Similarly, the space $\mathcal{E}_T(F) \cap D(T)$ is isomorphic to the space ${}_0\mathcal{O}_c(W,D(T^2))$ (by Lemma 3.3 (4)). Therefore, to prove the bijectivity of (3.8) it suffices to prove the bijectivity of $\zeta_V - T_{WV}$, when acting on the lower row of (3.7), which follows from Lemma 3.4. This shows that

$$\sigma(T,\mathcal{E}_T(F)) \subset F .$$

As we clearly have

$$\mathcal{E}_T(F) = \mathcal{E}_T(F \cap \sigma(T,X/Y)) ,$$

it follows from the above result that

$$\sigma(T,\mathcal{E}_T(F)) = \sigma(T,\mathcal{E}_T(F \cap \sigma(T,X/Y))) \subset F \cap \sigma(T,X/Y) ,$$

which completes the proof of the theorem.

3.8. REMARKS. 1° Theorem 3.7 is connected with another important observation of Apostol's. Namely, he proved directly that if X is a Banach space and $T \in L(X)$ has the SVEP, then there exists a holomorphic functional calculus with functions analytic in neighbourhoods of a given closed set $F \subset \mathbf{C}$, associated to the linear map $T|E_T(F)$ ([2], Theorem 2.10; see also [6], [7], [13], [22] for further development). It follows from Theorem 3.7 that if X is actually a Fréchet space and $T \in C(X)$, then $E_T(F) \in \mathrm{Inv}(T)$ and $\sigma(T, E_T(F)) \subset F$ for each closed $F \subset \mathbf{C}_\infty$. Hence the existence of a holomorphic functional calculus for $T|E_T(F)$ (as well as its consequences) can also be obtained from the general theory of Fréchet space operators (see [16], Section III.3). The Fréchet space structure of $E_T(F)$ and the spectral inclusion $\sigma(T, E_T(F)) \subset F$ (with respect to this structure) has been first noticed in [14] (when $E_T(F)$ is supposed to be closed in X, the assertion goes back to [3]).

2° If $T \in P(X/Y)$ has the SVEP, then the assignment $F \to E_T(F)$ provides a map with the properties (i) and (ii) from (1.1) (with $\mathbf{C}$ replaced by $\mathbf{C}_\infty$). If for every open cover $\{G_j\}_{j=1}^n$ of $\mathbf{C}$ one has

$$X/Y = E_T(\bar{G}_1) + \ldots + E_T(\bar{G}_n),$$

then the operator T is decomposable, via Theorem 3.7.

Conversely, we shall see that every decomposable operator has the SVEP and its spectral capacity is uniquely determined and coincides with the natural one, given by Lemma 3.6.

3° As one might expect (see Corollary 2.4), if $T \in P(x/Y)$ and $\infty \notin \sigma(T, X/Y)$, then $T \in L(X/Y)$. This assertion is obtained in [22]. For the convenience of the reader, we shall sketch its proof. Let $U = \mathbf{C}_\infty \setminus \sigma(T, X/Y)$ and let $\xi = x + Y \in X/Y$. Take $\phi \in {}_0O(U, D(T))$ such that $(\zeta_U - T_U)\phi = \xi$. If $f \in {}_0O(U, D_0(T))$ is in the coset ϕ and $g \in {}_0O(U, X)$ is in the coset $T_U\phi$, then $\zeta f - g - x \in O(U, Y)$. This shows that $x \in D_0(T)$. Therefore $D(T) = X/Y$.

3.9. LEMMA. *Let $T \in L(X/Y)$ be such that $\infty \notin \sigma(T, X/Y)$ and let $U \subset \mathbf{C}$ be open, $U \supset \sigma(T, X/Y)$. Then the operator*

$$\zeta_U - T_U : O(U, X/Y) \to O(U, X/Y)$$

is injective.

PROOF. Let $\phi \in O(U, X/Y)$ be such that $(\zeta_U - T_U)\phi = 0$. If $V = U \setminus \sigma(T, X/Y)$, then $\phi|V = 0$. In other words, if $f \in O(U, X)$ is in the coset ϕ, then $f|V \in O(V, Y)$. If

$\Delta \supset \sigma(T,X/Y)$ is a Cauchy domain such that $\overline{\Delta} \subset U$, and Γ is the boundary of Δ, then the Cauchy formula

$$g(z) = (2\pi i)^{-1} \int_\Gamma (w - z)^{-1} f(w)dw, \quad z \in \Delta ,$$

defines a function $g \in \mathcal{O}(\Delta, Y)$. On the other hand, since $f \in \mathcal{O}(U,X)$, we must have $f|\Delta = g|\Delta$. Therefore $f \in \mathcal{O}(U,Y)$, that is $\phi = 0$.

3.10. LEMMA. *Let* $T \in P(X/Y)$ *and let* $Z_j = X_j/Y \in \mathrm{Inv}(T)(j = 0,1,2)$ *be such that either* Z_1 *or* Z_2 *is in* $D(T)$ *and* $X/Y = Z_1 + Z_2$. *If* $\sigma(T,Z_o) \cap (\sigma(T,Z_2) \cup \sigma(T,Z_1 \cap Z_2)) = \emptyset$, *then* $Z_o \subset Z_1$.

PROOF. Let $\theta : Z_o \to X_2/(X_1 \cap X_2)$ be the operator given by the composite of the canonical map $X/Y \to X/X_1$, restricted to Z_o, and the natural isomorphism from X/X_1 onto $X_2/(X_1 \cap X_2)$ (induced by the decomposition $X/Y = Z_1 + Z_2$). We shall show that $\theta = 0$, which clearly implies our assertion.

Let $\xi \in Z_o$ and let $U = \rho(T,Z_o)$. Then there exists a section $\phi \in {}_o\mathcal{O}(U,Z_o \cap D(T))$ such that $(\zeta_U - T_U)\phi = \xi$. Let

$$\theta_o : Z_o \cap D(T) \to (X_2 \cap D_o(T))/(X_1 \cap X_2)$$

be the restriction of θ (note that $X_1 \cap X_2 \subset D_o(T)$ from the hypothesis). Then θ_o and θ induce, respectively, the operators

$$\theta_1 : {}_o\mathcal{O}(U,Z_o \cap D(T)) \to {}_o\mathcal{O}(U,(X_2 \cap D_o(T))/(X_1 \cap X_2)),$$

$$\theta_2 : \mathcal{O}(U,Z_o) \to \mathcal{O}(U,X_2/(X_1 \cap X_2)) .$$

Moreover,

$$\theta_2(\zeta_U - T_U)\phi = (\zeta_U - \hat{T}_U)\theta_1\phi = \theta\xi \tag{3.9}$$

(see [18], Theorem 2.9), where $\hat{T}$ is the operator induced by T in $X_2/(X_1 \cap X_2)$.

Next, let $V = \mathbf{C}_\infty \setminus (\sigma(T,Z_2) \cup \sigma(T,Z_1 \cap Z_2))$. Then

$$\sigma(\hat{T},X_2/(X_1 \cap X_2)) \subset \sigma(T,X_2/Y) \cup \sigma(T,(X_1 \cap X_2)/Y) =$$
$$= \sigma(T,Z_2) \cup \sigma(T,Z_1 \cap Z_2) ,$$

by Lemma 3.2. Hence the operator

$$\zeta_V - \hat{T}_V : {}_o\mathcal{O}(V,(X_2 \cap D_o(T))/(X_1 \cap X_2)) \to \mathcal{O}(V,X_2/(X_1 \cap X_2))$$

is bijective and we can find a section

$$\phi_2 \in {}_o\mathcal{O}(V,(X_2 \cap D_o(T))/(X_1 \cap X_2))$$

such that $(\zeta_V - \hat{T}_V)\phi_2 = \theta\xi$. If $\phi_1 = \theta_1\phi$, let us observe that

$$(\zeta_{U \cap V} - \hat{T}_{U \cap V})(\phi_1 | U \cap V - \phi_2 | U \cap V) = 0,$$

by (3.9). Thus there is a section

$$\phi_o \in {}_o\mathcal{O}(U \cup V,(X_2 \cap D_o(T))/(X_1 \cap X_2))$$

such that $\phi_o | U = \phi_1$ and $\phi_o | V = \phi_2$. Moreover $(\zeta_{U \cup V} - \hat{T}_{U \cup V})\phi_o = \theta\xi$. But $U \cup V = \mathbf{C}_\infty$. Hence $\phi_o = 0$ (see the proof of Theorem 3.7 from [18]), and so $\theta\xi = 0$. Consequently $Z_o \subset Z_1$.

The next result is a sufficient condition which insures the SVEP (see also [7], Bemerkung I.2.3 for Fréchet space operators with bounded spectrum).

3.11. THEOREM. *Let* $T \in P(X/Y)$ *be such that for every open cover* $\{G_1,G_2\}$ *of* $\mathbf{C}_\infty$ *there are a quotient Fréchet space* Z_o, *an operator* $S \in P(Z_o)$ *and two subspaces* Z_1, $Z_2 \in \mathrm{Inv}(S)$ *such that* $Z_o = Z_1 + Z_2$, $\sigma(S,Z_j \cap Z_k) \subset G_j \cap G_k$ $(j,k = 1,2)$, $X/Y \in \mathrm{Inv}(S)$ *and* $S | (X/Y) = T$. *Then* T *has the* SVEP.

PROOF. It suffices to prove that the operator

$$\zeta_U - T_U : \mathcal{O}(U,D(T)) \to \mathcal{O}(U,X/Y)$$

is injective for every open disc $U \subset \mathbf{C}$. Let U, V be open discs such that $V \subset \bar{V} \subset U$. Then $G_1 = U$ and $G_2 = \mathbf{C}_\infty \setminus \bar{V}$ provide an open cover of $\mathbf{C}_\infty$. Let $Z_j = X_j/Y$ $(j = 0,1,2)$ and $S \in P(Z_o)$ be given by the hypothesis, with respect to the cover $\{G_1,G_2\}$.

The complex of quotient Fréchet spaces

$$0 \to Z_1 \cap Z_2 \xrightarrow{\alpha} Z_1 \times Z_2 \xrightarrow{\beta} Z_o \to 0$$

is exact, where $\alpha(\eta) = (\eta, -\eta)$ and $\beta(\eta_1,\eta_2) = \eta_1 + \eta_2$. According to Remark 3.1, the complex

$$(3.10) \quad 0 \to \mathcal{O}(U,Z_1 \cap Z_2) \xrightarrow{\alpha_U} \mathcal{O}(U,Z_1) \times \mathcal{O}(U,Z_2) \xrightarrow{\beta_U} \mathcal{O}(U,Z_o) \to 0$$

is also exact, where $\alpha_U(\phi) = (\phi,-\phi)$ and $\beta_U(\phi_1,\phi_2) = \phi_1 + \phi_2$. The exactness of (3.10) shows that $\mathcal{O}(U,Z_o) = \mathcal{O}(U,Z_1) + \mathcal{O}(U,Z_2)$ and that $\mathcal{O}(U,Z_1) \cap \mathcal{O}(U,Z_2) = \mathcal{O}(U,Z_1 \cap Z_2)$. Therefore $\mathcal{O}(U,Z_o)/\mathcal{O}(U,Z_2)$ is isomorphic to $\mathcal{O}(U,Z_1)/\mathcal{O}(U,Z_1 \cap Z_2)$, which in turn is isomorphic to $\mathcal{O}(U,X_1/(X_1 \cap X_2))$. Let

$$\theta : \mathcal{O}(U,D(T)) \to \mathcal{O}(U,X_1/(X_1 \cap X_2))$$

be the composite of the canonical map

$$\mathcal{O}(U,Z_o) \to \mathcal{O}(U,Z_o)/\mathcal{O}(U,Z_2)$$

and the above isomorphism. We shall prove that the diagram

$$(3.11)\qquad \begin{array}{ccc} \mathcal{O}(U,D(S)) & \xrightarrow{\theta} & \mathcal{O}(U,X_1/(X_1 \cap X_2)) \\ \Big\downarrow \zeta_U - S_U & & \Big\downarrow \zeta_U - \hat{S}_U \\ \mathcal{O}(U,Z_o) & \xrightarrow{\theta} & \mathcal{O}(U,X_1/(X_1 \cap X_2)) \end{array}$$

is commutative, where $\hat{S}$ is induced by S in $X_1/(X_1 \cap X_2)$. First of all note that

$$\sigma(\hat{S},X_1/(X_1 \cap X_2)) \subset \sigma(S,Z_1) \cup \sigma(S,Z_1 \cap Z_2) \subset U$$

by Lemma 3.2 and the hypothesis. Since $\infty \notin U$ we must have $\hat{S} \in L(X_1 \cap X_2))$, by Remark 3.8.3°. If $\phi \in \mathcal{O}(U,D(S))$, we can write $\phi = \phi_1 + \phi_2$, with $\phi_j \in \mathcal{O}(U,Z_j)$ $(j = 1,2)$. As we have $\infty \notin \sigma(S,Z_1) \subset U$, then $Z_1 \subset D(S)$, as above. Hence $\phi_1 \in \mathcal{O}(U,D(S))$, and so $\phi_2 = \phi - \phi_1 \in \mathcal{O}(U,D(S))$. Therefore

$$(\zeta_U - S_U)\phi = (\zeta_U - S_U)\phi_1 + (\zeta_U - S_U)\phi_2 ,$$

and $(\zeta_U - S_U)\phi_j \in \mathcal{O}(U,Z_j)$ $(j = 1,2)$. Consequently

$$\theta(\zeta_U - S_U)\phi = (\zeta_U - \hat{S}_U)\theta\phi = (\zeta_U - \hat{S}_U)\theta\phi_1 ,$$

showing that (3.11) is commutative.

Now, let $\phi \in \mathcal{O}(U,D(T)) \subset \mathcal{O}(U,D(S))$ be such that $(\zeta_U - T_U)\phi = 0$, and let $\phi = \phi_1 + \phi_2$ be a decomposition of ϕ as above. Therefore, by the commutativity of (3.11),

$$0 = \theta(\zeta_U - S_U)\phi = (\zeta_U - \hat{S}_U)\theta\phi_1 .$$

According to Lemma 3.9, the operator $\zeta_U - \hat{S}_U$ is injective. Hence $\phi_1 \in \mathcal{O}(U,Z_1 \cap Z_2) \subset \subset \mathcal{O}(U,Z_2)$, and so $\phi = \phi_1 + \phi_2 \in \mathcal{O}(U,Z_2)$. Since $\sigma(S,Z_2) \cap V = \emptyset$, it follows that $\phi | V = 0$. As $V \subset \overline{V} \subset U$ is arbitrary, we must have $\phi = 0$.

3.12. REMARK. When $Y = \{0\}$ and therefore X, Z_j $(j = 0,1,2)$ are Fréchet spaces, then the requirement $\sigma(S,Z_j \cap Z_k) \subset G_j \cap G_k$ $(j,k = 1,2)$ from Theorem 3.11 may be replaced by weaker one $\sigma(S,Z_j) \subset G_j$ $(j = 1,2)$, provided Z_1,Z_2 are *closed* subspaces of Z_o (as stated in [7]). Indeed, in this case, if $U \subset \mathbf{C}$ is an open disc (more generally a simply connected open set) and $\sigma(T,Z_1) \subset U$, then $\sigma(S,Z_1 \cap Z_2) \subset U$, which suffices for the proof of Theorem 3.11. Nevertheless, if $Z_1 \cap Z_2 \in \mathrm{Lat}(Z_1)$ is not closed in Z_1, then the inclusion $\sigma(S,Z_1 \cap Z_2) \subset U$ may not be true, as simple examples show.

For operators with bounded spectrum, the condition from Theorem 3.11 is necessary too, modulo similarities (see also [7]).

3.13. PROPOSITION. *Let* $T \in L(X/Y)$ *have the* SVEP *and assume that*

$\infty \notin \sigma(T,X/Y)$. *Then for every open cover* $\{G_1,G_2\}$ *of* $\sigma(T,X/Y)$ *there are a quotient Fréchet space* Z_o, *an injective operator* $\theta : X/Y \to Z_o$, *an operator* $S \in L(Z_o)$ *and two subspaces* $Z_1,Z_2 \in \mathrm{Inv}(S)$ *such that* $Z_o = Z_1 + Z_2$, $\sigma(S,Z_j) \subset G_j$ $(j = 1,2)$ *and* $S|E_o = T_o$, *where* $T_o = \theta T\theta^{-1}$ *and* $E_o = \theta(X/Y)$.

PROOF. For every open and bounded set $U \subset \mathbf{C}$ we define the quotient Fréchet space

$$F_T(U) = O(U,X/Y)/(\zeta_U - T_U)O(U,X/Y)$$

(see [13] for Fréchet space operators). It is easily seen that T_U and ζ_U induce the same action on $F_T(U)$. Moreover, $\sigma(\zeta_U,F_T(U)) \subset \bar{U}$.

Now, let $\{U_1,U_2\}$ be an open cover of $\sigma(T,X/Y)$ such that $U_j \subset \bar{U}_j \subset G_j$, and with $\bar{U}_j$ compact in $\mathbf{C}$ $(j = 1,2)$. We define the quotient Fréchet space

$$Z_o = F_T(U_1) \times F_T(U_2)$$

and the operator $\theta : X/Y \to Z_o$ given by $\theta\xi = ([\xi]_1,[\xi]_2)$, where $[\xi]_j$ is the coset of ξ in $F_T(U_j)$ $(j = 1,2)$. Since T has the SVEP, the operator θ is clearly injective. We also set $Z_1 = F_T(U_1) \times \{0\}$, $Z_2 = \{0\} \times F_T(U_2)$ and $S \in L(Z_o)$ given by

$$S(\phi_1,\phi_2) = (\zeta_{U_1}\phi_1,\zeta_{U_2}\phi_2), \qquad (\phi_1,\phi_2) \in Z_o .$$

Having these objects defined, our assertions follow easily.

We are now in the position to prove the uniqueness of the spectral capacity attached to a decomposable operator in our sense (see also [9], [10], [16], [17], [22], etc.).

3.14. THEOREM. *Let* $T \in P(X/Y)$ *be decomposable and let* E *be a spectral capacity attached to* T. Then T *has the* SVEP *and* $E(P) = E_T(F)$ *for all closed* $F \subset \mathbf{C}_\infty$.

PROOF. That T has the SVEP clearly follows from Theorem 3.11. If $F = \bar{F} \subset \mathbf{C}_\infty$ is fixed and $\xi \in E(F)$, then the section

$$\phi = (\zeta_U - (T|E(F))_U)^{-1}\xi \in {}_oO(U,D(T))$$

satisfies the equation $(\zeta_U - T_U)\phi = \xi$, where $U = \mathbf{C}_\infty \setminus F$. Therefore $\xi \in E_T(F)$.

Conversely, let $\{G_1,G_2\}$ be an open cover of $\mathbf{C}_\infty$ such that $F \subset G_1$ and $\bar{G}_2 \cap F = \emptyset$, and let $Z_j = E(\bar{G}_j)$ $(j = 1,2)$. Since E is a spectral capacity attached to T, we have

$$\sigma(T,Z_2) \cup \sigma(T,Z_1 \cap Z_2) \subset \bar{G}_2 \cup (\bar{G}_1 \cap \bar{G}_2) = \bar{G}_2 .$$

Then it follows from Lemma 3.10 that $E_T(F) \subset E(\bar{G}_1)$.

If $\{G_{1,n}\}_{n=1}^{\infty}$ is a family of open sets such that each $G_{1,n}$ shares the properties of G_1 and $\cap \{\overline{G}_{1,n}; n \geq 1\} = F$, it results from (1.1) that

$$E_T(F) \subset \bigcap_{n=1}^{\infty} E(\overline{G}_{1,n}) = E(F) .$$

Consequently $E_T(F) = E(F)$ for each closed $F \subset \mathbf{C}_\infty$, and the proof of the theorem is completed.

3.15. REMARK. One can also settle in this context the problem solved in [15] for Banach space operators (see also [7], [13], [16], [22], etc.). Namely, let $T \in P(X/Y)$ have the SVEP. Then for every open $U \subset \mathbf{C}_\infty$ one can define the quotient Fréchet space

$$F_T(U) = O(U,X/Y)/(\zeta_U - T_U)_o O(U,D(T)) .$$

The assignment $U \to F_T(U)$, which is a presheaf with respect to the natural restrictions, corresponds to the sheaf model of a Fréchet space operator introduced in [13] (see also [11] for sheaf theoretical results). Since both $U \to O(U,X/Y)$ and $U \to {}_oO(U,D(T))$ (and therefore $U \to (\zeta_U - T_U)_o O(U,D(T)))$ are acyclic sheaves (this fact follows from the proof of Proposition 3.5 from [18]), one can derive that $U \to F_T(U)$ is actually a sheaf.

Next, if $F \subset U$ is a closed set, then the natural operator $E_T(F) \to F_T(U)$ is injective. Moreover, the image of $E_T(F)$ via this operator consists of those sections from $F_T(U)$ whose support is in F. If, in addition, T is assumed to be 2-decomposable (i.e. condition (iii) from (1.1) is valid only for $n \leq 2$), then one can show that T is decomposable, via the fact that the sheaf $U \to F_T(U)$ is, in this case, soft, as done in [13], Section 4. We omit the details (the case $T \in L(X/Y)$ with $\infty \notin \sigma(T,X)$ is treated in [22]).

3.16. EXAMPLES. 1° Let K be a compact subset of the real line and let $X = A'(K)$ be the Fréchet space of all analytic functionals carried by K (see, for instance, [12]). If $T \in L(X)$ is the operator induced by the multiplication with the independent variable, then T is decomposable and the spectral capacity of T is given by

$$E_T(F) = A'(F \cap K), \qquad F = \overline{F} \subset \mathbf{C}_\infty$$

(see [17] for details). We note that $E_T(F) \in \mathrm{Lat}(X)$ but, in general, $E_T(F)$ is not a closed subspace of X.

2° Let K_o, K be compact subsets of the real line, $K_o \subset K$, and let $X = A'(K)$, $Y = A'(K_o)$. Then X/Y is a quotient Fréchet space. If $T \in L(X/Y)$ is the strict operator induced by the multiplication with the independent variable, then T is decomposable [22]. The spectral capacity of T is given by

$$E_T(F) = (A'(K_o) + A'(F \cap K))/A'(K_o), \qquad F = \overline{F} \subset \mathbf{C}_\infty .$$

In particular, the strict operator induced by the multiplication with the independent variable in spaces of hyperfunctions on the real line [12] (which are quotient Fréchet spaces) is decomposable.

3° To get a genuine "unbounded" decomposable operator in a quotient Fréchet space, it suffices to take a direct sum of an operator of the previous type and, say, an unbounded selfadjoint operator.

REFERENCES

1. **Albrecht, E.; Vasilescu, F.-H.** : On spectral capacities, *Rev. Roumaine Math. Pures Appl.* **19** (1974), 701-705.

2. **Apostol, C.** : Spectral decompositions and functional calculus, *Rev. Roumaine Math. Pures Appl.* **13** (1968), 1481-1528.

3. **Colojoară, I.; Foiaș, C.**: *Theory of generalized spectral operators, Gordon and Breach*, New York, 1968.

4. **Dixmier, J.**: Etude sur les variétés et les opérateurs de Julia, *Bull. Soc. Math. France* **77** (1949), 11-101.

5. **Douady, R.** : Produits tensoriels topologiques et espaces nucléaires, *Asterisque*, Rev. Soc. Math. France, 1974, Exposé I.

6. **Eschmeier, J.** : Local properties of Taylor's analytic functional calculus, *Invent. Math.* **68** (1982), 103-116.

7. **Eschmeier, J.** : *Analytische Dualität und Tensorprodukte in der meherdimensionalen Spektraltheorie, Habilitationsschrift*, Münster, 1986.

8. **Foiaș, C.** : Spectral maximal spaces and decomposable operators in Banach spaces, *Arch. Math.* **14** (1963), 341-349.

9. **Foiaș, C.** : Spectral capacities and decomposable operators, *Rev. Roumaine Math. Pures Appl.* **13** (1968), 1539-1545.

10. **Frunză, Ș.** : The Taylor spectrum and spectral decompositions, *J. Functional Analysis*, **19** (1975), 390-421.

11. **Godement, R.** : *Topologie algèbrique et théorie des faisceaux, Hermann, Paris*, 1958.

12. **Hörmander, L.** : *The analysis of linear partial differential operators I*, Springer-Verlag, Berlin, 1983.

13. **Putinar, M.** : Spectral theory and sheaf theory I. *Operator Theory: Advances and Applications* Vol. **11**, pp. 283-297, Birkhäuser-Verlag, Basel, 1983.

14. **Putinar, M.; Vasilescu, F.-H.** : The local spectral theory needs Fréchet spaces, *Preprint Series in Mathematics* No. **16** (1982), INCREST, București.

15. **Radjabalipour, M.** : On equivalence of decomposable and 2-decomposable operators, *Pacific J. Math.* **77** (1978), 243-247.

16. **Vasilescu, F.-H.** : *Analytic functional calculus and spectral decompositions*, Editura Academiei and D. Reidel Publishing Company, Bucharest and Dordrecht, 1982.

17. **Vasilescu, F.-H.** : Analytic operators and spectral decompositions, *Indiana Univ. Math. J.* **34** (1985), 705-722.

18. **Vasilescu, F.-H.** : Spectral theory in quotient Fréchet spaces I, *Rev. Roumaine Math. Pures Appl.* **32** (1987), 561-579.

19. **Vasilescu, F.-H.** : Spectral theory in quotient Fréchet spaces II, Preprint, 1987.

20. **Waelbroeck, L.** : Le calcul symbolique dans les algèbres commutatives, *J. Math. Pures Appl.* **33** (1954), 143-186.

21. **Waelbroeck, L.** : Quotient Fréchet spaces, Preprint, 1984, (to appear in: *Rev. Roumaine Math. Pures Appl.*).

22. **Zhang Haitao** : *Generalized spectral decompositions* (in Romanian), Dissertation, University of Bucharest, 1987.

Department of Mathematics, INCREST
Bd. Păcii 220, 79622 Bucharest
Romania.

Operator Theory:
Advances and Applications, Vol. 32

Dedicated to the memory of Constantin Apostol

A NOTE ON QUASIDIAGONAL OPERATORS

Dan Voiculescu

Let $\mathcal{H}$ be a separable complex Hilbert space of infinite dimension and let $\mathcal{L}(\mathcal{H})$ and $\mathcal{K}(\mathcal{H})$ denote the bounded and respectively the compact operators on $\mathcal{H}$. An operator $T \in \mathcal{L}(\mathcal{H})$ is called quasidiagonal ([5]) if $T = D+K$ where $K \in \mathcal{K}(\mathcal{H})$ and D is block-diagonal, i.e. $D = D_1 \oplus D_2 \oplus \ldots$ for some decomposition $\mathcal{H} = \mathcal{H}_1 \oplus \mathcal{H}_2 \oplus \ldots$ where $\dim \mathcal{H}_j < \infty$ $(j = 1,2,\ldots)$.

This note deals with two questions: quasidiagonal operators with non-nuclearly embedded C^*-algebra and quasidiagonal operators which are not quasidiagonal relative to the Mačaev ideal.

From the general facts on nuclear C^*-algebras ([1],[2], [7],[15]) we derive the existence of a block-diagonal operator $T \in \mathcal{L}(\mathcal{H})$ such that for $C^*(T)$ the C^*-algebra of T, the inclusion $C^*(T) \hookrightarrow \mathcal{L}(\mathcal{H})$ is not nuclear. It follows in particular that such a block-diagonal operator is not a norm-limit of operators with finite-dimensional C^*-algebra (or equivalently a norm-limit of n-normal operators). Thus the result, we prove, easily implies the existence of such non-approximable quasidiagonal operators, for which Szarek [8], in response to a question of Herrero [16], had given a quite ingenious ad hoc argument.

An operator $T \in \mathcal{L}(\mathcal{H})$ is quasidiagonal relative to a normed ideal of compact operators ([9]) if $T = D+K$ where D is block-diagonal and K is in that normed ideal. Using our results on quasicentral approximate units for normed ideals ([12]) we exhibit a quasidiagonal operator which is not quasidiagonal relative to the Mačaev ideal. We don't know whether there are quasidiagonal operators which are not quasidiagonal with respect to a normed ideal strictly larger than the Mačaev ideal. The operator we construct also provides a counter-example to a question appearing in [17].

The present note has three sections. The first two sections deal with the two previously mentioned questions. The third section consists of concluding remarks. We discuss the class of nuclearly embedded C^*-algebras which appear to be the relevant class in connection with nonapproximable quasidiagonals. We also mention some of the related open problems concerning quasidiagonal operators.

The author would like to thank E. G. Effros and Z. J. Ruan for bibliographical help.

1. BLOCK-DIAGONAL OPERATORS WITH NON-NUCLEARLY EMBEDDED C^*-ALGEBRA

A completely positive map $\Phi: M \to N$, where M,N are C^*-algebras, is nuclear [1] if there is a net $(\tau_i, \sigma_i, P_i)_{i\in I}$ where the P_i are finite-dimensional C^*-algebras and $\tau_i: M \to P_i$, $\sigma_i: P_i \to N$ are completely positive maps such that Φ is the limit of the $\sigma_i \circ \tau_i$ in point-norm convergence. If $M = \overline{\bigcup_{j\in J} M_j}$ for an increasing net $(M_j)_{j\in J}$ of C^*-subalgebras, then Φ is nuclear iff the restrictions $\Phi|M_j$ are nuclear. A C^*-algebra M is nuclear [7] if id_M is nuclear [2].

<u>1.1 Proposition</u>. *There exists a block-diagonal operator* $T \in \mathcal{L}(\mathcal{H})$ *such that the inclusion* $j: C^*(T) \to \mathcal{L}(\mathcal{H})$ *is not a nuclear map.*

<u>Proof</u>. Fix a decomposition $\mathcal{H} = \mathcal{H}_1 \oplus \mathcal{H}_2 \oplus \dots$ with $\dim \mathcal{H}_j < \infty$ $(j \in \mathbb{N})$ and $\lim \dim \mathcal{H}_j = \infty$. Let P_j be the orthogonal projection of $\mathcal{H}$ onto $\mathcal{H}_j$ and let $A = \{T \in \mathcal{L}(\mathcal{H}) \mid [P_j,T]=0,\ j \in \mathbb{N}\}$. By [15] the von Neumann algebra A is not nuclear. Consider $E: \mathcal{L}(\mathcal{H}) \to A$ to be the conditional expectation $E(T) = \Sigma P_j T P_j$ and let $i: A \to \mathcal{L}(\mathcal{H})$ be the inclusion. Since $E \circ i = id_A$ and A is not nuclear we infer that i is not nuclear. Moreover A being the union of the net of its finitely generated C^*-subalgebras $(C^*(X))_{X\in\mathcal{G}}$, where $\mathcal{G}$ is the set of finite subsets of A, it follows that the inclusion $i_X: C^*(X) \to \mathcal{L}(\mathcal{H})$ is nonnuclear for some $X \in \mathcal{G}$.

To get a singly generated C*-algebra instead of $C^*(X)$ we apply some standard tricks of passing to matrices. Without loss of generality assume $X = \{T_1, \ldots, T_p\}$ with $T_j = T_j^*$, $\|T_j\| < 1$ $(j = 1, \ldots, p)$. Let $F \in \mathcal{L}(\mathbb{C}^p)$ be such that $F = -F^*$ and $(\exp F)e_k = e_{k+1}$ $(1 \leqslant k \leqslant p-1)$, $(\exp F)e_p = e_1$ where $e_1, \ldots, e_p$ is the canonical basis of $\mathbb{C}^p$. We define $T \in \mathcal{L}(\mathcal{H} \otimes \mathbb{C}^p)$ by

$$T = \sum_{j=1}^{p} (T_j + 2jI) \otimes Q_j + I \otimes F$$

where $Q_j \in \mathcal{L}(\mathbb{C}^p)$ is defined by $Q_j e_k = \delta_{j,k} e_j$. Clearly T acting on $\mathcal{H} \otimes \mathbb{C}^p = (\mathcal{H}_1 \otimes \mathbb{C}^p) \oplus (\mathcal{H}_2 \otimes \mathbb{C}^p) \oplus \ldots$ is block-diagonal. The C*-algebra of T contains

$$T + T^* = 2 \sum_{j=1}^{p} (T_j + 2jI) \otimes Q_j$$

and
$$T - T^* = 2I \otimes F.$$

Since the spectra of the $T_j + 2jI$ are pairwise disjoint we infer that the $I \otimes Q_j$ are spectral projections for $T+T^*$ and are contained in $C^*(T)$. Also $I \otimes \exp F \in C^*(T)$ and hence $I \otimes \mathcal{L}(\mathbb{C}^p) \subset C^*(T)$. It follows easily that $C^*(T) = C^*(T_1, \ldots, T_p) \otimes \mathcal{L}(\mathbb{C}^p)$. The inclusion $C^*(T) \hookrightarrow \mathcal{L}(\mathcal{H} \otimes \mathbb{C}^p)$ is not nuclear because $C^*(T_1, \ldots, T_p) \to \mathcal{L}(\mathcal{H})$ is not nuclear. Indeed the inclusion $C^*(T_1, \ldots, T_p) \to \mathcal{L}(\mathcal{H})$ is the composition of $C^*(T_1, \ldots, T_p) \to C^*(T)$, $C^*(T) \to \mathcal{L}(\mathcal{H} \otimes \mathbb{C}^p)$ and $\mathcal{L}(\mathcal{H} \otimes \mathbb{C}^p) \to \mathcal{L}(\mathcal{H})$ where the last map takes S into $(I \otimes Q_1)S|\mathcal{H} \otimes e_1$. Q.E.D.

1.2 Proposition. *Let* $T \in \mathcal{L}(\mathcal{H})$ *be a norm-limit of operators* T_k $(k \in \mathbb{N})$ *such that* $\dim C^*(T_k) < \infty$. *It follows that the inclusion* $C^*(T) \to \mathcal{L}(\mathcal{H})$ *is nuclear.*

Proof. Since $\dim C^*(T_k) < \infty$ there exist conditional expectations E_k of $\mathcal{L}(\mathcal{H})$ onto $C^*(T_k)$ (these are completely positive maps, $E_k(\mathcal{L}(\mathcal{H})) = C^*(T_k)$, $E_k^2 = E_k$, $\|E_k\| = 1$, $E_k(XSY) = XE_k(S)Y$ if $S \in \mathcal{L}(\mathcal{H})$, $X, Y \in C^*(T_k)$). Further let $j: C^*(T) \to \mathcal{L}(\mathcal{H})$ be the inclusion and let $\Omega = \{a \in C^*(T) \mid \lim_{n\to\infty} \|(E_n \circ j)(a) - j(a)\| = 0\}$. We shall prove that $\Omega = C^*(T)$. Since E_n and j are contractive

Ω is closed. Moreover Ω is a vector space and since $E_n(S^*) = E_n(S)^*$ we have $\Omega^* = \Omega$. We have $T \in \Omega$. Thus in order to conclude the proof it will be sufficient to prove that $x,y \in \Omega \Longrightarrow xy \in \Omega$. We have:

$$\begin{aligned}\|(E_n \circ j)(xy) - j(xy)\| &\leqslant \|E_n(j(xy) - E_n(j(x))E_n(j(y)))\| \\ &\quad + \|E_n(j(x))E_n(j(y)) - j(x)j(y)\| \\ &\leqslant 2\|E_n(j(x))E_n(j(y)) - j(x)j(y)\|\end{aligned}$$

which easily gives the desired conclusion. Q.E.D.

Combining the preceding two propositions we obtain, via a different route from Szarek [8], a negative answer to a question of Herrero in [16].

1.3 Corollary. *There exists a block-diagonal operator* $T \in \mathcal{L}(\mathcal{H})$ *which is not a norm-limit of operators with finite-dimensional* C*-*algebra.*

2. A QUASIDIAGONAL OPERATOR NON-QUASIDIAGONAL WITH RESPECT TO THE MAČAEV IDEAL

We shall denote the Mačaev ideal [4] by $\mathcal{C}_\infty^-$. It is the norm-closure of finite rank operators with respect to the norm

$$|X|_\infty^- = \sum_{k\geqslant 1} k^{-1}\mu_k$$

where $\mu_1 \geqslant \mu_2 \geqslant \ldots$ are the eigenvalues of $(X^*X)^{\frac{1}{2}}$. An operator T is quasidiagonal relative to the Mačaev ideal if $T = D+K$ where D is the block-diagonal and $K \in \mathcal{C}_\infty^-$. This is equivalent to

$$qd_\infty^-(T) = \liminf_{P \in \mathcal{P}} |[P,T]|_\infty^- = 0$$

where $\mathcal{P}$ is the set of finite rank self-adjoint projections in $\mathcal{H}$ with its natural order. In ([11],[12]) we considered $k_\infty^-(T)$ defined by

$$k_\infty^-(T) = \liminf_{A \in \mathcal{R}_1^+} |[A,T]|_\infty^- \quad .$$

where $\mathcal{R}_1^+$ is the set of finite rank operators A on $\mathcal{H}$ such that $0 \leqslant A \leqslant I$ endowed with its natural order. This number measures the obstruction to the existence of a quasicentral approximate unit relative to $\mathcal{C}_\infty^-$ for T and we have $k_\infty^-(T) \leqslant qd_\infty^-(T)$. The existence of a quasidiagonal operator which is not quasidiagonal relative to the Macaev ideal will follow from the following fact.

2.1 Proposition. *There exists a quasidiagonal operator* $T \in \mathcal{L}(\mathcal{H})$ *such that* $k_\infty^-(T) > 0$.

Proof. The idea is to use the pair of unitaries for which their joint k_∞^- does not vanish, exhibited in [12], in order to construct T.

Let $G = \mathbb{Z} * \mathbb{Z}$ be the free group on two generators and $C^*(G)$ its full C^*-algebra. Let further $u_1, u_2 \in C^*(G)$ be the unitaries corresponding to the generators of G. It is a known fact that we can find finite-dimensional representations $\sigma_n : C^*(G) \to \mathcal{L}(\mathcal{H}_n)$ $n \in \mathbb{N}$, $\dim \mathcal{H}_n < \infty$ such that $\sigma = \bigoplus_n \sigma_n$ is a faithful representation of $C^*(G)$. In addition we may also assume $\sigma(C^*(G)) \cap \mathcal{K}(\oplus \mathcal{H}_n) = 0$. Let further λ be the left regular representation of $C^*(G)$ on $\ell^2(G)$ and let $\mu = \lambda \oplus \sigma$. We shall take $\mathcal{H} = \ell^2(G) \oplus \bigoplus_n \mathcal{H}_n$ and $T = \mu(a)$ where

$$a = \phi(u_1 + u_1^*) + i\phi(u_2 + u_2^*)$$

with $\phi(t) = -1 + \frac{2}{\pi} \arccos \frac{t}{2}$. In view of the author's non-commutative Weyl-von Neumann type theorem [10], T is unitarily equivalent to a compact perturbation of $\sigma(a)$ and hence quasidiagonal.

On the other hand, by [11],

$$\begin{aligned} k_\infty^-(T) &\geqslant \max(k_\infty^-(\lambda(a)), k_\infty^-(\sigma(a))) = k_\infty^-(\sigma(a)) \\ &\geqslant k_\infty^-(\sigma(\phi(u_1 + u_1^*)), \sigma(\phi(u_2 + u_2^*))) \end{aligned}$$

where $k_\infty^-(X_1, \ldots, X_n) = \liminf_{A \in R_1^+} \max_{1 \leqslant k \leqslant n} (|[A, X_k]|_\infty)$.

Let $v_j = \exp(\pi i\phi(u_j + u^*))$ $(j = 1,2)$. Since

$$|[\sigma(v_j), A]|_\infty^- = |[\exp(\pi i\sigma(\phi(u_j + u_j^*))), A]|_\infty^-$$
$$\leq \pi|[\sigma(\phi(u_j + u_j^*)), A]|_\infty^- \exp(\pi\|\sigma(\phi(u_j + u_j^*))\|)$$

we see that $k_\infty^-(T) > 0$ will follow if we prove that $k_\infty^-(\sigma(v_1), \sigma(v_2)) > 0$. By ([11] 1.4) we have $k_\infty^-(\sigma(v_1), \sigma(v_2)) = k_\infty^-(\lambda(v_1), \lambda(v_2))$.

Let $\xi \in \ell^2(G)$ be the trace-vector $\xi(g) = \delta_{g,e}$ $(g \in G)$ and let $\tau(x) = \langle \lambda(x)\xi, \xi\rangle$ be the corresponding trace on $C^*(G)$. Note that the measures determined by τ on the spectra of u_1, u_2 coincide with the Haar measure on $\{z \in \mathbb{C} \mid |z| = 1\}$. It is easily seen that v_1, v_2 have the same property. On the other hand, the C^*-algebras of u_1 and u_2 form a free pair of subalgebras in $C^*(G)$ with respect to τ, in the sense of [13]. This implies that the C^*-algebras of v_1 and v_2 also form a free pair of subalgebras in $C^*(G)$ with respect to τ. Let $\alpha: C^*(G) \longrightarrow C^*(G)$ be the endomorphism such that $\alpha(u_j) = v_j$ $(j = 1,2)$. Since a state is completely determined on an algebra generated by two subalgebras, forming a free pair with respect to that state, by the restrictions to those subalgebras, we infer that $\tau\circ\alpha = \tau$. Let $X = \overline{\lambda(\alpha(C^*(G)))\xi}$. Since the representations of $C^*(G)$ associated with the states τ and $\tau\circ\alpha$ are λ and $(\lambda\circ\alpha)|X$ we infer from $\tau = \tau\circ\alpha$ the fact that these representations are unitarily equivalent. It follows in particular that $k_\infty^-(\lambda(u_1), \lambda(u_2)) = k_\infty^-((\lambda(v_1)|X), (\lambda(v_2)|X))$. This implies in view of ([11] 1.4) and ([12], 3.2) that

$$k_\infty^-(\lambda(v_1), \lambda(v_2)) \geq k_\infty^-(\lambda(u_1), \lambda(u_2)) > 0 \quad . \qquad \text{Q.E.D.}$$

<u>2.2 Corollary</u>. *There exists a quasidiagonal operator which is not quasidiagonal relative to the Macaev ideal.*

For the Schatten-von Neumann class consider the numbers $qd_p(T)$ and $k_p(T)$ ([9],[11]) defined by

$$qd_p(T) = \liminf_{P \in \mathcal{P}} |[P,T]|_p$$

$$k_p(T) = \liminf_{A \in \mathcal{R}_1^+} |[A,T]|_p$$

2.3 Corollary. *There exists an operator* $T \in \mathcal{L}(\mathcal{H})$, *such that* T *cannot be expressed in the form* $T = \bigoplus_{j \in \mathbb{N}} S_j + K$ *with* $qd_p(S_j) < \infty$ *and* $K \in \mathcal{C}_p$ *for any* $p < \infty$.

Proof. Assume $T = \bigoplus_{j \in \mathbb{N}} S_j + K$, $qd_p(S_j) < \infty$, $K \in \mathcal{C}_p$ for some $p < \infty$. Without loss of generality we may assume $1 < p < \infty$. Since $qd_p(S_j) < \infty$ we infer $k_p(S_j) < \infty$ which by ([12] 2.4) implies $k_p(S_k) = 0$. Hence by ([11] 1.4 and 1.3) we would have $k_p(T) = 0$. Clearly an operator T with $k_\infty^-(T) > 0$ provides a counterexample. Q.E.D.

The preceding corollary for p=2 provides a negative answer to one of the questions in problem 4.37 of [16].

3. CONCLUDING REMARKS AND OPEN PROBLEMS

The result in section 1 suggests that it may be of interest to consider the class of *nuclearly embedded* C^*-*algebras*.

3.1 Definition. A C^*-algebra M is *nuclearly embedded* if there exists a C^*-algebra N and an injective homomorphism $j: M \to N$ which is nuclear.

Clearly a C^*-subalgebra of a nuclear C^*-algebra is nuclearly embedded, but we don't know whether every nuclearly embedded C^*-algebra is a C^*-subalgebra of a nuclear C^*-algebra. Also, E. G. Effros suggested to us that it may be of interest to compare the class of nuclearly embedded C^*-algebras to the classes contained in [3].

3.2 Proposition. *If* M *is a nuclearly embedded algebra, then every completely positive map* $\phi: M \to \mathcal{L}(\mathcal{X})$ *is nuclear.*

Proof. Since M is the union of the increasing net of its finitely generated C^*-subalgebras it is sufficient to prove the proposition for separable M. Also, clearly, M may be assumed to be nuclear and because of the Stinespring dilation theorem the proof reduces to the case when ϕ is a unital *-homomorphism. Under these new assumptions there is no loss of generality in replacing ϕ by $\phi \oplus (\pi \otimes I_{\mathcal{X}})$ where π is a representation on some separable Hilbert space $\mathcal{H}$ such that $\operatorname{Ker} \pi = 0$ and $\pi(M) \cap \mathcal{K}(\mathcal{H}) = 0$. This new ϕ is unitarily equivalent to a direct sum $\bigoplus_{i\in I} \phi_i$ where $\phi_i : M \to \mathcal{L}(\mathcal{H})$ are unital *-homomorphisms with $\operatorname{Ker} \phi_i = 0$ and $\mathcal{K}(\mathcal{H}) \cap \phi_i(M) = 0$.

Since M is nuclearly embedded there is a faithful *-representation of M on a Hilbert space which is nuclear. Every subrepresentation of this representation and every multiple of this representation being nuclear, we can find a representation $\rho: M \to \mathcal{L}(\mathcal{H})$ which is nuclear $\operatorname{Ker} \rho = 0$ and $\mathcal{K}(\mathcal{H}) \cap \rho(M) = 0$.

In view of the author's non-commutative Weyl-von Neumann type theorem [10], there are unitary operators $U_{n,i} \in \mathcal{L}(\mathcal{H})$ such that

$$\lim_{n\to\infty}(\sup_{i\in I}\|U_{n,i}\,\rho(a)U_{n,i}^{*} - \phi_i(a)\|) = 0$$

for all $a \in A$. Hence the nuclearity of $\bigoplus_{i\in I}\rho_i$ where ρ_i are copies of ρ, implies the nuclearity of $\bigoplus_{i\in I}\phi_i$. Q.E.D.

We conclude with a problem concerning the role of the Macaev ideal in quasidiagonality questions.

In [14] we have shown that for every n-tuple of operators τ we have:

$$k_{\Phi}(\tau) = \liminf_{A\in \mathcal{R}_1^{+}} |[A,\tau]|_{\Phi} = 0$$

whenever the normed ideal $\mathfrak{S}_{\Phi}^{(0)}$ (see [4]) is strictly larger than the Macaev ideal $\mathcal{C}_{\infty}^{-}$. It is therefore natural to ask the following question about quasidiagonality relative to $\mathfrak{S}_{\Phi}^{(0)}$.

<u>3.3 Problem</u>. If $\mathfrak{S}_{\Phi}^{(0)}$ is a normed ideal strictly larger than the Macaev ideal $\mathcal{C}_{\infty}^{-}$, does it follow that every quasidiagonal operator is also quasidiagonal relative to $\mathfrak{S}_{\Phi}^{(0)}$?

3.4 Remark. The analogue of the preceding problem for quasitriangularity is also open, but we should note that up to now the largest normed ideal, for which examples of quasitriangular operators, which are not quasitriangular relative to that normed ideal, are known, is the Hilbert-Schmidt class. Hence in quasitriangularity ideals larger than the Hilbert-Schmidt class should be considered.

REFERENCES

1. M.D.Choi, E.G.Effros: The completely positive lifting problem for C*-algebras, Annals of Mathematics **104** (1976), 585-609.

2. M.D.Choi, E.G.Effros: Nuclear C*-algebras and the approximation property, Amer. J. Math. **100** (1978), 61-79.

3. E.G.Effros, U.Haagerup: Lifting problems and local reflexivity for C*-algebras, Duke Math. J. (1985).

4. I.T.Gohberg, M.G.Krein: Introduction to the theory of non-selfadjoint operators (Russian), Moscow (1965).

5. P.R.Halmos: Ten problems in Hilbert space, Bull. Amer. Math. Soc. **76** (1970), 887-933.

6. D.A.Herrero, S.J.Szarek: How well can a n×n matrix be approximated by reducible ones?, preprint.

7. E.C.Lance: On nuclear C*-algebras, J. Functional Analysis **12** (1973), 157-176.

8. S.J.Szarek: A quasidiagonal operator which is not a limit of M-normals, preprint.

9. D.Voiculescu: Some extensions of quasitriangularity, Rev. Roumaine Math. Pures Appl. **18** (1973), 1303-1320.

10. D.Voiculescu: A non-commutative Weyl-von Nuemann theorem, Rev. Roumaine Math. Pures Appl. **21** (1976), 97-113.

11. D.Voiculescu: Some results on norm-ideal perturbations of Hilbert space operators, J. Operator Theory **2** (1979), 3-37.

12. D.Voiculescu: Some results on norm-ideal perturbations of Hilbert space operators II, J. Operator Theory **5** (1981), 77-100.

13. D.Voiculescu: Symmetries of some reduced free product C*-algebras, in *Operator Algebras and Their Connections with Topology and Ergodic Theory*, Lecture Notes in Math. vol. 11321 (Springer-Verlag, 1985), 556-588.

14. D.Voiculescu: On the existence of quasicentral approximate units relative to normed ideals, in preparation.

15. S.A.Wassermann: On tensor products of certain group C*-algebras, J. Functional Analysis **23** (1976), 239-254.

16. E.G.Effros, editor: A selection of problems, in *Operator Algebras and K-Theory* (American Mathematical Society, 1982).

17. V.P.Havin, S.V.Hruščev and N.K.Nikolskii, editors: *Linear and Complex Analysis Problem Book*, Lecture Notes in Math. vol. 1043 (Springer-Verlag).

Department of Mathematics
University of California
Berkeley, CA 94720 USA